Carola Richter, Melanie Radue, Christine Horz-Ishak, Anna Litvinenko, Hanan Badr, Anke Fiedler (eds.)
Cosmopolitan Communication Studies

Media Studies | 118

Carola Richter (Dr. phil.) is a professor for international communication at Freie Universität Berlin. In her research, she focuses on media systems and communication cultures in the MENA region (Middle East and North Africa), foreign news coverage, media and migration as well as on public diplomacy. She is the co-founder of AREACORE, the Arab-European Association of Media and Communication Researchers, and director of the Center for Media and Information Literacy (CeMIL) at Freie Universität Berlin.
Melanie Radue (Dr. rer. pol.) is a research fellow at the Chair of Development Politics at the Universität of Passau and works on topics of international communication with a specification in de-Westernization. She completed her doctorate at Friedrich-Alexander-Universität Erlangen-Nürnberg on the topic of "Media Systems in Non-Western Contexts. Freedom of the Media in Southeast Asia". Her research interest lies in media freedom, comparative media systems and social movements. Beyond her academic work she is engaged in the design, implementation, and management of media assistance projects.
Christine Horz-Ishak (Dr. phil.) is a professor of transcultural media communication at Technische Hochschule Köln. Her research and teaching focus on media, migration and diaspora, as well as on the diversity and participation of minorities in traditional media and in the digital context. She is a scholarship holder of the Initiativprogramm of Technische Hochschule Köln and a founding member of the DFG-Network Cosmopolitan Communication Studies (KosmoKW).
Anna Litvinenko (Dr. phil.) is a researcher at the Digitalization and Participation department at the Institute for Media and Communication Studies at Freie Universität Berlin, and an associated researcher at the Weizenbaum Institute for the Networked Society. After receiving her PhD in 2007 and until 2015, she was an associate professor at Saint Petersburg State University. There, she headed the German-Russian Centre of Journalism, as well as the Office for International Scientific Exchange. Her research focuses on political communication in authoritarian regimes, internet governance, as well as on comparing mediated communication across different contexts.
Hanan Badr (Dr. phil.) is a full professor and head of the Public Spheres and Inequalities Unit at the Department of Communication Studies at Paris Lodron Universität Salzburg, Austria. Her academic socialization has evolved between European and Arab universities where she studied journalism and international communication in Cairo and Erfurt. Her work focuses on global inequalities and communication, journalism and media coverage, comparing media systems, activism and transformation.
Anke Fiedler (Dr. phil.) is a researcher and lecturer at the Universität of Greifswald. Her research focuses on media structures in (post-)conflict and transitional states, media and migration, and the role of media development in democratic transitions. Her research has taken her to countries in North and Sub-Saharan Africa and the Middle East. She is a founding member of the DFG network Cosmopolitan Communication Studies (KosmoKW).

Carola Richter, Melanie Radue, Christine Horz-Ishak, Anna Litvinenko,
Hanan Badr, Anke Fiedler (eds.)

Cosmopolitan Communication Studies

Toward Deep Internationalization

[transcript]

This publication has been supported by Deutsche Forschungsgemeinschaft (DFG, the German Research Foundation) and the Open Access Publication Fund of Freie Universität Berlin.

Bibliographic information published by the Deutsche Nationalbibliothek
The Deutsche Nationalbibliothek lists this publication in the Deutsche Nationalbibliografie; detailed bibliographic data are available in the Internet at https://dnb.dnb.de

transcript Verlag | Hermannstraße 26 | D-33602 Bielefeld | live@transcript-verlag.de

Cover design: Kordula Röckenhaus
Cover illustration: Image by soundset from Pixabay
Printing: Elanders Waiblingen GmbH, Waiblingen
https://doi.org/10.14361/9783839476772
Print-ISBN: 978-3-8376-7677-8 | PDF-ISBN: 978-3-8394-7677-2
ISSN of series: 2569-2240 | eISSN of series: 2702-8984

Printed on permanent acid-free text paper.

Contents

Cosmopolitan communication studies

Toward deep internationalization. An introduction

Carola Richter, Melanie Radue, Christine Horz-Ishak, Anna Litvinenko, Hanan Badr & Anke Fiedler

Introduction

This book aims to contribute to a "deep internationalization" of media and communication studies by offering insights and guidance on how to integrate a cosmopolitan perspective in our discipline.[1] Building on the debates on de-Westernization and cosmopolitanism in the decades since the 2000s, the book advocates for the inclusion of both global and local perspectives and context-led approaches in communication studies. We argue that acknowledging and incorporating epistemologies, topics, and methodologies from diverse regions, contexts, and backgrounds will enhance the comprehensiveness and relevance of our discipline and foster a more inclusive and meaningful understanding in communication studies.

This book is the main outcome of the research network Cosmopolitan Communication Studies, which was set up in 2019 and was supported by the Deutsche Forschungsgemeinschaft (DFG, German Research Foundation) from 2021–2025. It comprises both established and emerging scholars located mainly in Germany who are campaigning for a deep internationalization. The network has provided a mapping of the research and teaching landscape in communication studies in Germany, pointing to its lack of internationalization. It has published policy and debate papers and organized several conferences and PhD workshops.[2] This book is the product of multiple rounds of discussions among network members and associated authors, aiming to contribute to the debate

1 In the following, we consistently use the term "communication studies" as we consider it an umbrella term for what is in most parts of the world referred to as "media and communication studies." In Germany, however, there is a somewhat rigid separation between (more literature and theater-oriented) media studies and (more mass media-oriented) communication studies (see also Richter et al., 2024).

2 For more information about the network and its output, see: https://www.polsoz.fu-berlin.de/en/kommwiss/arbeitsstellen/internationale_kommunikation/Projekte-und-Publikationen/Cosmopolitan-Communication-Studies/index.html?ts=1685960329

about internationalization and advocating a cosmopolitan turn in communication studies. Although the network was set up within a German-speaking context, we recognize that the topics and challenges addressed in this book go far beyond the German-speaking research community. Thus, the primary aim of this work is to foster the connection of research communities from different parts of the world to advance the cosmopolitanization of communication studies.

What do we mean by cosmopolitan communication studies? Attempting a definition

The call for de-Westernizing communication studies started in the late 1990s. James Curran and Myung-Jin Park's book *De-Westernizing Media Studies* (2000) positioned itself as "part of a growing reaction against the self-absorption and parochialism of much Western media theory" (p. 1). Several books and articles have been published since then that have addressed this subject. Daya Thussu's edited volume titled *Internationalizing Media Studies* (2009) brought together scholars from different parts of the world, claiming that "theories of globalization . . . have failed to globalize imagination that is to retrieve and disseminate theories of the global from non-Western and non-metropolitan centres" (p. 17). The book *Internationalizing "International Communication"* that was edited by Chin-Chuan Lee (2015) also collected the voices of scholars who had been calling for more global perspectives. Multiple conferences and publications aimed to push forward a less hegemonic approach that presents multiple and critical perspectives on what is understood as international and inter-/transcultural in our discipline (for an overview, see Glück, 2018).

In a special issue of *Communication Theory* in 2014, Silvio Waisbord and Claudia Mellado produced a "reassessment" of the de-Westernization of communication studies. Indeed, thus far, many scholars have injected knowledge that was, for a long time, seen as marginal and, therefore, irrelevant to mainstream communication studies. These scholars include Winston Mano and viola milton (2021) and their Afrokology approach, Daya Thussu (2013) with his emphasis on "Chindia" as a new global player, and several scholars from Latin America showing the merits of a critical decolonial approach to communication and media (Ganter & Ortega, 2019).

At the same time, ongoing debates about White Western dominance in communication studies, emphasized in the #CommunicationSoWhite campaign (Chakravartty et al., 2018), continuous calls for decolonization in social sciences (Badr & Ganter, 2021; Carpentier et al., 2020), and observations about the glaring lack of international and transcultural perspectives in German communication studies' curricula made by the editors of this volume (Badr et al., 2020; Richter et al., 2023), have shown that there is still a need to truly de-Westernize and to become more cosmopolitan. Waisbord and Mellado (2014) have argued in their reassessment that "the subject of study, the body of evidence, analytical frameworks, and academic cultures" (p. 363) as a whole need to be critically reviewed and subjected to de-Westernization.

In this introduction, we continue this debate and call for a cosmopolitan turn in communication studies. We aim to identify the actual obstacles and constraints that will need to be overcome to achieve a more cosmopolitan perspective. Yet, we do not simply want to stress the deficits. Instead, in this book, we strive to make concrete suggestions on how meaningful cosmopolitan approaches could look like in the multiple subfields that constitute communication studies. In doing so, we rely on Waisbord's (2015) assessment that "cosmopolitan scholarship is not reduced to being hospitable to 'international' research. Instead, it is a globalized perspective that critically considers world differences to probe theoretical arguments and define empirical questions" (pp. 185–186). In other words, it is not enough to be aware of and acknowledge research and knowledge produced in non-Western regions but to systematically incorporate it into dominant (Western) approaches and understandings of communication processes. Moreover, such an approach implies including inter- and transdisciplinary perspectives and struggling with the problems arising therefrom, or, as Carpentier et al. (2020) have put it, "cosmopolitan researchers look across and share disciplinary, cultural, geographical, linguistic, and structural borders, accepting the challenges that this imposes" (p. 292). A mere diversification of personnel and approaches would hence be an abridging of a "deep internationalization" and a "deep transculturalization" in the aforementioned sense. However, a true enhancement of academic cosmopolitanism calls for an epistemic transformation through theoretical and methodological openness and an examination of the contextual and structural reasons for the invisibility of some voices (Badr & Ganter, 2021).

To pursue this endeavor, we have embraced new perspectives on cosmopolitan research. These perspectives may have arisen from "below" and from "within" communication studies. First, cosmopolitanism from below seeks to identify and take seriously the perspectives stemming from non-privileged contexts. Without a context sensitive cosmopolitanization from below, a Eurocentric view, with its inherent projection of an occidental superiority, is the most likely outcome (Gunaratne, 2010; McQuail, 2000; Wang, 2011). Waisbord and Mellado (2014) referred to this as a call to review the "subject of study," "body of evidence," and "analytical frameworks" of communication research. They demanded a "shift in the analytical mindset" (p. 365). They argued that researchers should be "curious about the applicability of concepts, theories, and arguments across settings, aware of the impact of particular conditions on academic production, and modest about the generalizability of conclusions" (p. 365). Second, a cosmopolitanization from within communication studies demands non-coercive and egalitarian communication studies. Here, awareness of one's own (Occidental) identity and positionality plays a decisive role in the perception of other contexts (Hantrais, 1999; Said, 1996) because "scientists, like other observers, hold a myriad of preconceptions and biases about the way the world operates" (Kim, 2007, p. 280). With reference to Beck and Sznaider (2006), a cosmopolitanization from within also means including perspectives from superdiverse societies that have emerged in the West. Ultimately, this is a call to change academic cultures and their dominant ideas of knowledge production (Alves & Medeiros, 2021, pp. 4–7).

In terms of a definition, cosmopolitan communication studies are the result of a "deep internationalization" that goes beyond a simple gathering of case studies around the world. A cosmopolitan approach in communication studies is characterized by valuing the "common bonds" that shape communication phenomena at various times and places, albeit "recognizing differences and shared conditions" (Waisbord, 2016, p. 880). Cosmopolitanism is meant to "really benefit from the interconnectedness that globalization provides communication studies with" (Alves & Medeiros, 2021, p. 5). It can enrich the discipline with new angles of knowledge through epistemic diversity (Mutsvairo et al., 2021). Also, it is a way to expand analytical frameworks by incorporating comparative expertise. A cosmopolitan approach in communication studies is thus a normative horizon of inclusive knowledge production from all world regions.

Why do we lack cosmopolitanism in communication studies? A deficit analysis

The Western gaze

When browsing through the prominent and most highly ranked journals and textbooks in communication studies, the obvious finding is the predominance of the Western gaze in research and a pronounced US- and Eurocentrism. Not only do the theories and concepts primarily come from the West, but samples and examples almost exclusively stem from Western Europe, North America, and Australia (de Albuquerque, 2021; Ganter & Ortega, 2019; Hanitzsch, 2019, p. 215). Even if research from non-Western scholars or on non-Western examples is included in the dominant journals, most of such research focuses on building cumulative knowledge and generalizability, hence the replication and application of well-established typologies, indicators, and categories predominantly developed in and for Western contexts. Despite the rich academic traditions in regions such as the Middle East, post-Soviet states, South America, and East Asia, English-speaking communication studies have largely overlooked academic production schools from these areas (Demeter, 2017; Kim, 2009).

We should question approaches that rely on US- and Eurocentrism, not only but also because they prevent us from finding more convincing explanations. Referring to the economic crises and the rise of populism in the West that took many researchers by surprise, the South African scholar Herman Wasserman argued: "I think there's a fundamental realization that still has to dawn on many people in the Global North that you have to find the answers somewhere else. And I think that's the work, that this sort of internationalization has to do: to shift that perception" (Grüne et al., 2024, p. 14).

However, the current political economy of academic systems and the strong Western funding schemes are part of the problem rather than part of the solution. They often foster short-lived research according to sociopolitical requirements of Western countries to deal with current crises. Research on countries beyond the Western world often follows a kind of a "market" and "crisis" logic. For example, when there is political attention in the West toward the South because of a revolution (such as during the so-called Arab Spring in 2011), a war or militant conflict in which the West is involved (as in Ukraine in 2022, Iraq in 2003, or Afghanistan in 2001 and 2021), or when a health crisis or similar emergency occurs in which Western countries might be affected (as in the case of Ebola or crises that stimulate migration toward the North)—in these cases, academic interest also increases. This results in research foci that are less critical than required and help security agencies and law enforcement authorities to gain information about minorities or countries with vulnerable political systems rather than foci that promote to study transcultural innovations and networks and cross-national communication in the sense of cosmopolitanism (Ashwell & Croucher, 2018). These funding initiatives and organizations also foster an administrative research line that serves the national interests of the donor states and/or funding organizations, such as the EU or the World Bank (e.g., St. Clair, 2006). Thus, the "global impact of western theories cannot be explained only by their intrinsic merits, but as the result of the socialization of scholars from all parts of the world in western educational institutions, and the networks built around them" (de Albuquerque, 2021, p. 180).

Globalization is leading to an increase in the economic, political, and cultural interdependencies of all world regions, mostly driven by a neoliberal logic that impacts our knowledge production. Therefore, communication studies are still shaped by US- and Eurocentrism and an instrumental understanding of learning about the world instead of taking a truly cosmopolitan view.

Lack of contextualization

A typical problem that prevents a cosmopolitan turn in communication studies is that analyses and interpretations of communication processes and media phenomena in non-Western regions often lack adequate contextualization. Contextual knowledge is critical in understanding specific phenomena and actions and attributing meaning to them. Contextualization requires first and foremost scholars with language and cultural expertise about the country and region in focus instead of "parachute" scholars who are not familiar with the media systems and communication cultures they study. The particular characteristics of social, political, economic, cultural, and historical contexts are often not properly acknowledged, and instead, social concepts and categories from other contexts are imposed (de Albuquerque, 2021). Consequently, analyzing media environments, most notably in under-researched regions, requires consideration of context-specific characteristics (Chakravartty & Roy, 2013; Kuo & Chew, 2009, p. 423; Willnat & Aw, 2009) without the assumption of exceptional uniqueness or otherness. Therefore, contextualization and being aware of one's own academic perspective on and relation to the objects of analysis are the basis for cosmopolitan research in communication studies (Hantrais, 1999, pp. 103–104). Cosmopolitan research must thus be designed inductively to be versatile but concrete enough to capture the contexts in question with its inherent values and philosophies (e.g., Miike, 2002). Such a research approach contrasts with the current practice of placing mass communication contexts in predefined pigeonholes, where we shoehorn them into improper models regardless of their actual characteristics (see, for example, the debate about Hallin and Mancini's, 2004 media system typology and its applicability to the Global South in Hallin and Mancini, 2011). Contextualization generally challenges the use of typologies of hegemonic approaches. Consequently, "context" should be more widely acknowledged.

Epistemic violence

The lack of contextualization is not only a problem because it does not allow us to understand properly. It can also lead to epistemic violence, including constant (implicit or explicit) othering and silencing voices in mainstream (Western) research (Spivak, 1988). This concerns approaches and methodologies at the same time. Othering creates concepts of social reality that reproduce the normative dichotomies of civilized vs. uncivilized worlds (Hall, 1994). The unreflected (or even conscious) use of terminologies that have the power to label and reproduce labels is an essential component of epistemic violence and othering. For example, seemingly "normal" geographic labels carry certain assumptions. The term "post-Soviet" was introduced as a descriptor for countries emerging from the Soviet Union's collapse, initially viewed as a transitional label (Buyandelgeriyn, 2008). However, more than 35 years later, this label no longer accurately reflects the current realities of these countries (Sagatienė, 2023). For example, Ukraine's ongoing fight for independence, particularly its rejection of ties to Russian imperialism and the Soviet legacy, suggests that continuing to use the term "post-Soviet" may overlook the aspirations of these nations. Another such label is the term "Middle East," referring to a heterogeneous region from a geopolitical point describing a region emanating from a Eurocentric gaze. An intuitive counter-question is: Middle of what, and East of what? (El Houri, 2024). Similarly, the term "Global South" is a useful tool for discussing global inequalities, but it too must be reconsidered each time it is used (Haug, 2021). The countries grouped under this term represent a broad diversity of experiences, histories, and political realities. Rather than applying it as a blanket label or as "a fancy equivalent for the idea of 'rest of the world'" (de Albuquerque, 2021, p. 186), scholars should reflect critically on the countries they are describing, taking into account the varied dynamics at play in different regions. The "Global South" label, while convenient, risks simplifying this diversity if not continuously reflected upon and reassessed.

Also, an emphasis on normative and strongly deterministic approaches, such as the "public sphere," "democratic participation," "good governance," "media freedom," or "journalistic objectivity," may result in a devaluation of what is happening in many parts of the world instead of understanding it within its particular context (Ganter & Badr, 2022). In the case of "journalistic professionalism," which, for example, mostly refers to the paradigm of objectivity, a differentiated perspective is needed that takes other values of journalism into account, such as advocacy (Standaert et al., 2021). Waisbord (2016) referred to "translation" as a way to confront "the clash between dogmatism and difference, language slips and gaps, and the possibility of (mis)understandings" (p. 871) that are inherent in academic knowledge production about the "other." Yet, translation in a cosmopolitan sense is not easy to achieve because the underlying foundation for the above described phenomena is the power asymmetries in knowledge production and knowledge transfer.

After two decades wherein there has been a demand for de-Westernization in knowledge production, one still has to conclude that communication studies are dominated by White researchers located in the West (Chakravartty et al., 2018; Ganter & Ortega, 2019; Alves & Medeiros, 2021). One reason for this is a highly unbalanced knowledge production and transfer. Research has proven the power of geopolitical position in the production of academic knowledge. For example, the location of scientific journals matters regarding their visibility; whether a journal is located in the USA or Europe or elsewhere makes a difference (de Albuquerque, 2021). The limited or even (im)mobility of scholars from the Global South due to visa restrictions and a lack of funding prevents them from gaining equal access to participation in international conferences or scholarship programs (Axyonova et al., 2022). In addition, as in many disciplines, English is seen as the lingua franca for publishing in communication studies—yet, it is questionable whether its almost exclusive use in science enables the free circulation of knowledge. Suzina (2021) described the hegemony of English as a "sterilization of scientific work" (p. 171). "English only" prevents many scholars from describing their ideas and local concepts with specific wording, which in turn, leads to the fact that literature reviews often only consider some relevant texts because they are published in English. Yet, it is precisely through language that an alternative reality to mainstream English-based communication research can be created (Demeter et al., 2022). What is more, many researchers in the Global South lack the means of paying for professional English copy editing, which often results in a blowback in the peer review process of the most important and highly ranked journals that are exclusively in the hands of European and US-American publishing houses (Goyanes, 2020). Despite a transformation trend toward more editorial diversity in gatekeeper positions and "mindful inclusiveness," the road toward true epistemic diversity is still long (Rao, 2019). Here again, the commodification of publishing comes into play: publishing houses' main customers are financially potent Western universities for whom they want to produce allegedly relevant knowledge, that is, local Western knowledge with which consumers can easily identify (de Albuquerque, 2021). However, instead of consumer-centric knowledge production for Western scholars, we need more context-based knowledge to broaden the scope.

Restricting methodologies

Research methodologies often fail to reflect the realities of the Global South because they are not sufficiently adapted to the specific contexts and challenges of conducting research in these regions. Traditional methodologies, largely developed in the West, assume the availability of reliable and comprehensive statistical data, such as media outreach, circulation figures, the number of journalists and media outlets, or media ownership structures. However, in the Global South, this data may frequently be unavailable, unreliable, or incomplete due to a range of structural problems, including limited resources for data collection, weak institutional frameworks, and political constraints. Questions about access to archives, open science, security, and safety of the research subjects, or even questions concerning basic infrastructure such as electricity or the Internet, rarely appear in communication methods training. As a result, methodologies that rely on such data are ill-suited to these contexts, as they fail to account for local complexities or to adapt to the limitations of data collection in these regions (Moyo, 2020).

In addition, due to the highly commercialized publication system described above, scholars are not encouraged to do research in peripheral regions, as established and dominant methods are difficult to apply. Ultimately, these methodological limitations restrict our understanding of media systems in the Global South, leaving vast areas of the world inadequately studied and poorly understood.

However, context sensitive research can be achieved through qualitative methods. Research has shown that outside the West, such research is more present. For example, in Brazil, mostly qualitative approaches were applied to study communication phenomena during the COVID-19 pandemic, in contrast to the overall use of quantitative methods (Barreto de Souza et al., 2023). Methods must also reflect the possibility that different contexts might follow different realities. Therefore, for example, a historical perspective that accounts for the transformative dynamics of communication can be helpful in embracing the possibility of the plurality of path dependencies (e.g., Roudakova, 2011).

How to achieve cosmopolitanism? A tentative guideline

The obstacles to increased cosmopolitanism in communication studies are thus structural and related to a West–South asymmetry in knowledge production, but they also stem from an ignorance that is entrenched in the comfortable position of Western scholars. That ultimately means that we as scholars should challenge ourselves to rethink our epistemic comfort zones and dare to create epistemic discomfort by pushing the boundaries, thus enabling a more inclusive, context sensitive understanding of media and communication phenomena and thereby slowly changing the structural imbalances.

While aiming for true and deep internationalization, we should not make the mistake of essentializing the West or instrumentalizing the call for de-Westernization to push nationalist or identity-focused explanations. Waisbord and Mellado (2014) warned us of "academic commissars patrolling the borders of legitimate knowledge" (p. 368), and Ray (2012) called upon us to de-Westernize, "but with a critical edge" (p. 238). Despite the structural and epistemic problems outlined above, we should also not forget that there are several developments that we can build on when aiming to achieve cosmopolitanism in communication studies.

Incorporate the flow of people and ideas into knowledge production

There is a historically grown and constant transnational flow of people in academia—voluntarily or by force (see Axyonova et al., 2022)—that helps to bring in new perspectives and shake up established explanations. Instead of mainstreaming their ideas and approaches according to "our" academic systems, communication scholars in the West could gain a more cosmopolitan perspective by placing the knowledge of "others" at the center of their studies (Richter et al., 2024).

At the same time, tokenism and a false understanding of representation through identity attribution should be avoided. A local scholar is not per se better equipped to explain certain developments than a foreign one, nor do they necessarily represent the heterogeneity of explanations that might exist in a specific context. Once engaging in cooperation with and while seeking the inclusion of non-Western voices, scholars should critically reflect on whether they select only those voices that fit their expectations and confirm their pre-established explanations. To achieve true cosmopolitanism, Western scholars might have to allow conflicting voices and resistance to prevalent ideas (Richter et al., 2024; Volk, 2021). Cosmopolitan academic environments thus need to be open to voices that critically reflect seemingly taken for granted perspectives and standards from "other" points of view.

This call for acknowledging and engaging with different ways of thinking is connected to the need for a much stronger inclusion of the flows of ideas circulating around the world. These flows of ideas might not always be easily accessible because of a lack of translations or because they are not published by the dominant publishing houses. Yet, Western scholars should strive to gain this knowledge by expanding their scope when conducting literature reviews by including cooperation partners with respective language access, channeling funds into translation and publishing open access in different languages, and making creative use of tools of artificial intelligence. The call for "mindful inclusion" (Rao, 2019, p. 698) through deliberate and responsible citation practices aims at fostering justice and adding visibility to sources from marginalized groups who have credible and authentic expertise. As citation—in the neoliberal publishing scene—serves as academic capital through metrics, we need to shift to inclusive citational practices as criteria for excellence and diversity.

Context matters: Engaging in comparison and interdisciplinarity

As communication studies build bridges to other disciplines, such as political science, sociology, area, or cultural studies, drawing on more open, transnational, and postcolonial flows of ideas in these disciplines can advance a cosmopolitan understanding of communication studies as well. Hafez and Grüne (2022) showed, for example, how such an interdisciplinary perspective helps us to understand communication phenomena on the macro and micro levels. They borrowed from political science and area studies to explore broader systemic contexts, and at the same time, included sociological and cultural perspectives to understand everyday life communication.

Here, we underline the need for comparative research in particular. Comparisons serve as a central epistemological strategy for uncovering and understanding conditioning contexts (Thomaß, 2016, p. 51). Working with inductive comparisons can provide scholars with starting points for understanding the configurations of media environments and avoiding a Eurocentric bias by using a hermeneutic analysis of latent structures and processes (McLeod & Blumler, 1987, pp. 314–316). Through comparison, scholars can reflect on normative ideals and not only learn about different research realities but also potentially change how they perceive and produce analytical categories beyond hegemonic concepts and theories. Comparison creates an inclusive matrix of knowledge, especially when designing research across the most similar and most different designs. Yet, as de Albuquerque (2021) has warned us, when engaging in comparison, we should refrain from defining "our analytical subjects in the function of their negative relation with regard to the Anglophone western standards" (p. 185).

What seems to be most important for deep internationalization and true cosmopolitanism is seeking (local) context knowledge before jumping to conclusions. For example, the digital age has given us unprecedented access to vast amounts of data, particularly from social media and online platforms. However, the sheer volume of data does not necessarily lead to deeper knowledge. The risk with big data research is that it can become decontextualized, overlooking the cultural, political, and social contexts that shape these data points. The assumption that more data means better generalizability, scalability, and predictive power often overlooks that smaller-scale, context-led research provides a better nuanced, grounded understanding (Powers & Vera-Zambrano, 2018).

An example from transcultural communication for the need for a solid understanding of context is the research on media productions of minority groups in Western societies, particularly if these groups come from countries and regions from the Global South. We need to acknowledge that the migration situation influences people's political expressions about their home countries. Yet, we cannot assume that they reconstruct different political camps within the migration situation nor that they seek a separation from the host country's society, as is often done in simplified explanations. Rather, we have to take into account transcultural communication flows, traditional methods of political critique, and cultural expressions to understand minority communication from a cosmopolitan perspective (Horz, 2014). This shows how much context knowledge matters.

Adapting methodologies and research foci

Diversifying contexts and research objects also demands an adaptation of methodologies and research foci. A one-instrument-fits-all methodological approach is doomed to fail in a cosmopolitan setting. In order to understand transnational and global patterns and common experiences, comparative studies should take center stage, aiming to examine phenomena across countries, regions, or social groups. In particular, context-focused research, for example, that uses ethnographic and participatory methods should be considered, that is, methods that "explore and influence social reality in partnership" (von Unger, 2014, p. 1). This means that research is no longer just about people but "with" people. This leads to a double objective: the participation of social actors as co-researchers, as well as measures for individual and collective empowerment (von Unger, 2014, p. 1). In particular, the targeted inclusion of marginalized groups of people and actors in the research processes can bring about a redistribution of power relations in the shaping of society, result in the reduction of marginalization, and thus strengthen social cohesion (Hamidi & Mielke Möglich, 2021). Another important approach is not to dismiss research in closed or invisible contexts just because research is difficult as many gatekeepers, that is, reviewers, editors, and funding entities, do (Badr, 2024). An intersectional perspective is conducive to analyzing the different variables, such as gender, race, religion, or ethnicity, that create social inequalities (Crenshaw, 1989). Critical approaches, such as feminist or postcolonial studies, also expand our research with their distinct focus on power imbalances and social inequalities.

Immersive approaches, such as participatory action research (PAR), could be a technique to counter the abovementioned lack of data that scholars may be confronted with—in global contexts but also in researching transcultural communication phenomena situated in Western superdiverse societies (Vertovec, 2021). Participatory designs can produce situated knowledge by collaborative analysis (Cornish et al., 2023). This means the ability first to analyze structures of inequality and second to understand and include standpoints of marginalized and vulnerable research subjects (Harding, 1991).

Moreover, research should not focus solely on digital or mediatized communication, as many vital social and communicative processes occur outside the digital realm, and even beyond mediatized publics. A more comprehensive approach to research must include these offline communication practices, providing a fuller picture of the social dynamics at play. A self-reflexive approach must be taken throughout the research process to reflect on the power structures underlying the research—including institutional power, historical or (post)colonial legacies, gender or ethnic disparities, unequal distribution of resources, etc.—and how they may influence one's own research.

What is the contribution of this book? Incorporating cosmopolitanism into the subfields of communication studies

After outlining the challenges and potential avenues for achieving a cosmopolitan turn in communication studies, the subsequent chapters aim to facilitate the realization of this turn across various subfields within the discipline. Many scholars and students based in the West feel the urgency to de-Westernize or adopt a cosmopolitan approach, yet they often lack concrete guidelines beyond general and abstract propositions. In response, we have chosen to illuminate various subfields of communication studies, with the aim of demonstrating how a cosmopolitan perspective can be applied within these domains. While it was beyond the scope of this book to address every subfield, we have tried to be as comprehensive as possible by including areas such as journalism research and education, political communication, media ethics, media governance, media system research, crisis communication, war and conflict coverage, interpersonal communication, media development, and media diversity studies. Each chapter seeks to illustrate how a cosmopolitan perspective can enhance our understanding of both global and local communication processes, which are ultimately interconnected. The frameworks presented provide a clearer and more concrete understanding of how the challenges outlined earlier can be addressed. These chapters frequently draw on best practice examples and engage with relevant literature to facilitate a seamless transition from a Western-centric to a cosmopolitan approach for all interested scholars in the field of communication studies.

Barbara Thomass deals with **media ethics**, one of the crucial subfields of communication studies. Building on media ethics, one can analyze communication and media practices in light of the norms and values that ought to guide them. Before being able to do this, however, these norms and values need to be defined and agreed upon. But are norms and values universal? How can we deal with universalism from a cosmopolitan perspective that should be sympathetic to and acknowledge the differences of possibly contradicting norms and values? In her chapter, Barbara Thomass takes us on a historical journey to understand the idea of universalism and its underlying normativity. She raises the question of how to strike the balance between normative universalism and contextualizing cosmopolitanism. She argues that we must consider different perspectives while avoiding epistemic violence and othering in defining ethical norms.

Christine Horz-Ishak guides us into the subfield of **media diversity** studies. She explains the need for "unboxing" the diversity paradigm in order to achieve a cosmopolitan approach. Picking up critical debates about diversity as the structural production of difference, Horz-Ishak demonstrates that media diversity research often tends to be instrumental, tokenistic, and aligned with official politics, thus lacking a genuinely cosmopolitan perspective. Drawing on the cosmopolitan critique, she develops a heuristic framework for media diversity research, which is useful for unveiling emancipatory communication phenomena of minorities as well as power structures and inequalities within media organizations. This framework offers a foundation for conducting more systematic and contextualized analyses of diversity within the media in the future.

Kai Hafez and Anne Grüne address in their chapter several subfields of communication studies, such as **interpersonal communication**, **media sociology**, and **political communication**. They focus on non-mediated political and social communication from an international comparative perspective as a means to deliberately counteract a technocentric trend in communication studies. They argue that areas of non-mediated forms of communication are crucial to understanding party and government politics but also interactions in everyday lifeworlds. Their approach to cosmopolitanism advocates for a polycentric and interdisciplinary perspective on communication beyond mediatization, aiming to understand global transformations and the role of non-mediated forms of communication within them.

"Context matters" is the message that Melanie Radue, Johanna Mack, and Carola Richter emphasize in their chapter on **media system research.** This is one of the key subfields in communication studies, in which scholars attempt to understand why the media is performing and organized as it is in various countries by analyzing the underlying structures and conditions. The authors argue that current research, however, is often shaped by Western norm-driven deficit analyses that ignore the respective contexts. Context-led approaches that focus on historical path dependencies, power imbalances, and relations are presented as remedies by the authors.

The subfield of **media governance** concerns both the informal and formal processes and practices that determine the framework in which media perform. Similar to the chapter on media systems, Sarah Anne Ganter emphasizes the importance of carefully studying political, economic, and cultural contexts to avoid falling into the trap of hegemonic analyses of media policies. She highlights the significant shortcoming of relying heavily on Western case studies while neglecting research from the Global South. Ganter argues that excluding the Global South leads to overlooking diverse realities and perspectives crucial for understanding media governance, including debates on what media freedom, independence, or privacy rights mean to people and decision-makers in different regions of the world.

The subsequent chapter delves into the subfield of **risk and crisis communication**. Risk and crisis communication are specialized fields of media practice designed to help people understand potential hazards, minimize harm, and maintain trust during and after critical events. In their effort to reimagine risk and crisis communication research through a cosmopolitan lens, Pauline Gidget Estella, Martin Löffelholz, and Yi Xu specifically critique the enduring US- and Eurocentrism in the production of knowledge about crisis communication. They argue that, without a cosmopolitan perspective, risk and crisis communication cannot be effectively practiced or researched in increasingly multicultural societies, especially given the transnational nature of most crises.

In the next four chapters, we refer mainly to **journalism studies** as one of the central subfields of communication studies, including journalism cultures, journalistic coverage and data-driven journalism.

Melanie Radue, Thomas Eckerl, Oliver Hahn, and Beate Illg discuss the need to evaluate different **journalistic cultures** and role models against the specific contexts in which they appear. By highlighting some of what are often referred to pejoratively as prefix journalisms, they explain how it is essential to overcome constructed boundaries in mainstream understandings of journalism research. By highlighting the cosmopolitan potential arising from a review of different types of journalism and a critical analysis of academic typologies, this chapter serves as a guide for conducting a meaningful, context sensitive analysis of journalism cultures.

War and conflict reporting is a distinct area of journalism, one in which othering, biased knowledge production, and blind spots of coverage can be detected, as if under a magnifying glass. The same applies for research on war coverage, which often appears to align with an instrumental Western agenda. Kathrin Schleicher and Aynur Sarısakaloğlu therefore call for empowering non-Western narratives in research on war coverage. They explain why war and conflicts become visible in journalistic coverage and why they do not. Both authors criticize the often biased portrayal of wars and give recommendations for adopting a more cosmopolitan perspective while conducting research on war coverage.

In another chapter, Aynur Sarısakaloğlu examines the **digitalization of journalism** through the lens of algorithm-driven journalism. Automation and artificial intelligence are increasingly shaping newsrooms, although this shift occurs unevenly across the globe, leading to what the author terms an "AI-driven divide." This mirrors the long-standing digital divide in the unequal distribution of technologies between the West and the Global South. However, it is not only about unequal access—issues such as technology-inscribed biases and algorithmic colonialism also require attention. The author argues that addressing these challenges necessitates a cosmopolitan mindset and offers a conceptual framework for advancing cosmopolitan research on algorithm-driven journalism, focusing on ethical and governance questions.

The next chapter then turns to the topic of data-driven journalism, and it explores the subject of **fact-checking in journalism**. Regina Cazzamatta introduces the work of Latin American fact-checkers and the ways in which they combat misinformation. In examining the distinctive features of the Latin American context, she addresses geographic disparities in knowledge production about fact-checking and journalism. Latin American fact-checkers frequently lack access to infrastructure, financial resources, information, and capacity building opportunities, somewhat similar to their counterparts in the West. However, to assess their practices, it is essential to comprehend the specific contexts in which they operate, as their methodologies may diverge from those of their Western colleagues. Cazzamatta posits that universalizing approaches to media systems offer only limited insights into diverse fact-checking practices.

The following two chapters address the subfields of media development (or media assistance) and journalism education, the latter most often being included in media development initiatives, too. In these two subfields, international and cross-cultural cooperation is inscribed per se. Nevertheless, it is crucial to reflect on power asymmetries in knowledge production and transfer, as well as conceptual and methodological Eurocentrism. A cosmopolitan approach offers potential remedies for these problems.

Concerning the subfield of **media development**, Susanne Fengler and colleagues present a critical review of "The industry known as 'media development.'" The authors have been involved in the MEDAS 21 project, a postgraduate program focused on investigating the media's role in peacebuilding, journalism training, and women journalists' perspectives in the Global South. Building on the project's findings, they apply a critical lens to current media assistance practices, particularly the dependencies between donor and recipient countries. In light of shifting geopolitical realities and growing awareness of power asymmetries in the Global South, the authors argue that a cosmopolitan approach to media assistance is essential.

Finally, Mira Keßler with Kefa Hamidi and Beate Illg discuss the challenges and possibilities of **international journalism training**. Western concepts of journalism are typically regarded as the standard for "good" journalism and are emphasized in journalism education globally. However, inequality, as well as context-specific norms, practices, and the daily realities of journalists, may necessitate alternative understandings of journalistic roles. The authors draw on examples from Nepal and Afghanistan to demonstrate that with a cosmopolitan mindset, a hierarchy-free exchange of knowledge between research and practice could be established, making journalism education more meaningful.

These 12 chapters, each addressing specific subfields of communication studies, are framed by two key chapters. The first provides a deeper understanding of the history—and the notable absence—of cosmopolitanism in communication studies. The final chapter explores the future of cosmopolitan communication studies, emphasizing the potential for progress through carefully planned international cooperation.

The chapter on the **historical trajectories** of entanglement and ignorance looks at the factors influencing the process of deep internationalization in the academic field of communication studies more generally. Delving into the history of German, French, and Brazilian communication studies, Stefanie Averbeck-Lietz, Lisa Bolz, and Otávio Daros unfold the historical developments that show why entangled research traditions exist in some communities, while in other communities the variety of research traditions seem to be completely ignored. For the first time, the authors reveal how German newspaper studies during the Nazi era attempted a colonial-style internationalization—efforts that were rightly ignored by other research communities, including the Germans themselves after World War II. In contrast, the French and Brazilian academic communities engaged in extensive exchanges. This historical chapter thus not only helps explain why certain scientific communities are more or less cosmopolitan, but it also highlights that seemingly peripheral communities may offer broader perspectives on media and communication phenomena that deserve recognition.

The concluding chapter addresses the challenges and solutions involved in assembling and working within **international research teams**. Sophia Volk argues that genuinely international and inclusive research teams are crucial for producing meaningful, context sensitive studies. Drawing on extensive research—including systematic reviews of English-language journals, anecdotal reflections, and a qualitative study on comparative communication scholarship—Volk emphasizes that the inclusion of diverse perspectives is key to fostering a cosmopolitan approach in communication studies. As previously discussed, collaboration with scholars from various regions and backgrounds is essential for achieving this. However, such collaborations often face challenges, including misunderstandings, differing work routines, "culture clashes," and power asymmetries, particularly in terms of resources. Volk critically examines these typical issues and offers solutions and best practice examples to help overcome these obstacles.

The chapters in this book present a variety of approaches to a cosmopolitan turn in communication studies, both from "below" and from "within." They not only highlight the challenges we face in expanding our regional and epistemological perspectives but also offer guidelines and examples for addressing these issues. As this book emerged from fruitful discussions within a network of scholars with diverse perspectives and experiences, we hope to extend this dialogue to a broader academic community. We aim for this book to serve as a catalyst for further research, a guide for fostering greater cosmopolitanism in communication studies, and an invitation to join us in an ongoing conversation about the deep internationalization of the field.

Acknowledgements

The editors would like to particularly thank Huyen My Nguyen who contributed tremendously to this book by pre-structuring, formatting, and editing. In addition, we thank the Deutsche Forschungsgemeinschaft (DFG, German Research Foundation) for funding the research network that provided the basis for this book.

References

Alves, C. N. R., & Medeiros, D. (2021). Towards cosmopolitanism in German academia? Shedding light on colonial underpinnings of communication research in a globalized world. *Global Media Journal – German Edition, 11*(1). https://doi.org/10.22032/dbt.49167

Ashwell, D., & Croucher, S. M. (2018). Communication and the Global South. In M. Powers (Ed.), *Oxford research encyclopedia of communication*. Oxford University Press. https://doi.org/10.1093/acrefore/9780190228613.013.649

Axyonova, V., Kohstall, F., & Richter, C. (Eds.). (2022). *Academics in exile: Networks, knowledge exchange and new forms of internationalization*. transcript.

Badr, H. (2024). Researching closed contexts: What we can learn from analyzing so-called constrained, inaccessible and invisible media contexts. In S. Ramasubramanian & O. O. Banjo (Eds.), *The Oxford handbook of media and social justice* (pp. 131–140). Oxford University Press.

Badr, H., & Ganter, S. A. (2021). Towards cosmopolitan media and communication studies: Bringing diverse epistemic perspectives into the field. *Global Media Journal – German Edition, 11*(1). https://doi.org/10.22032/dbt.49164

Badr, H., Behmer, M., Fengler, S., Fiedler, A., Grüne, A., Hafez, K., Hahn, O., Hamidi, K., Hanitzsch, T., Horz, C., Illg, B., Litvinenko, A., Löffelholz, M., Radue, M., Richter, C., Thomaß, B., & Töpfl, F. (2020). Kosmopolitische Kommunikationswissenschaft: Plädoyer für eine "tiefe Internationalisierung" des Fachs in Deutschland: Ein wissenschaftliches Positionspapier. *Publizistik, 65*(3), 295–303. https://doi.org/10.1007/s11616-020-00576-6

Barreto de Souza Martins, F., Yu, J., & Domahidi, E. (2023). A global health crisis with divided research traditions? A comparative review of Brazilian and international research in communication on the COVID-19 pandemic. *Annals of the International Communication Association*, 47(4), 479–496. https://doi.org/10.1080/23808985.2023.2242817

Beck, U., & Sznaider, N. (2006). Unpacking cosmopolitanism for the social sciences: A research agenda. *British Journal of Sociology*, 57(1), 1–23. https://doi.org/10.1111/j.1468-4446.2006.00091.x

Buyandelgeriyn, M. (2008). Post-post-transition theories: Walking on multiple paths. *Annual Review of Anthropology*, 37, 235–250. https://doi.org/10.1146/annurev.anthro.37.081407.085214

Carpentier, N., Ganter, S. A., Ortega, F., & Torrico, E. (2020). A debate on post-colonialism and de-coloniality: Latin American and European perspectives on change and hope. In F. O. Paulino, G. Kaplún, M. V. Mariño, & L. Custódio (Eds.), *Research traditions in dialogue: Communication studies in Latin America and Europe* (pp. 275–293). Media XXI.

Chakravartty, P., & Roy, S. (2013). Media pluralism redux: Towards new frameworks of comparative media studies "beyond the West". *Political Communication*, 30(3), 349–370. https://doi.org/10.1080/10584609.2012.737429

Chakravartty, P., Kuo, R., Grubbs, V., & McIlwain, C. (2018). #CommunicationSoWhite. *Journal of Communication*, 68(2), 254–266. https://doi.org/10.1093/joc/jqy003

Cornish, F., Breton, N., Moreno-Tabarez, U., Delgado, J., Rua, M., de-Graft Aikins, A., & Hodgetts, D. (2023). Participatory action research. *Nature Reviews Methods Primers*, 3, Article 34. https://doi.org/10.1038/s43586-023-00214-1

Crenshaw, K. (1989). Demarginalizing the intersection of race and sex: A black feminist critique of antidiscrimination doctrine, feminist theory and antiracist politics. *University of Chicago Legal Forum*, 1989(1), Article 8. http://chicagounbound.uchicago.edu/uclf/vol1989/iss1/8

Curran, J., & Park, M.-J. (Eds.). (2000). *De-Westernizing media studies*. Routledge. https://doi.org/10.4324/9780203981764

de Albuquerque, A. (2021). The institutional basis of anglophone western centrality. *Media, Culture & Society*, 43(1), 180–188. https://doi.org/10.1177/0163443720957893

Demeter, M. (2017). The core-periphery problem in communication research: A network analysis of leading publication. *Publishing Research Quarterly*, 33(4), 402–420. https://doi.org/10.1007/s12109-017-9535-2

Demeter, M., Goyanes, M., Navarro, F., Mihalik, J., & Mellado, C. (2022). Rethinking de-Westernization in communication studies: The Ibero-American movement in international publishing. *International Journal of Communication*, 16, 3027–3046. https://ijoc.org/index.php/ijoc/article/view/18485

El Houri, W. (2024, September 11). *From MENA to WANA: Why terminologies matter*. Global Voices. https://globalvoices.org/2024/09/11/from-mena-to-wana-why-terminologies-matter/

Ganter, S. A., & Badr, H. (Eds.). (2022). *Media governance: A cosmopolitan critique*. Palgrave Macmillan. https://doi.org/10.1007/978-3-031-05020-6

Ganter, S. A., & Ortega, F. (2019). The invisibility of Latin American scholarship in European media and communication studies: Challenges and opportunities of de-Westernization and academic cosmopolitanism. *International Journal of Communication, 13*, 68–91. https://ijoc.org/index.php/ijoc/article/view/8449

Glück, A. (2018). De-Westernization and decolonization in media studies. In M. Powers (Ed.), *Oxford encyclopedia of communication*. Oxford University Press. https://doi.org/10.1093/acrefore/9780190228613.013.898

Goyanes, M. (2020). Editorial boards in communication sciences journals: Plurality or standardization? *International Communication Gazette, 82*(4), 342–364. https://doi.org/10.1177/1748048518825322

Grüne, A., Wasserman, H., de Albuquerque, A., & Kraidy, M. M. (2024). From islandization to joint internationalization: A conversation about a "deep" internationalization in communication studies between Anne Grüne, Herman Wasserman, Afonso de Albuquerque and Marwan M. Kraidy. *Global Media Journal – German Edition, 14*(1). https://doi.org/10.60678/gmj-de.v14i1.300

Gunaratne, S. A. (2010). De-Westernizing communication/social science research: Opportunities and limitations. *Media, Culture & Society, 32*(3), 473–500. https://doi.org/10.1177/0163443709361159

Hafez, K., & Grüne, A. (2022). *Foundations of global communication: A conceptual handbook*. Routledge.

Hall, S. (1994). The West and the rest: Discourse and power. In S. Hall & B. Gieben (Eds.), *Formations of modernity* (pp. 184–227). Polity Press. (Original work published 1992)

Hallin, D. C., & Mancini, P. (2004). *Comparing media systems: Three models of media and politics*. Cambridge University Press.

Hallin, D. C., & Mancini, P. (2011). *Comparing media systems beyond the Western World*. Cambridge University Press. https://doi.org/10.1017/CBO9781139005098

Hamidi, K., & Mielke Möglich, A. (2021). Kommunikation für Sozialen Wandel. *Publizistik, 66*(3–4), 565–588. https://doi.org/10.1007/s11616-021-00679-8

Hanitzsch, T. (2019). Journalism studies still needs to fix Western bias. *Journalism, 20*(1), 214–217. https://doi.org/10.1177/1464884918807353

Hantrais, L. (1999). Contextualization in cross-national comparative research. *International Journal of Social Research Methodology, 2*(2), 93–108. https://doi.org/10.1080/136455799295078

Harding, S. (1991). *Whose science? Whose knowledge? Thinking from women's lives*. Cornell University Press. https://www.jstor.org/stable/10.7591/j.ctt1hhfnmg

Haug, S. (2021, September 28). What or where is the 'Global South'? A social science perspective. *LSE Blogs*. https://blogs.lse.ac.uk/impactofsocialsciences/2021/09/28/what-or-where-is-the-global-south-a-social-science-perspective/

Horz, C. (2014). *Medien – Migration – Partizipation: Eine Studie am Beispiel iranischer Fernsehproduktion im Offenen Kanal*. transcript.

Kim, M.-S. (2007). The four cultures of cultural research. *Communication Monographs, 74*(2), 279–285. https://doi.org/10.1080/03637750701393063

Kim, M.-S. (2009). Cultural bias in communication science: Challenges of overcoming ethnocentric paradigms in Asia. *Asian Journal of Communication, 19*(4), 412–421. https://doi.org/10.1080/01292980903293338

Kuo, E. C. Y., & Chew, H. E. (2009). Beyond ethnocentrism in communication theory: Towards a culture-centric approach. *Asian Journal of Communication, 19*(4), 422–437. https://doi.org/10.1080/01292980903293361

Lee, C.-C. (Ed.). (2015). *Internationalizing "International Communication"*. University of Michigan Press. https://doi.org/10.2307/j.ctv65sxh2

Mano, W., & milton, v. c. (2021). Afrokology of media and communication studies: Theorising from the margins. In W. Mano & v. c. milton (Eds.), *Routledge handbook of African media and communication studies* (pp.19–42). Routledge.

McLeod, J. M., & Blumler, J. G. (1987). The macrosocial level of communication science. In C. R. Berger & S. H. Chaffee (Eds.), *Handbook of communication science* (pp. 271–322). Sage.

McQuail, D. (2000). Some reflections on the Western bias of media theory. *Asian Journal of Communication, 10*(2), 1–13. https://doi.org/10.1080/01292980009364781

Miike, Y. (2002). Theorizing culture and communication in the Asian context: An assumptive foundation. *Intercultural Communication Studies, 6*(1), 1–22.

Moyo, L. (2020). *The decolonial turn in media studies in Africa and the Global South*. Palgrave Macmillan. https://doi.org/10.1007/978-3-030-52832-4

Mutsvairo, B., Borges-Rey, E., Bebawi, S., Márquez-Ramírez, M., Mellado, C., Mabweazara, H. M., Demeter, M., Głowacki, M., Badr, H., & Thussu, D. (2021). Ontologies of journalism in the Global South. *Journalism & Mass Communication Quarterly, 98*(4), 996–1016. https://doi.org/10.1177/10776990211048883

Powers, M., & Vera-Zambrano, S. (2018). The universal and the contextual of media systems: Research design, epistemology, and the production of comparative knowledge. *International Journal of Press/Politics, 23*(2), 143–160. https://doi.org/10.1177/1940161218771899

Rao, S. (2019). Commentary: Inclusion and a discipline. *Digital Journalism, 7*(5), 698–703. https://doi.org/10.1080/21670811.2019.1634482

Ray, T. (2012). To de-Westernize, yes, but with a critical edge: A response to Gunaratne and others. *Media, Culture & Society, 34*(2), 238–249. https://doi.org/10.1177/0163443711432414

Richter, C., Grüne, A., Hafez, K., Fiedler, A., Behmer, M., Horz-Ishak, C., Badr, H., Litvinenko, A., Hahn, O., Radue, M., Sarısakaloğlu, A., Löffelholz, M., Fengler, S., Illg, B., Hamidi, K., Hanitzsch, T., & Thomaß, B. (2023). Die "tiefe Internationalisierung" der deutschen Kommunikationswissenschaft? Eine Evaluation der Personal- und Forschungsstrukturen sowie der Lehrprogramme deutscher Hochschulen. *Global Media Journal – German Edition, 13*(1). https://doi.org/10.22032/dbt.57904

Richter, C., Litvinenko, A., Priyadharma, S., & Badr, H. (2024). Internationalizing German communication studies: Learning from the world. A conversation between Carola Richter, Anna Litvinenko, Subekti Priyadharma and Hanan Badr. *Global Media Journal – German Edition, 14*(1). https://doi.org/10.60678/gmj-de.v14i1.299

Roudakova, N. (2011). Comparing processes: Media, "transitions," and historical change. In D. C. Hallin & P. Mancini (Eds.), *Comparing media systems beyond the Western world* (pp. 246–277). Cambridge University Press. https://doi.org/10.1017/CBO9781139005098.014

Sagatienė, D. (2023, July 11). Challenging the 'Post-Soviet' label and colonial mindsets: NATO summit in Vilnius. *Verfassungsblog*. https://dx.doi.org/10.17176/20230711-231150-0

Said, E. (1996). From Orientalism. In P. Mongia (Ed.), *Contemporary postcolonial theory: A reader* (pp. 20–36). Routledge.

Spivak, G. C. (1988). Can the subaltern speak? In C. Nelson & L. Grossberg (Eds.), *Marxism and the interpretation of culture* (pp. 66–111). University of Illinois Press.

St. Clair, A. L. (2006). The World Bank as a transnational expertised institution. *Global Governance, 12*(1), 77–95. https://www.jstor.org/stable/27800599

Standaert, O., Hanitzsch, T., & Dedonder, J. (2021). In their own words: A normative-empirical approach to journalistic roles around the world. *Journalism, 22*(4), 919–936. https://doi.org/10.1177/1464884919853183

Suzina, A. C. (2021). English as lingua franca: Or the sterilisation of scientific work. *Media, Culture & Society, 43*(1), 171–179. https://doi.org/10.1177/0163443720957906

Thomaß, B. (2016). Der Vergleich als Metamethode in der Kommunikationswissenschaft. In S. Averbeck-Lietz & M. Meyen (Eds.), *Handbuch nicht standardisierte Methoden in der Kommunikationswissenschaft* (pp. 49–65). Springer VS. https://doi.org/10.1007/978-3-658-01656-2_4

Thussu, D. K. (2013). De-Americanising media studies and the rise of "Chindia". *Javnost – The Public, 20*(4), 31–44. https://doi.org/10.1080/13183222.2013.11009126

Thussu, D. K. (Ed.). (2009). *Internationalizing media studies*. Routledge.

Vertovec, S. (2021). The social organization of difference. *Ethnic and Racial Studies, 44*(8), 1273–1295. https://doi.org/10.1080/01419870.2021.1884733

Volk, S. C. (2021). *Comparative communication research: A study of the conceptual, methodological, and social challenges of international collaborative studies in communication science*. Springer VS. https://doi.org/10.1007/978-3-658-36228-7

von Unger, H. (2014). *Partizipative Forschung: Einführung in die Forschungspraxis*. Springer VS. https://doi.org/10.1007/978-3-658-01290-8

Waisbord, S. (2015). De-Westernization and cosmopolitan media studies. In C.-C. Lee (Ed.), *Internationalizing "International Communication"* (pp. 178–200). University of Michigan Press. https://doi.org/10.2307/j.ctv65sxh2.11

Waisbord, S. (2016). Communication studies without frontiers? Translation and cosmopolitanism across academic cultures. *International Journal of Communication, 10*, 868–886. https://ijoc.org/index.php/ijoc/article/view/3483/0

Waisbord, S., & Mellado, C. (2014). De-Westernizing communication studies: A reassessment. *Communication Theory, 24*(4), 361–372. https://doi.org/10.1111/comt.12044

Wang, G. (2011). Orientalism, Occidentalism and communication research. In G. Wang (Ed.), *De-Westernizing communication research: Altering questions and changing frameworks* (pp. 58–76). Routledge.

Willnat, L., & Aw, A. (2009). The big unknown: Conclusions about political communication in Asia. In L. Willnat & A. Aw (Eds.), *Political communication in Asia* (pp. 216–228). Routledge. https://doi.org/10.4324/9780203885680

Historical trajectories of entanglement and ignorance

German, French, and Brazilian communication studies in dialogue

Stefanie Averbeck-Lietz, Lisa Bolz & Otávio Daros

Introduction

International research is one of the major *buzzwords* in career strategies and research projects, but this does not mean that communication studies have therefore become more "cosmopolitan" in a normative sense of openness for different or even unknown traditions. Various factors such as (mutual) ignorance and isolation can—in part—explain dynamics that we observe today: when speaking about "international research," often only a certain type of international research is taken into consideration (i.e., comparative project-based research relying on third-party funding). International career paths work well within the Western European and US-American spheres, but hardly beyond. We therefore suggest taking a close look at some of the results of international ignorance and isolation as well as at the lack of transnational academic crossings. This article is a first attempt to write the history of the field of communication studies from a cosmopolitan perspective, while the communities under analysis have been and remain more or less disconnected from each other.

Our aim—not least based on personal experiences as researchers in all three contexts of German, French and Brazilian research—is to show that a concept such as "cosmopolitanism" challenges not only the epistemological and methodological perspectives of communication studies but also its social shape. Who did research with whom about what (e.g., in journalism research or media system research), and was this from a more national perspective or a more inter- or transnational one? What role do language barriers play (Simonson et al., 2022), what role national historical, political, and economic contexts of communication studies (Löblich & Scheu, 2011), not least the policies and politics regarding the universities as organizational and institutional bodies for a cosmopolitan turn?

The inter- and trans-nationalization of the field of communication studies will be regarded from the viewpoints of three national research communities: Germany, France, and Brazil. We also ask if there are relations between them. Again, we state that there exists a certain "German isolation" from France (Averbeck-Lietz et al., 2019) and Brazil

(Averbeck-Lietz, 2023; Ganter & Ortega, 2019; Richter et al., 2023). Until today, this German isolation seems to be deeply rooted in the historical and political context of German Zeitungswissenschaft (newspaper studies) during 1933 and 1945 and the difficult re-establishing of Publizistikwissenschaft (studies on public communication) after 1945 (Klein, 2006; Kutsch, 2010, 2023; Löblich, 2010; Scheu, 2012; Wiedemann, 2012).

Writing the history of the discipline is a kind of memory history (Scheu, 2023). Memory is selective and, in the sense of the sociology of knowledge, related to social, cultural, and generational positions—such are often nationally framed and, even more, related to one's own academic environment. Thus, our assumption is that remembering the history of communication studies in each of the countries under focus is somehow different and disconnected from each other, even if there may be common traditions with regard to the traveling of ideas and persons. We aim to reveal both the entanglements and the disconnections to support future international research, taking into consideration today's lack of mutual recognition and knowledge about diverse academic traditions, which often hinders international research teams (Volk, 2021).

Even though international comparisons have advantages in helping us to better understand differences, we do *not* want to impose a frame for comparison.[1] Instead, we want to emphasize and understand the (lack of) international and cosmopolitan dynamics within the broader field of communication studies. Why does "border crossing" (Wessler & Averbeck-Lietz, 2012) fail in the domain of research?

We want to contribute to the study of both the international perspectives and influences in Germany, France, and Brazil (the idea of the international), as well as the history and the structure of the academic discipline (the social configuration of international work; see Averbeck-Lietz & Löblich, 2017). Why focus on these three countries? While German communication studies have their roots in the first third of the twentieth century, the French and Brazilian academic disciplines were founded much later. French communication scholars are working internationally, but less within the Euro-American sphere, and at the same time, the structure of French academia favors French publications and career paths. Brazilian communication studies meanwhile have had a strong orientation toward international research, especially in its beginnings, gaining influence from including but not limited to German and French scholars. Despite a strong academic and intellectual exchange with France, Brazilian scholars still struggle today regarding international recognition on a horizontal level.

In respect to Germany, we highlight the roots of the early internationalization before 1945, which had an impact on the post-1945 period in the Federal Republic of Germany, also known as West Germany. France is a near neighbor of Germany and their interrelations are relevant to examine, especially as there seems to be a lack of broad understanding between those two countries in comparison with other neighboring countries, such as Switzerland, Austria, or the Netherlands. Brazil, the third country of our case study, has allowed us to have another perspective as there have been relatively few exchanges between Germany and Brazil after 1970, while there has been much more exchange and scientific influence between France and Brazil, countries that share some

1 For a comparative approach regarding concrete research fields under international comparison, such as public opinion research or reception analysis, see Averbeck-Lietz & Löblich (2017).

intellectual traditions. The three authors of this article come from German, French, and Brazilian backgrounds, while all have transnational knowledge in and about each of the three countries.

Our aim is to delve into international exchanges and transnational knowledge circulation that go beyond what is often perceived in Germany, which has a strong Anglo-Saxon approach to communication studies (Koivisto & Thomas, 2007) and therefore often excludes other perspectives. In the first part, we want to examine an unknown or forgotten part of German international communication history. Usually, the epistemological turn during the 1960s and then a new orientation toward American communication studies are put forward (Löblich, 2010), but there is still an undiscovered chapter of German international communication research before 1945 that overlapped with colonial studies. In the second part, we explain international research practices from a French point of view. In 2023, the USA-based journal *History of Media Studies* published a special section on (lacking) French–German communication research.[2] We want to explain why French scholars might lean toward certain international practices and cooperation rather than toward others. In the third and last part, we depict some intellectual exchanges and influences between Germany and Brazil as well as between France and Brazil—two academic interrelations that differ from each other. Taking into account a Brazilian perspective in international research is not common and underlines the limitations regarding the horizontalization of internationalization.

The authors of this article cannot argue from a systematic research project or commonly selected sources; this article has to be seen as a step in the direction of cosmopolitan research to understand entangled histories being connected and disconnected at the same time (Löblich & Averbeck-Lietz, 2016). Certainly, the German "case" is in fact not a case "for" but an exception: Nazi newspaper studies were supportive of the state and, at least in Germany, they were "forgotten" or even silenced, thus representing blind spots in the collective memory of the field of modern communication studies (Duchkowitsch et al., 2004).[3] To the best of our knowledge, we are able to show for the first time that beyond propaganda research, one of the research objects of National Socialist newspaper studies was indeed international media but it was structurally embedded into the Nazi ideology and its colonial aims.[4] This was the burdened background of establishing new connections between Germany and other countries after 1945.

---

2 See https://hms.mediastudies.press/volume-3---2023. The special section deals with the lacks and some entanglements in political communication, media system research, media and cultural studies, and journalism research. Lisa Bolz authored the chapter on journalism studies.

3 In March 2024, on the occasion of the National Congress of the German Communication Association (DGPuK) at the University of Erfurt, this Nazi past was again a topic for various reasons, not the least of which was to inform the youngest generation of researchers about and encourage further research on it.

4 This article focuses on media mapping on the African continent; for South Eastern European research embedded into the Nazi ideology, see Heinelt (2003) and Höpken (2021).

Germany: An internationalization before 1945 hindering international cooperation after World War II

The strong social science quantitative orientation, which has national German peculiarities (Löblich, 2010), particularly when compared with France and Brazil, and the organizational, theoretical, and methodological separation between *communication* and *media* studies in German universities (Wagner, 2023), was often not very compatible with international research during the twentieth century. One may think of cultural studies and their sluggish adaptation in Germany (Schwer, 2005; Tröger & van den Ecker, 2023), especially Latin American cultural studies, which remain largely unknown with their approach of empowerment and objects of study such as popular culture and inequality (Massmann, 2004; Rüdiger & Escosteguy, 2017). Such transnational lines only opened up from the 1990s onward with the adaptation of cultural studies in Germany (Hepp et al., 2015). In recent decades, critical (Güney et al., 2023) and normative approaches (Karmasin et al., 2013; Zillich et al., 2016) as well as inter- and transnational (Wessler & Brüggemann, 2012) and comparative research (Pfetsch & Esser, 2014; Thomaß, 2013) have been established in German-language communication studies, the latter predominantly in media system and journalism research and often based on standardized methodologies, while the comparative focus is (still) relatively restricted to Anglo-American, US-centered communication studies and research (Koivisto & Thomas, 2007; Richter et al., 2023).

If we want to understand the German case, we have to keep in mind the position of German Zeitungswissenschaft as an international outlaw after 1945, while the Nazi newspaper studies via certain actors' constellations built continuities during the 1950s and 1960s in German academia and journalism (Hachmeister, 1998; Kutsch, 2006, 2023). Recently, Francisco Rüdiger (2019) revealed the Nazi past of German newspaper studies to the Brazilian public through work with primary sources.

Again, more hidden, there is a history of newspaper and radio studies prefiguring international media system research. It is contextualized by a "combination" of colonial, imperial, and national socialist perspectives, again hindering neutral international perspectives after 1945 and cosmopolitan ones even more.

The "Africa Books" and articles of German newspaper scientists before 1945

In this chapter, the past of German communication studies can only be marked by a few first case studies analyzing crucial publications dealing with international topics by prominent German scholars. There is an urgent need for deeper research on this type of early international outreaching, but definitely *not* cosmopolitan research.[5] On the contrary, it is to reveal the "dark side" of internationalization combined with the normalization of racist and national socialist positions after 1933.

5 Until now, there is no systematic research on international media research during the Nazi times. The journal *Zeitungswissenschaft* (*Newspaper Studies*) as the leading journal in the field from 1926 through 1944 would be a first source.

Walter Hagemann's critique of "colonial methods" from the late 1920s and its reframing within Nazi-vocabulary

> Afrika ist kein Niemandsland mit einigen Millionen intelligenter Arbeitsmaschinen, das man nach Belieben aufteilen und 'entwickeln' kann. . . . Die Emanzipation des Schwarzen wird vielleicht einmal das Ende des weißen Kolonisators sein.[6] (Hagemann, 1929, pp. 17–24)
> Eine Weile schien es den fortschrittsgläubigen Europäern, als ob die Welt weiß geworden wäre und Europa der Kraftmittelpunkt der Weltkugel sein und bleiben werde. Beide Annahmen erwiesen sich als falsch.[7] (Hagemann, 1943, p. 5)

In 1929, Walter Hagemann (1900–1964), a German journalist, later a founding figure of German Publizistikwissenschaft wrote *The Revision of Colonial Methods in Africa*. It was a critical book on how colonialists handled and abused their power (Hachmeister, 1987, p. 139). Nevertheless, Hagemann's first two books (1928, 1929) on colonial action are not at all free from racial stereotyping, neither from the time typical N*word (he also used the terms *schwarze* [black] and *farbige* [colored] people), nor from asserting the idea of backwardness and "uncivilized" black people (Wiedemann, 2012, pp. 90–97). Hagemann's own attitudes were nationalistic (Hagemann, 1929, pp. 58–59), pro-Catholic and pro-missionary (pp. 31–35), and anti-communist (pp. 24–35).

Hagemann did not write a book on media or communication but about observations of "development" (Hagemann, 1929, p. 17): on black trade unions, racist and classist (in)justice, criminal land "reforms," unequal payment of white and black workers (p. 9), the separation of settlements (the so-called "color bar," p. 76), and on black people who had "no political rights" (p. 10). Hagemann denounced the "theory" of the inferiority of the black race as an ideology fostering abuse (p. 19). He proposed that white African leaders should consider the example of Brazil as a "eine friedliche Lösung der Rassenfrage" ("a peaceful solution for the question of race," p. 2), noting that Brazil was a country where black people had been allowed to vote since 1891.[8]

Hagemann denounced the European powers for using propaganda, structural lies, false promises, and the functional abuse of the Christian religion (Hagemann, 1929, pp. 19, 67), by which he meant the strategic communication, structural, and factual violence causing poverty, illness, and death. He explicitly mentioned the *Vernichtungspolitik* (annihilation politics) and *Ausrottung* (eradication) of the Herero people (p. 37). Ninety-two years later, the German Government declared the Genocide of the Herero a fact. Astonishingly, some pages later in his text, we can read Hagemann's impressions about Togo,

---

6 "Africa is not a no-man's land with a few million intelligent working machines that can be divided up and 'developed' as desired. . . . The emancipation of the black will perhaps one day be the end of the white colonizer" [Translation by the authors].

7 "For a while, it seemed to Europeans who believed in progress that the world had turned white and that Europe would be and remain the center of power in the world. Both assumptions turned out to be wrong" [Translation by the authors].

8 Hagemann did not mention at all that only literate male citizens who earned a certain income were allowed to vote at that time.

where—in his words—the majority of blacks were "zufrieden" (content) with the German rule, which overall had been "einwandfrei" (flawless) (p. 84).[9]

Concerning African media, Hagemann observed that there were "Eingeborenen-Zeitungen" (with a pejorative undertone: indigenous newspapers)[10] in many regions which were not at all read nor appreciated by the White populations (p. 10), while—referring to South Africa—museums, theatres, cinemas, and parks were forbidden for "colored" people. English dailies like the South African *Cape Times* and the newspapers of Southwest Africa and Kenya were sources used for Hagemann's books. Yet, it remains unclear how systematically he used the press as a source to develop his argumentation.

During the period of 1938 through 1943, Hagemann wrote another book on Africa, and in some paragraphs, he mirrored the racist and the "völkische" (ethnical) vocabulary of National Socialism (Hagemann, 1943, pp. 58, 105). Not a word on the genocide was included any longer; instead, Hagemann denounced the Herrero for riots. In Hagemann's words, the German colonies had been "robbed" by other European powers and the League of Nations (pp. 51, 144, 163–168). This was a typical but false German common sense at the time, also very visible in articles of the then leading journal in newspaper research *Zeitungswissenschaft* (Dresler, 1940, p. 522).

Hagemann was not known for using Nazi speech. On the contrary, the regime saw the Catholic, nationalist journalist as an "opponent" of it (Wiedemann, 2017, p. 943). Hagemann (1932) denounced the emerging National Socialist Movement as antisemitic and totalitarian. After the victory of the Allied Forces, Hagemann (1948) wrote the first book in German that analyzed the Nazi media system and its planned propaganda. Nevertheless, during the early 1940s, Hagemann's tonality changed: contrary to his first book about the African continent from 1929, in his new book (1943), colonialism was no longer opposed. Accordingly, Hagemann's biographer Wiedemann documented "opportunist" pro-Nazi statements in his writings around 1940. Two years earlier, Hagemann had lost his position as an editor in chief of the Catholic journal *Germania*. It is conceivable that he was trying to write in a more conformist mode to secure his possibilities to gain work (Wiedemann, 2012, pp. 136–146, 2017, pp. 951–954; see also Hachmeister, 1987, p. 157, for more detail).

Two other scholars who—like Hagemann—never became party members of the National Socialist Party (NSDAP) also wrote on Africa: Emil Dovifat[11] and Kurt Wagenführ.[12]

9 Which cannot be seen as a historical truth; see Habermas (2016) on German violence in Togo.

10 "Eingeborenen-Zeitung" was the time typical notion often used in the journal *Zeitungswissenschaft* (e.g., see d'Ester, 1937, p. 301).

11 Concerning the controversial debate about Dovifat's role after 1933, see Pfeiffer (2018). Based on archive sources, Averbeck (1999, pp. 361–369) argued that he tried to protect his assistant Hans Traub (1901–1943) who had a Jewish background from being expelled from the Berlin Institute of Newspaper Studies. Benedict (1986, pp. 12–14) mentioned that Dovifat was arrested after his speech at a regional Katholikentag (Catholic convention) denouncing the National Socialist race politics in 1934.

12 Karl d'Ester was a professor for newspaper studies at the University of Munich; he visited the African continent in 1936.

The media in Africa: Dovifat, Dietze, and Wagenführ's media mapping from 1942

The first systematic mapping of African media done by a German research team under derogatory terminologies such as "Eingeborenenblätter" (indigenous leaves) (Dovifat, 1942, p. 490) had been realized during the Nazi era by one well-established scholar and two younger researchers: Emil Dovifat (1890–1969), Roderich Eduard Dietze (1909–1960) and Kurt Wagenführ (1903–1987). Dietze wrote a chapter on broadcasting media co-authored by Wagenführ (Dietze & Wagenführ, 1942).[13] In 1933, the young Wagenführ was expelled by the Nazi administration for his liberal attitudes from his position at the radio station *Deutsche Welle*.[14] Some years later, he held courses at the Deutsche Hochschule für Politik (German School of Politics), a Nazi institution at the time. His topics were the organization and the effects of international broadcasting (Kutsch, 1985, pp. 63–64). After 1940, he taught at the University of Leipzig and its overtly Nazi-friendly institute for newspaper studies. Furthermore, he established a journal and a broadcasting research center affiliated with Dovifat's institute in Berlin (Bohrmann & Kutsch, 1976; Ravenstein, 1983). After World War II, Wagenführ worked for German print and public service media, taught courses in several German institutes for communication studies, and was involved in founding the prestigious Hans-Bredow-Institut of Hamburg (Kutsch, 1985, pp. 388–389; Lerg, 1987), but he was also denied a position in the German Public Broadcasting Station, the *NWDR*, by the British Controller in 1947 (Munzinger, n.d.).

Dovifat, Dietze, and Wagenführ mainly conducted desk research. However, at least Wagenführ had been on research stays in North Africa and the Middle East from 1933–1934 (Ravenstein, 1983, p. 99). Their database was fuzzy and consisted of statistics from the League of Nations, the *South African Broadcasting Company*, and the *BBC*. For Dovifat's chapter on the African printed press, a student, Gerhard Steinbrücker,[15] collected secondary data from the collection of the newspaper science institute at Berlin such as the *Handbuch der Weltpresse* (Handbook of the World Press, Bömer, 1937) and from some rare dissertations on press and journalism in "German East Africa" such as the one by Dietrich Redeker (1911–1979) dating from 1937 (Dovifat, 1942, pp. 483, 494–495). Today, Redeker is known as a former journalist for "Rassefragen" (racial issues), deeply rooted in Nazi ideology (Brieden, 2020). Other sources were Adolf Dresler's (1898–1971) articles on colonial media in Africa published in the journal *Zeitungswissenschaft* which served Dovifat without any critical hints as a basis to write his own article. Dovifat's reference author Dresler was the editor of the political journal *Die Oase. Feldzeitung des Deutschen Afrika-Korps*. In 1934, Dresler graduated with a dissertation supervised by Karl d'Ester (1881–1960) at the University of Munich on the Italian press.[16] At that time,

---

13 Dietze became a prominent sports reporter, during the 1940s, also working for the BBC as a German political correspondent, during the 1950s for the German Public Service Broadcasting. See German National Archive, http://www.bundesarchiv.de/oeffentlichkeitsarbeit/bilder_dokumente/01789/index-4.html.de.

14 See "Wagenführ, Kurt" in Munzinger Online/Personen – Internationales Biographisches Archiv, http://www.munzinger.de/document/00000004211.

15 To our knowledge, biographical data are not available.

16 Ten years earlier Dresler's book on Mussolini (1924) reflected differences between fascist and "völkisch"; in another book Dresler (1925) devalued the League of Nations as a capitalist, pro-Jewish organization supporting a "world-finance" system.

Dresler headed the press office of the NSDAP and constantly contributed articles on colonial media, such as the (fascist) Italian and the (Vichy) French press, to the journal *Zeitungswissenschaft* co-edited by d'Ester.

In their two chapters from 1942, Dovifat, Dietze, and Wagenführ wrote in a highly descriptive manner enriched by statistical information about the development of newspapers (languages, ownership, resources, publics) and the very first steps of broadcasting on and for the African continent, including their technical dimensions. Their chapters were edited by the *Kolonialpolitische Amt* (Colonial Political Office) of the NSDAP[17] under the guidance of Franz Ritter von Epp (1868–1946), a former high-ranking military officer in the *Schutztruppe Südwestafrika* (the so-called Protection Force of South West Africa). In this function von Epp had been involved in the murder of the Herero (Wächter, 1999; Zorn, 1959). Dovifat, Dietze, and Wagenführ's collaborating authors (not in their own chapters, but in the book) were situated at the *Reichspost*. The terminology of these authors was highly conformist to the Nazi ideology, categorizing three "types" of technical staff members involved in establishing African telegraphy, postal services, broadcasting, and press: European, Asian and Black workers. The Blacks were seen as "inferior" to all others and stereotyped in a highly dehumanizing way (see Obst, 1942, p. 26).

The two long book chapters of Dovifat, Dietze, and Wagenführ did not include any distancing from colonialism or the neo-colonial ambitions of the Nazi State and/or their editor in chief Franz Ritter von Epp or their collaborating authors from the *Reichspost*. Dovifat, Hagemann, Dietze, Wagenführ, and also d'Ester (1937) used the N*word in the time typical manner. Only Dovifat occasionally put quotation marks: "N*" (Dovifat, 1942, p. 498). Nevertheless, Dovifat confronted the Indigenous population with a common narrative of the "courageous" German colonial "pioneers" ("mutige deutsche Kolonialpioniere"; Dovifat, 1942, p. 493). Dovifat (and/or his student co-writing the article?) saw some African newspapers under "Jewish influence" operating propaganda against Germany (Dovifat, 1942, p. 498). Articles covering the growing South African press and news system while negatively stereotyping the Jewish owners of African media were also published by the journal *Zeitungswissenschaft* (Grünbeck, 1939).[18] In the same journal, Dresler (1940, pp. 524–527) highlighted von Epp's power position with regard to the emerging Nazi colonial politics. Without a doubt, the central journal in the field was read by Dovifat, Dietze, and Wagenführ.

---

17 Not only Dovifat and Wagenführ worked for the Colonial Political Office but also the former doctoral student of d'Ester, Josef Hardy Krumbach (1910–1972) who personally knew Hagemann and Dresler. Krumbach wrote on African topics serving as an editor in chief of the Nazi journal *Kolonie und Heimat* (Averbeck, 1999, pp. 449–452, 562–563). It remains unclear what impact the personal relationships between the newspaper scholars had for their careers in the context of the Colonial Political Office and related organizations.

18 The Nazi past of Max Grünbeck (1907–1984)—another student of d'Ester—mayor of the town of Friedrichshafen from 1951 through 1977 (Vierhaus, 2011) became a public subject recently. See https://www.gruene-fraktion-fn.de/home/artikel/antrag-zur-vergangenheit-von-max-gruenbeck/ and https://www.friedrichshafen.de/buerger-stadt/nachrichten-videos-bilder/alle-nachrichten/detailseite/nachrichten/vor-75-jahren-wahl-des-buergermeisters-max-gruenbeck/.

Why remember the colonialist view in German newspaper studies?

The abovementioned examples show that we cannot ignore this German past when we want to understand the history of internationalization of German communication studies, which started much earlier than 1945.

Our remembering of the past is selective, ignorant, and even isolated from other European communities of research. We know only a little about their memories of newspaper studies under German occupation when German professors came to build the new discipline in the Czech Republic and Austria (Charvát et al., 2023; Jírak & Köpplová, 2017; Kniefacz, 2008). In France, which had been occupied by the Germany from 1940 to 1944, a discipline of communication studies formally evolved during the 1970s much later than in the Czech Republic or Austria, but research topics such as newspaper history, press law, and journalism research existed since the 1930s, rooted in academic disciplines like history and law (Averbeck-Lietz, 2010; Boure, 2002). Similar roots are found in the academic field of newspaper studies in the Netherlands and Belgium (Hemels, 2017; Van den Bulck & Van den Bulck, 2017). Each of the countries had been under German occupation. Why would have Austrian, Czech, Polish, Danish, Dutch, Belgian, and French researchers welcomed their German counterparts with open arms after 1945? Even if there were some (famous) mediating door openers for German research to get back into the international field during the 1950s and 1960s, such as Henk Prakke (1900–1964), Kurt Baschwitz (1886–1986), Alphons Silbermann (1909–2000), and Paul F. Lazarsfeld (1901–1976), this was not common.

France: International academic milieus beyond the Anglo-Saxon tradition of communication studies

Despite being neighboring countries, France and Germany are quite different regarding communication studies and international perspectives and collaboration within this field of research. German scholars are, for example, very visible within the International Communication Association (ICA) and are frequently published in well-known English-speaking journals. When working internationally, most French scholars are cooperating in other international networks and milieus. In the following, we will take a look at the early days of the International Association for Media and Communication Research (IAMCR) and the often forgotten history of this association, which was founded in France, before considering international barriers as well as international French cooperation, especially toward French-speaking countries and Brazil.

International cooperation and international research within French-speaking countries since the founding of UNESCO's Department of Mass Communication and the early days of IAMCR

Although French communication studies were officially founded in 1974, research within this field in France had been done previously (Pélissier & Demers, 2014). UNESCO founded its "International Institute of the Press and Information, designed to promote the training of journalists and the study of press problems throughout the world" (Hamelink & Nordenstreng, 2016) in 1946, setting two primary goals: "establishing training centers for journalists and founding an international organization for the promotion of scientific research on mass communication," (Hamelink & Nordenstreng, 2016) and it additionally published the series *Reports and Papers on Mass Communication*. Important for the further development of the institutionalization of research regarding communication and journalism was the 1956 conference at the UNESCO headquarters in Paris (Hamelink & Nordenstreng, 2016), where a committee was created that paved the way for the later IAMCR. Fernand Terrou (1905–1976), the founding father and first director of the Institut Français de Presse (IFP) at the University of Paris, valued international exchanges and was part of the interim committee that convened the IAMCR founding conference at UNESCO in December 1957. He then became the first president of the IAMCR. It is notable that the early French journalism and communication research was internationally oriented, with Terrou involved in international associations and UNESCO's Department of Mass Communication (Bolz, 2023).

Even though today many French scholars do not participate in the big international conferences for different reasons (financial, career strategies, different methodological approaches, etc.), there are laboratoires de recherche (research laboratories) and individual scholars who actively look for international affiliations, as seen, for example, with the ICA French Chapter, established by four research units in 2024: Crem (University of Lorraine), Cimeos (University Bourgogne Franche-Comté), Cresat (University of Haute Alsace), and Elliadd (University of Franche-Comté, Technological University of Belfort-Montbéliard) (see Raichvarg, 2024).

Various reasons prevent French scholars, especially young scholars, from participating in big international conferences or publishing internationally and therefore being visible in the Anglo-Saxon community. Aspects such as language, concepts, and career (strategies) function as barriers: French is still the major publication language, and within the French communication studies community, French communication journals are more visible than English-speaking journals. When aiming for an academic career in France, it is therefore more important to have published in French journals than in English-speaking journals. Furthermore, methodology is another major difference compared with German communication studies. The Anglo-Saxon approach in communication studies is mostly theory-oriented with a quantitative approach. French communication studies have their roots in literary studies, with Robert Escarpit (literary studies scholar), Jean Meyriat (documentation scholar), and Roland Barthes being the founding fathers (Averbeck-Lietz, 2010, pp. 353–364; Jeanneret & Ollivier, 2004). The methodological approach is mainly qualitative and less influenced by some major theories. These differences can be complementary, but it is often complicated for French scholars to get into conferences such as ICA or into major English-speaking journals due to specific selection criteria that rarely ever tolerate different approaches. Accordingly, only a few French scholars take part in international associations (ICA, ECREA, IAMCR). In addition, French academia is structured differently in comparison to that of Germany. Whereas, in Germany, scholars meet annually at a national conference and at thematic section conferences, French communication studies organized in the

Société Française des Sciences de l'Information et de la Communication (SFSIC) are less structured around thematic sections that are comparable to ICA sections.

Due to the different histories and cultures of the academic discipline, different languages, and different cultural heritages, some authors who are internationally often read and cited are less known and cited in France and vice versa. To give some examples from the French–German context, Niklas Luhmann's work has considerably shaped a part of German communication studies, but only some of his texts have been translated into French (Bolz, 2023). Whereas Pierre Bourdieu's theory on cultural and economic capital seems to be more cited by German communication scholars, his text on television (Bourdieu, 1996) is more read in France (Benson & Neveu, 2005; Krämer, 2023; Pélissier & Demers, 2014; Schäfer, 2004). In general, critical theory is more read within French communication studies than in the German discipline (despite major authors being German). Such different concepts within both academic disciplines lead to different perceptions of certain research objects, with journalism being one example that is seen slightly differently in both countries (Bolz, 2019).

These dynamics create different citation milieus (Averbeck-Lietz et al., 2019) regarding certain authors, theories, and concepts, and they oblige, for example, young scholars to decide in which community they want to be more visible. Even though it is not easy to get a permanent academic position in France, French academia is still more appealing for young French scholars than other university systems where scholars have to work for several years on non-permanent contracts before being able to apply for one of the sought-after professorships—the German system is particularly unforgiving as many talented postdocs have to leave academia and change careers in their 40s. At French universities, the *maître de conférences* positions are permanent and internationally comparable to associate professorships. "Internationalization" and "international profiles" might officially be selection criteria during recruitment processes, but in reality, the candidates have spent a lot of time in French academia before being appointed to an academic position.

France and Germany, despite being neighboring countries, are far from each other when it comes to cooperation and co-working in communication studies (Averbeck-Lietz et al., 2019). The international relations of French communication scholars are more oriented toward some French-speaking countries, especially Belgium and Canada, as well as some other countries, depending on personal connections, relationships, and institutional funding opportunities. The opportunities provided by the Agence Nationale de la Recherche (ANR) shape the research landscape. Big research projects such as Médias19 and ANR's Numapresse, in which Canadian journalism history researchers were involved, are quite visible in a research landscape that is not (yet) exclusively structured by large, funded research projects. The absence of a language barrier enables easy international cooperation, especially between France and Belgium or between France and French-speaking Canada (even with binational calls for projects).

Even though there are strong links between France and other French-speaking countries such as Belgium or Canada, only limited cooperation exists between France and French-speaking African countries within communication studies (the UNESCO chair Pratiques journalistiques et médiatiques. Entre mondialisation et diversité culturelle (Journalistic and media practices. Between globalization and cultural diversity) situated at the University of Strasbourg regularly works with scholars from African countries), the influence is rather unidirectional. The (history of the) entanglement of the media field—with a French impact in the past on the media system in North Africa—is not mirrored by academic interests. France has a quite different history with regard to (de-)colonialization than Germany, and much stronger relations to the Global South, especially Maghrebian Universities (Hammami, 2005; Idelson, 2023). Concepts in racism studies and "alterité" were relevant earlier than in Germany (Averbeck-Lietz, 2018), which did not for a considerable time understand itself as an "immigration society."

The gap between French and German communication studies has been mentioned and explored by several researchers from both countries (Averbeck-Lietz et al., 2019; Bolz, 2019; Hubé, 2020). Some rare French–German cooperation exists within this field but remains the result of the personal commitment of certain scholars, such as the members of the organization committee of the French-German-Suisse PhD congress held at Mulhouse (France) and Basle (Switzerland) in 2019, which was organized by the three national research organizations: the DGPuK, SFSIC, and SGKM. At the same time, there are other more stable international milieus within communication studies that work, such as the exchanges between France and Brazil, in this field of research.

International cooperation beyond language barriers: The case of French-Brazilian academic exchanges

International exchanges and collaboration are quite diverse within French communication studies; thus, we want to closely examine these French–Brazilian academic exchanges. Even though there is no significant exchange between the two countries, some individuals have had a huge impact on French–Brazilian knowledge transfer, and the mutual influence is worthwhile to look at.

Regarding Latin America, there are epistemological and milieu connections, some of which were via forced emigration from Latin American dictatorships to Paris during the late 1960s and 1970s, with Armand Mattelart and Eliséo Verón being two highly influential scholars who came from Latin America to France (Averbeck-Lietz, 2010, pp. 414–446; Massmann, 2004). In 2018, an issue of the French journal *Communication & langages* was even dedicated to Verón and his work "between the worlds" (Gomez-Mejia et al., 2018). Much stronger is the link the reverse: in Brazil, for example, many leading professors in communication studies were "trained" by Michel Maffesoli at Sorbonne University, such as Vera Veiga França (Universidade Federal de Minas Gerais), André Lemos (Universidade Federal da Bahia), Juremir Machado da Silva (Pontifícia Universidade Católica do Rio Grande do Sul), and Luiz Claudio Martino (Universidade de Brasília).

Beyond personal connections, there are common milieus between France and Brazil. Critical approaches are valued much more in France and Latin America, especially as Brazilian communication studies have been influenced by the Frankfurt School and French theories (Rüdiger & Escosteguy, 2017). Major French sociologists, such as Jean Baudrillard, Edgar Morin, and Dominique Wolton, are important in Brazilian research and teaching. However, the French–Brazilian connections go beyond shared theoretical references. Today, many young Brazilian scholars do their PhDs in France and thus contribute to a growing French–Brazilian academic milieu. The personal interest and investment of communication studies professors, such as Nicole D'Almeida, Denis

Ruellan, and Fabio Henrique Pereira, remain indispensable for the exchanges and collaboration in this field of research. But today, too, French–Brazilian exchanges are common in different research areas, such as organizational communication research (Paris Rego de Souza, 2017) or journalism studies with the trilingual journal *Sur le Journalisme, About Journalism, Sobre Jornalismo*, or the Brazil–France-Francophone Belgium Journalism Research Network (Bolz, 2023).

These examples underline that beyond the Anglophone tradition, there are other cultures of exchange and other international citation milieus that shape the field of communication studies. Often, the most important links have been made by influential individuals, but looking at today's research landscape, it is clear that the French–Brazilian milieu goes well beyond the work of some outstanding professors. Even though the language barrier remains for many, the gain within such an active international community is to learn about different traditions and value them for themselves to put cosmopolitan communication studies into reality.

Brazil and Latin America: Diversification of references in the formation of the regional field and challenges of internationalization

France and Germany were the academic traditions that served as a parameter for the creation of Brazilian universities in the first half of the twentieth century. Although the institutional organization of the Universidade de São Paulo (1934) took the Universität zu Berlin as a reference, the Universidade do Rio de Janeiro (1920) followed the Napoleonic model of higher education. Both Brazilian institutions were formed intellectually, notably by the French intelligentsia (Paula, 2002). Among the professors who were part of the so-called French missions at the Universidade de São Paulo were the historian Fernand Braudel (1902–1985), the anthropologist Claude Lèvi-Strauss (1908–2009), and the sociologist Roger Bastide (1898–1974) who worked in Brazil for nearly two decades and was recognized for his study of Afro-Brazilian religions (Bastide, 1960/1978).

When they returned to Europe and the generation of Brazilian academics trained by them came to replace them in their chairs, the area of communication studies did not yet exist in Brazil. Therefore, despite the marked Franco-European influence on the development of social sciences and philosophy in the country (Merkel, 2022), it cannot be said that the Brazilian field of communication was based on the same foundations. From the end of the 1940s into the beginning of the 1960s, there were a few undergraduate courses in journalism that operated in an improvised way within the philosophy faculties, without forming their own intellectual project. In any case, they served as embryos for the experimental schools of communication, when they opened in universities such as São Paulo, Rio Janeiro, and Brasília, in the mid-1960s (Marques de Melo, 1974).

Communication studies emerged in Brazil and other parts of the world largely as a result of the international action of the United Nations Educational, Scientific and Cultural Organization (UNESCO) in the bipolar context of the Cold War (Wagman, 2016). In South America, there were times of coups and military dictatorships that, in addition to censorship, torture, and other forms of violence, imposed changes on the State. In Brazil, one of these reforms took place in 1968 in the context of higher education, when the US-American university organization model began to be progressively adopted (Paula, 2002). Paradoxically, however, on an intellectual level, the period was marked by waves of ideological criticism against North American imperialism, making Marxism a larger current within the humanities, including in communication and journalism studies (Rüdiger & Daros, 2022).

CIESPAL and a common project for communication education in Latin America

Created as annexes to the philosophy faculties, the first undergraduate journalism programs in Brazil not only lacked infrastructure for professional training but also didactic material for teaching basic subjects about the press. Books of a theoretical or practical nature on journalism by both Brazilian authors and by translators of foreign authors into Portuguese were rare. Meanwhile, in Spanish America, the translation of Emil Dovifat's work, *Periodismo* (1937/1959), was already circulating and free from any ideological discussions about his role in Nazi newspaper studies.

In fact, journalism schools in Latin American countries under Spanish colonization were greatly informed by Spain, a country with its own theoretical tradition of journalism. Spanish-speaking countries have historically shared many references, including in terms of bibliography. Brazil, on the other hand, did not benefit from its linguistic proximity to Portugal in this sense, as the field of communication developed there late (Ribeiro, 2011).

The change in this development framework in Brazil and other Latin American countries occurred through the Centro Internacional de Estudios Superiores de la Comunicación para América Latina (CIESPAL), which first promoted the area of journalism and then guided its conversion to communication. The center's headquarters were established in the city of Quito, Ecuador in 1959, and for decades, it was the main meeting point for Latin American scholars (Daros, 2023). For these and other reasons, it can be speculated that the emergence of the field of communication in Latin America as a whole occurred in a more unified way compared to other parts of the world (Waisbord, 2014).

In the first years, the center basically operated under the following dynamic: hiring North American and European professors to teach Latin American journalists and academics, who spent time in Quito usually as UNESCO scholarship holders. In this initial phase, which can be called pre-critical, the Ford Foundation was also an important financier of the actions, under the mediation of Raymond B. Nixon (1903–1997), who replaced the French Fernand Terrou as president of the IAMCR. This articulation reinforced the tendency to benefit visiting professors from the USA and the translation of works into Spanish by such scholars as Wilbur Schramm, Paul J. Deutschmann, Ralph O. Nafziger, and David Manning White. Among the French, the spokesperson was Jacques Kayser (1900–1963), deputy director of the IFP and also a member of the IAMCR.

However, from the 1970s onward, there was a progressive departure from the North American zone of influence, as it began to count on substantial support from the German social-democratic foundation Friedrich-Ebert-Stiftung (FES), as well as eventually from Konrad-Adenauer-Stiftung (KAS) and the Radio Netherlands Training Centre (RNTC). The assumptions of mass communication research became the target of objection among Latin American scholars, who, before claiming their own tradition of thought, made great use of German critical theory and French semiology to denounce the strategies of North American cultural and media imperialism, as did Luis Ramiro Beltrán, Armand Mattelart, and others linked to CIESPAL (Daros, 2023; Massmann, 2004).

Among the Brazilians who arrived at the center as lecturers in the 1960s and 1970s, names such as Danton Jobim, Luiz Beltrão, and Décio Pignatari stand out, all of whom were pioneering figures in the field of communication and information in Brazil. Many others would find opportunities for specialization there before becoming or consolidating themselves as regional leaders in the area: José Marques de Melo, Cremilda Medina, Christa Berger, Doris Fagundes Haussen, Raquel Paiva, etc. Ultimately, however, generations of academics who helped to highlight the emergence of critical awareness and the change in the role played by CIESPAL over the decades have moved from North–South cooperation to the promotion of South–South cooperation, despite the numerous challenges faced by countries in the region.

Intellectual influences from France and Germany on Brazilian scholarship

While intellectual exchange with the USA tended to be labyrinthine, marked by ups and downs (Daros & Rüdiger, 2022; Simonson et al., 2022), a friendly and stable relationship was established with France and its intelligentsia. One of the milestones of the Franco–Brazilian exchange occurred in the context of the Brazilian military dictatorship (1964–1985), a time when doctoral programs were rare in Brazil, but there was a demand for qualified personnel to fill teaching vacancies in schools and departments that were opening in universities across the country.

France was the priority destination of an intelligentsia that, as a rule, rejected the USA due to its association with South American military governments. As a critic of the regime, the case of José Freitas Nobre is representative: a journalist and politician, in addition to being one of the first communication professors at Universidade de São Paulo, he received a doctorate in law and information economics from Université de Paris, under the supervision of the French professor Terrou. His departure from Brazil to Europe was, not by chance, at the end of the 1960s, the period of greatest repression and violation of human rights by the Brazilian regime.

Whether for political or other reasons, a considerable part of the current list of senior researchers in communication and cultural studies in Brazil migrated at that time to Paris for training: Renato Ortiz completed both his master's and doctorate degrees at the École des Hautes Études en Sciences Sociales (EHESS), under the supervision of Edgar Morin and Roger Bastide; Antonio Fausto Neto also received his doctorate at EHESS, but under the supervision of Jacques Perriault; and Muniz Sodré and José Luiz Braga completed both their master's and doctorate degrees at Université de Paris, with Maurice Mouillaud as their advisor.

Exchanges with French academia remained high even after Brazil's re-democratization, resulting in the construction of strong and long-lasting Brazilian–French milieus. This helps explain why philosophers and critical theorists, such as Guy Debord, Jean Baudrillard, Jean-François Lyotard, Paul Ricœur, Roland Barthes, Gilles Deleuze, Michel Foucault, Pierre Bourdieu, Edgar Morin, Bruno Latour, Dominique Wolton, and others, were more present in communication studies in Brazil than North American social scientists, from the tradition of empirical mass communication research (Daros, 2021).

The figure who most probably brought together Brazilian researchers for training was the sociologist Michel Maffesoli, disciple of Gilbert Durand and professor at the Université de Paris. Under his supervision were some of the academics who have played a leading role in the process of consolidation and expansion of communication studies in Brazil over the last three decades, including Vera Veiga França, Luiz Claudio Martino, Juremir Machado da Silva, and André Lemos. These scholars helped disseminate many of the key concepts in Maffesoli's work, such as everyday life, imaginary, and sociality.

This intellectual collective, in general terms, sought to oppose the Frankfurt School's legacy of criticism of ideology and the culture industry, introduced in Latin America by the works of communication scholars such as Antonio Pasquali (Sanchez Narvarte & Komissarov, 2019), but particularly in Brazil, largely thanks to the initiative of Gabriel Cohn (1971). However, with notable exceptions, the main one being Francisco Rüdiger (2004), most Brazilian communication theorists who adhered to German critical theory separated themselves from the tradition over time.

The main example is that of Ciro Marcondes Filho, who was supervised by Dieter Prokop at Goethe-Universität Frankfurt, who was notable for studying the phenomena of communication and journalism under the categories of commodity, ideology, and domination in the 1980s, replacing them with problematics of postmodernity in the 1990s. After 2000, he reinvented himself as a communication theorist based on the philosophy of Edmund Husserl and his phenomenological followers: Emmanuel Levinas, Maurice Merleau-Ponty, and others (Daros, 2022).

On the other hand, Francisco Rüdiger (2019, 2020) not only maintained his interest in the theoretical tradition represented by the Frankfurt School but also expanded and diversified his themes of study based on new connections with German scholarship. As an intellectual historian, he rescued the heritage of *Zeitungs- und Publizistikwissenschaft*, presenting an unprecedented systematization on the topic to the Brazilian public. Also, among German journalism theorists, the best-known name in Brazilian academia has been Otto Groth (1875–1975), whose work was translated by Liriam Sponholz (Groth, 1960–1972/2011).

Another theoretical link between Germany and Brazil was created by Norval Baitello Junior, who received his doctorate at the Freie Universität Berlin under the supervision of Ivan Bystrina and became a specialist in cultural semiotics and media theory in Brazil, being an interlocutor of the work of German professor Harry Pross (Baitello, 2005). Still, the main reference for semiotics among Brazilians is probably Charles Sanders Peirce, whose main interpreter in Brazil among communication scholars is Lucia Santaella, in many of her works in partnership with the German linguist and semiologist Winfried Nöth (Nöth & Santaella, 2007).

Horizontalization of internationalization and its limitations

It can be said that in recent years, there has been a process of horizontalization of international relations. If, in the initial decades of the field, foreign professors came to Brazil to teach and Brazilian researchers went abroad to obtain training, this dynamic, without having changed completely, is much more complex today. This is because projects in the form of partnerships have increasingly emerged aiming at mutual collaboration between scholars of different nationalities, with common research interests. An example previously mentioned is *Sur le Journalisme, About Journalism, Sobre Jornalismo*, a multilingual journal that brings together editors from France and Brazil, as well as Canada, Belgium, and Argentina.

There are large comparative global projects such as the Worlds of Journalism Study (WJS), which, although based in Europe and the USA, have been developed with a concern for inclusion of researchers of as many nationalities as possible. However, it must be admitted that, exceptions aside, Brazilians and other scholars from developing countries are practically absent from the boards of projects that claim to be global and from journals that claim to be international. This organizational aspect implies less representation in the production of knowledge, marginalization of non-Western theoretical frameworks, low acceptance of works that are not based on Anglo-Saxon sources, etc. (Ganter & Ortega, 2019; Goyanes, 2020).

In Brazil, advances in the institutionalization of internationalization have historically occurred through the federal government's funding agencies, such as the Coordenação de Aperfeiçoamento de Pessoal de Nível Superior (CAPES) and the Conselho Nacional de Desenvolvimento Científico e Tecnológico (CNPq), which have programs in partnership with the Deutscher Akademischer Austauschdienst (DAAD). But we see that there is much to be done to achieve effective results and to have true equity in intellectual exchange between countries, that is, for Brazilian and Latin American research to have an impact on international scholarship commensurate with the influence that Western Europe and the USA have had on the development of the global field (see also Cazzamatta in this book).

Conclusion and outlook

In Germany, international perspectives in research extend back to the early days of communication studies—but they were not at all cosmopolitan. This is a memory that has been forgotten or even silenced today, like National Socialism was and still is in so many parts of German society and academia. It took a long time for German research to come back to international science milieus, which in our estimation must be taken into account if we regard German communication studies today. Currently, German communication studies is based on a methodological turn in the 1960s, when North American approaches had an impact on German communication research and when empirical research methods replaced the former normative approaches (Löblich, 2010). International publications and careers are important to those who want to succeed in German academia, but today's perception of "international" is often a limited one and considers mainly the German-speaking countries, the USA, the UK, Australia, the Netherlands, and Scandinavia. Big international conferences of the ICA or the European Communication and Research Association (ECREA) are valued higher than smaller ones—regarding the nationalities that are represented in such conferences, German academics are usually among the top three (Scharkow & Trepte, 2024, p. 24) and publications that are not English or German are very rarely cited within German academia. When promoting international research and cooperation, one must keep in mind that the German international perspective is only one among other perspectives (Richter et al., 2023).

Another perspective on international research is the French one, as French communication studies are more oriented toward the French-speaking community, especially those of Canada and Belgium, with scattered international cooperation regarding certain topics. The structure of French academia (French being the major language in terms of recruitment process, citations milieus, etc.) and the perspective of being able to get a permanent position in France often lead young scholars to stay in France, and strict selection criteria that do not apply to the French academic culture of communication studies often prevent scholars from attending big international conferences such as ICA or getting into the most well-known English-speaking journals within the field of communication studies.

International work and international perspectives in research projects are often qualities that are sought after, but a closer look reveals that the different international perspectives need to be close enough to each other methodologically and theoretically that certain kinds of international visions are not neglected or excluded. The methodological differences, the different citation milieus, and the language barrier between France and Germany are significant factors that make a regular French–German exchange difficult that goes beyond the personal commitment of a small number of scholars.

The exchanges between France and Brazil are more common, and the French–Brazilian milieu is an interesting case study. Even though the influence of French scholars is more important in Brazil than vice versa, some Brazilian scholars have had a huge impact on French academia within the field of communication studies. Today, there are regular exchanges between France and Brazil and scholars who travel between both countries, as well as established French–Brazilian academic projects such as conference cycles or journals.

These positive outcomes and exchanges cannot and should not mask the fact that many research projects take into account data and information from Brazil or other countries from the "Global South," but without Brazilian scholars, for example, being on important boards or editorial committees and the like. This underlines the differences between decolonization and cosmopolitan approaches. Working internationally should mean knowing, accepting, and integrating different points of view and different ways of working (Miike & Yin, 2022). When reflecting on the different dynamics regarding international perspectives, this could mean questioning one's own (imagined) "international community" or the constraints in terms of recruitment processes.

Cosmopolitanism is thus conceptually developed in this chapter in the context of the history of communication studies in order to question diverse traditions, their connected and disconnected paths, and their (normative) views. It is assumed that science per se functions transnationally ("Traveling of ideas," Said, 1983), but the reality shows that this is not arbitrary and takes place along disciplinary, social, historical, economic, (science-)cultural, and (science-)political conditions (Averbeck-Lietz & Löblich, 2017; Löblich & Scheu, 2011; Simonson & Park, 2016). Cosmopolitanism needs openness but also knowledge about other (diverging) perspectives and, not least, an awareness of the lack of cosmopolitanism. This article aims to strengthen such an awareness on the basis of mutual dialogue between different traditions and communities of scholars.

References

Averbeck, S. (1999). *Kommunikation als Prozess: Soziologische Perspektiven in der Zeitungswissenschaft 1927–1935*. LIT.

Averbeck-Lietz, S. (2010). *Kommunikationswissenschaft in Frankreich: Der epistemologische Diskurs der Sciences de l'information et de la communication 1975–2005*. Avinus.

Averbeck-Lietz, S. (2018). Intercultural communication in France. In Y. Y. Kim (Ed.), *The international encyclopedia of intercultural communication* (pp. 1093–1101). Wiley Blackwell. http://dx.doi.org/10.1002/9781118783665.ieicc0082

Averbeck-Lietz, S. (2023). On (missing) links between German, Latin American, and French mediatization research: Reflections on diverse research milieus and their traditions. *Matrizes*, *17*(3), 241–272. http://dx.doi.org/10.11606/issn.1982-8160.v17i3p241-272

Averbeck-Lietz, S., & Löblich, M. (2017). Kommunikationswissenschaft vergleichend und transnational: Eine Einführung. In S. Averbeck-Lietz (Ed.), *Kommunikationswissenschaft im internationalen Vergleich: Transnationale Perspektiven* (pp. 1–29). Springer VS. https://doi.org/10.1007/978-3-531-18950-5_1

Averbeck-Lietz, S., Bonnet, F., Cordonnier, S., & Wilhelm, C. (2019). Communication studies in France: Looking for a "Terre du milieu"? *Publizistik*, *64*(3), 363–380. https://doi.org/10.1007/s11616-019-00504-3

Baitello, N., Jr. (2005). *A Era da Iconofagia: Ensaios de Comunicação e Cultura*. Hacker.

Bastide, R. (1978). *The African religions of Brazil: Toward a sociology of the interpenetration of civilizations* (H. Sebba, Trans.). Johns Hopkins University Press. (Original work published 1960)

Benedict, K.-U. (1986). *Emil Dovifat: Ein katholischer Hochschullehrer und Publizist*. Mathias-Grünwald-Verlag.

Benson, R., & Neveu, E. (2005). *Bourdieu and the journalistic field*. Polity Press.

Bohrmann, H., & Kutsch, A. (1976). Rundfunkwissenschaft im Dritten Reich. *Studienkreis Rundfunk und Geschichte*, *2*(2), 14–20. https://rundfunkundgeschichte.de/assets/RuG_02_2.pdf

Bolz, L. (2019). Recherches sur le journalisme en France et en Allemagne, un dialogue impossible? Regards croisés sur des méthodologies et des développements divergents. *Revue française des sciences de l'information et de la communication*, *18*. https://doi.org/10.4000/rfsic.7702

Bolz, L. (2023). Journalism studies and journalism education in France and in Germany. *History of Media Studies*, *3*. https://doi.org/10.32376/d895a0ea.b732c99f

Bömer, K. (Ed.). (1937). *Handbuch der Weltpresse: Eine Darstellung des Zeitungswesens aller Länder* (3rd. rev. ed.). Armannen.

Bourdieu, P. (1996). *Sur la télévison: Suivi de l'emprise de journalisme*. Raisons d'agir.

Boure, R. (Ed.). (2002). *Les origines des sciences de l'information et de la communication : Regards croisés*. Presse Universitaires de Septentrion. https://doi.org/10.4000/communicationorganisation.2797

Brieden, H. (2020, September 6). *Dr. Dietrich Redeker: Vom Fachjournalisten für Rassefragen zum Heimatschriftsteller*. Radioflora. https://radioflora.de/dr-dietrich-redeker-vom-fachjournalisten-fuer-rassenfragen-zum-heimatschriftsteller/

Charvát, M., Jírak, J., & Köpplová, B. (2023). Studying media in the Czech Republic: Past and presence of Czech media studies. *Publizistik*, *68*(1), 109–130. https://doi.org/10.1007/s11616-022-00768-2

Cohn, G. (Ed.). (1971). *Comunicação e Indústria Cultural*. Nacional.

d'Ester, K. (1937). Streiflichter auf afrikanisches Nachrichten- und Pressewesen. *Zeitungswissenschaft*, *12*(5), 288–307.

Daros, O. (2021). French theoretical and methodological influences on Brazilian journalism research. *Media, Culture & Society*, *43*(8), 1553–1564. https://doi.org/10.1177/0163443721999936

Daros, O. (2022). A critical tribute to Ciro Marcondes' new communication theory. *Communication Theory*, *32*(1), 173–178. https://doi.org/10.1093/ct/qtab004

Daros, O. (2023). CIESPAL and the development of education and research in communication in Latin America. *Studies in Communication Sciences*, *23*(1), 105–120. https://doi.org/10.24434/j.scom2023.01.3334

Daros, O., & Rüdiger, F. (2022). Paradigm shift in mid-twentieth century Brazilian journalism: A negative dialectics of decoloniality? *Journalism Studies*, *23*(13), 1703–1720. https://doi.org/10.1080/1461670X.2022.2083004

Dietze, E. R., & Wagenführ, K. (1942). Das Rundfunkwesen in Afrika. In E. Obst (Ed.), *Post- und Fernmeldewesen: Das Rundfunkwesen. Das Zeitungswesen* (pp. 435–482). De Gruyter Mouton.

Dovifat, E. (1942). Das Zeitungswesen in Afrika. In E. Obst (Ed.), *Post- und Fernmeldewesen. Das Rundfunkwesen: Das Zeitungswesen* (pp. 483–518). De Gruyter Mouton.

Dovifat, E. (1959). *Periodismo* (F. Blanco, Trans.). UTEHA. (Original work published 1937)

Dresler, A. (1924). *Mussolini*. Hammer.

Dresler, A. (1925). *Der Völkerbund der Weltfinanz: Ein Warnruf an Deutschland*. Hammer.

Dresler, A. (1940). Die deutschen Kolonien und die Presse. *Zeitungswissenschaft*, *15*(10), 514–528.

Duchkowitsch, W., Hausjell, F., & Semrad, B. (2004). *Die Spirale des Schweigens: Zum Umgang mit der nationalsozialistischen Zeitungswissenschaft*. LIT.

Ganter S. A., & Ortega, F. (2019). The invisibility of Latin American scholarship in European media and communication studies: Challenges and opportunities of de-Westernization and academic cosmopolitanism. *International Journal of Communication*, *13*, 68–91. https://ijoc.org/index.php/ijoc/article/view/8449

Gomez-Mejia, G., Le Marec, J., & Souchier, E. (2018). Verón entre les mondes. *Communication & Langages*, *196*(2), 9–26. https://doi.org/10.3917/comla1.196.0009

Goyanes, M. (2020). Editorial boards in communication sciences journals: Plurality or standardization? *International Communication Gazette*, *82*(4), 342–364. https://doi.org/10.1177/1748048518825322

Groth, O. (2011). *O poder cultural desconhecido: Fundamentos da ciência dos jornais* (L. Sponholz, Trans.). Vozes. (Original work published 1960–1972)

Grünbeck, M. (1939). Union von Südafrika: Auflösung des Schlesinger Konzerns. *Zeitungswissenschaft, 14*(5), 331–333.

Güney, S., Hille, L., Pfeiffer, J., Porak, L., & Theine, H. (Eds.). (2023). *Eigentum, Medien, Öffentlichkeit: Verhandlungen des Netzwerkes Kritische Kommunikationswissenschaft*. Westend. https://doi.org/10.53291/9783949925092

Habermas, R. (2016). *Skandal in Togo: Ein Kapitel deutscher Kolonialherrschaft*. S. Fischer.

Hachmeister, L. (1987). *Theoretische Publizistik: Studien zur Geschichte der Kommunikationswissenschaft in Deutschland*. Volker Spiess.

Hachmeister, L. (1998). *Der Gegnerforscher: Die Karriere des SS-Führers Franz Alfred Six*. C.H. Beck.

Hagemann, W. (1928). *Gestaltwandel Afrikas*. Germania.

Hagemann, W. (1929). *Die Revision der Kolonialmethoden in Afrika*. Aschendorff.

Hagemann, W. (1932). Faschismus als europäisches Problem. *Zeitschrift für Politik, 31*, 306–318.

Hagemann, W. (1943). *Uns ruft Afrika! Reiseskizzen*. Deutsche Verlagsgesellschaft.

Hagemann, W. (1948). *Publizistik im Dritten Reich – ein Beitrag zur Methodik der Massenführung*. Hansischer Gildeverlag.

Hamelink, C., & Nordenstreng, K. (2016). Looking at history through the International Association for Media and Communication Research (IAMCR). *IAMCR website*. https://iamcr.org/node/3578

Hammami, S. (2005). Les sciences de l'information et de la communication dans le monde arabe : Réflexions sur les difficultés d'émergence d'une discipline. *Revue Tunisienne de communication, 1*(45), 7–42. https://archivesic.ccsd.cnrs.fr/sic_00001746

Heinelt, P. (2003). *'PR-Päpste.' Die kontinuierlichen Karrieren von Carl Hundhausen, Albert Oeckel und Franz Ronneberger*. Karl Dietz.

Hemels, J. (2017). Die Anerkennung und Entwicklung der Kommunikationswissenschaft in den Niederlanden. In S. Averbeck-Lietz (Ed.), *Kommunikationswissenschaft im internationalen Vergleich: Transnationale Perspektiven* (pp. 53–92). Springer VS.

Hepp, A., Krotz, F., Lingenberg, S., & Wimmer, J. (Eds.). (2015). *Handbuch Cultural Studies und Medienanalyse*. Springer VS. https://doi.org/10.1007/978-3-531-19021-1

Höpken, W. (2021). *Wissenschaft – Politik – Biografie: Die deutsche Südosteuropaforschung und ihre Akteure am Beispiel von Franz Ronneberger*. De Gruyter Oldenbourg.

Hubé, N. (2020). À la recherche d'une universalité du journalisme: La *Journalistik* allemande. *Revue Française des Sciences de l'information et de la communication, 19*. https://doi.org/10.4000/rfsic.9269

Idelson, B. (2023). Les sciences de l'information et de la communication en terrains indianocéaniques: Entre déterminisme digital et approche distanciée des concepts. *Communication, 40*(1). https://doi.org/10.4000/communication.17088

Jeanneret, Y., & Ollivier, B. (2004). Les SIC en perspective. *Hermes, 38*(1), 86–88. https://doi.org/10.4267/2042/9429

Jírak, J., & Köpplová, B. (2017). Zur Entwicklung der Kommunikationswissenschaft in der tschechischen Universität und Gesellschaft. In S. Averbeck-Lietz (Ed.), *Kommunika-*

tionswissenschaft im internationalen Vergleich: Transnationale Perspektiven (pp. 239–271). Springer VS. https://doi.org/10.1007/978-3-531-18950-5_9

Karmasin, M., Rath, M., & Thomaß, B. (Eds.). (2013). *Normativität in der Kommunikationswissenschaft*. Springer VS. https://doi.org/10.1007/978-3-531-19015-0

Klein, P. (2006). *Henk Prakke und die funktionale Publizistik: Über die Entgrenzung der Publizistik- zur Kommunikationswissenschaft*. LIT.

Kniefacz, K. (2008). *Zeitungswissenschaft in Wien 1900–1945: Die Institutionalisierung im Kontext der deutschsprachigen Fachentwicklung* [Diploma thesis, University of Vienna]. https://doi.org/10.25365/thesis.1046

Koivisto, J., & Thomas, P. (2007). *Mapping communication and media research: Germany* (Research reports 6). University of Helsinki.

Krämer, B. (2023). How German communication research discovered Bourdieu but missed his potential for the study of (populist) political communication. *History of Media Studies, 3*. https://doi.org/10.32376/d895a0ea.7ec13efb

Kutsch, A. (1985). *Rundfunkwissenschaft im Dritten Reich*. K.G. Saur.

Kutsch, A. (2006). Verdrängte Vergangenheit: Darstellungstechniken und Deutungen der Fachgeschichte im "Dritten Reich" in den Personalien der Publizistik. In C. Holtz-Bacha, A. Kutsch, W. Langenbucher, & K. Schönbach (Eds.), *50 Jahre Publizistik* (pp. 73–112). VS Verlag für Sozialwissenschaften.

Kutsch, A. (2010). Die Entstehung des Deutschen Zeitungswissenschaftlichen Verbandes. *Jahrbuch für Kommunikationsgeschichte, 12*, 121–144. https://www.jstor.org/stable/20852787

Kutsch, A. (2023). Die drei Anläufe zur Gründung der Publizistik. *Publizistik, 68*(4), 519–529. https://doi.org/10.1007/s11616-023-00817-4

Lerg, W. B. (1987). Kurt Wagenführ und die Rundfunkkunde. *Studienkreis Rundfunk und Geschichte, 13*(4), 393–398.

Löblich, M. (2010). *Die empirisch-sozialwissenschaftliche Wende in der Publizistik- und Zeitungswissenschaft*. Herbert von Halem.

Löblich, M., & Averbeck-Lietz, S. (2016). The transnational flow of ideas and histoire croisée with attention of the cases of France and Germany. In P. Simonson & D. W. Park (Eds.), *The international history of communication study* (pp. 25–46). Routledge.

Löblich, M., & Scheu, A. M. (2011). Writing the history of communication studies: A sociology of science approach. *Communication Theory, 21*(1), 1–22. https://doi.org/10.1111/j.1468-2885.2010.01373.x

Marques de Melo, J. (1974). *Contribuições para uma pedagogia da comunicação*. Paulinas.

Massmann, A. (2004). Von "Kommunikation für Entwicklung" zur "cultura mediática". *Publizistik, 49*(3), 275–291. https://doi.org/10.1007/s11616-004-0072-6

Merkel, I. (2022). *Terms of exchange: Brazilian intellectuals and the French social sciences*. University of Chicago Press.

Miike, Y., & Yin, J. (Eds.). (2022). *The handbook of global interventions in communication theory*. Routledge.

Munzinger. (n.d.). *Kurt Wagenführ*. https://www.munzinger.de/search/go/document.jsp?id=00000004211

Nöth, W., & Santaella, L. (2007). Die Relevanz der Peirceschen Semiotik des Vagen für die Theorie der Kommunikation. In S. J. Schmidt (Ed.), *Die (Un)Möglichkeit der Kommunikation* (pp. 73–96). Peter Lang.

Obst, E. (Ed.). (1942). *Post- und Fernmeldewesen: Das Rundfunkwesen. Das Zeitungswesen*. De Gruyter Mouton. https://doi.org/10.1515/9783111435800

Paris Rego de Souza, A. Z. (2017). Une communauté scientifique en construction: La recherche en Communication Organisationnelle, France et Brésil. *Communication & Organisation*, 52, 35–46. https://doi.org/10.4000/communicationorganisation.5640

Paula, M. F. C. (2002). USP e UFRJ: A influência das concepções alemã e francesa em suas fundações. *Tempo Social*, 14(2), 147–161. https://doi.org/10.1590/S0103-20702002000200008

Pélissier, N., & Demers, F. (2014). Recherches sur le journalisme: Un savoir dispersé en voie de structuration. *Revue française des sciences de l'information et de la communication*, 5. https://doi.org/10.4000/rfsic.1135

Pfeiffer, J. (2018). Emil Dovifat. In M. Meyen & T. Wiedemann (Eds.), *Biografisches Lexikon der Kommunikationswissenschaft*. Herbert von Halem. http://blexkom.halemverlag.de/emil-dovifat/

Pfetsch, B., & Esser, F. (2014). Political communication in comparative perspective: Key concepts and new insights. In C. Reinemann (Ed.), *Political communication* (pp. 87–105). De Gruyter Mouton. https://doi.org/10.1515/9783110238174.87

Raichvarg, D. (2024, March 1). REACH-ing out: Bienvenue ICA France. *ICA Newsletter*. https://www.icahdq.org/blogpost/1523657/498383/REACH-ing-Out-Bienvenue-ICA-France

Ravenstein, M. (1983). Kurt Wagenführ 80 Jahre. *Publizistik*, 28(1), 98–99.

Ribeiro, N. (2011). The rise of a new field: Researching communication history in the Iberian countries. *Medien & Zeit*, 26(3), 7–16.

Richter, C., Grüne, A., Hafez, K., Fiedler, A., Behmer, M, Horz-Ishak, C., Badr, H., Litvinenko, A., Hahn, O., Radue, M., Sarısakaloğlu, A., Löffelholz, M., Fengler, S., Illg, B., Hamidi, K., Hanitzsch, T., & Thomaß, B. (2023). Die "tiefe Internationalisierung" der deutschen Kommunikationswissenschaft? Eine Evaluation der Personal- und Forschungsstrukturen sowie der Lehrprogramme deutscher Hochschulen. *Global Media Journal – German Edition*, 13(1). https://doi.org/10.22032/dbt.57904

Rüdiger, F. (2004). *Theodor Adorno e a crítica à indústria cultural: Comunicação e teoria crítica da sociedade* (3rd ed.). EDIPUCRS.

Rüdiger, F. (2019). *Síntese de história da Publicística: Estágios reflexivos da ciência da comunicação pública alemã*. Insular. https://doi.org/10.1590/1982-25532019344198

Rüdiger, F. (2020). *Origens do pensamento acadêmico em jornalismo: Alemanha, União Soviética, Itália e Japão* (2nd ed.). Insular.

Rüdiger, F., & Daros, O. (2022). Marxist thinking and journalism theory in Brazil. *Rethinking Marxism*, 34(4), 538–557. https://doi.org/10.1080/08935696.2022.2139079

Rüdiger, F., & Escosteguy, A. C. (2017). Brazilian research in communication: Historical synopsis and reflexive trends of academic work in an emerging country. In S. Averbeck-Lietz (Ed.), *Kommunikationswissenschaft im internationalen Vergleich: Transnationale Perspektiven* (pp. 359–383). Springer VS. https://doi.org/10.1007/978-3-531-18950-5_14

Said, E. (1983). *The world, the text, and the critic*. Harvard University Press.

Sanchez Narvarte, R. E., & Komissarov, S. (2019). Antonio Pasquali y la Escuela de Frankfurt: Lecturas y apropiaciones (1966–1970). *Chasqui. Revista Latinoamericana de Comunicación*, *141*, 257–274. https://doi.org/10.16921/chasqui.v0i141.3850

Schäfer, S. (2004). Journalismus als soziales Feld: Das relationale Denken Pierre Bourdieus als Grundlage für eine Journalismustheorie. In M. Löffelholz (Ed.), *Theorien des Journalismus: Ein diskursives Handbuch* (pp. 321–334). VS Verlag für Sozialwissenschaften. https://doi.org/10.1007/978-3-663-01620-5_15

Scharkow, M., & Trepte, S. (2024). National diversity at conferences of the International Communication Association. *Annals of the International Communication Association*, *48*(1), 17–36. https://doi.org/10.1080/23808985.2023.2261018

Scheu, A. M. (2012). *Adornos Erben in der Kommunikationswissenschaft: Eine Verdrängungsgeschichte*. Herbert von Halem.

Scheu, A. M. (2023). Fachgeschichte und Erinnerungsforschung: Die Beziehung von Fachgeschichte und Fachgedächtnis am Beispiel der deutschen Kommunikationswissenschaft. In C. Pentzold & C. Lohmeier (Eds.), *Handbuch kommunikationswissenschaftliche Erinnerungsforschung* (pp. 391–406). De Gruyter. https://doi.org/10.1515/978 3110629743-018

Schwer, K. (2005). "Typisch deutsch?" Zur zögerlichen Rezeption der Cultural Studies in der deutschen Kommunikationswissenschaft. *Münchener Beiträge zur Kommunikationswissenschaft*, *2*. https://doi.org/10.5282/ubm/epub.521

Simonson, P., & Park, D. W. (Eds.). (2016). *The international history of communication study*. Routledge.

Simonson, P., Park, D. W., & Pooley, J. (2022). Exclusions/exclusiones: The role for history in the field's reckoning. *History of Media Studies*, *2*. https://doi.org/10.32376/d895a0ea .ed348e03

Thomaß, B. (Ed.). (2013). *Mediensysteme im internationalen Vergleich* (2nd rev. ed.). UVK. https://doi.org/10.36198/9783838539324

Tröger, M., & van den Ecker, M. (2023). Where's the critique? On the dearth of critical theory in German communication research. *Media Theory*, *7*(1), 257–276. https://journalcontent.mediatheoryjournal.org/index.php/mt/article/view/888

Van den Bulck, J., & Van den Bulck, H. (2017). Communication sciences in flanders: A history. In S. Averbeck-Lietz (Ed.), *Kommunikationswissenschaft im internationalen Vergleich: Transnationale Perspektiven* (pp. 93–113). Springer VS.

Vierhaus, R. (2011). Max Grünbeck. In R. Vierhaus (Ed.), *Deutsche Biographische Enzyklopädie Online*. K. G. Saur. https://www.degruyter.com/database/DBE/entry/dbe.16-4054/html?lang=de

Volk, S. C. (2021). *Comparative communication research: A study of the conceptual, methodological and social challenges of international collaborative studies in communication science*. Springer VS. https://doi.org/10.1007/978-3-658-36228-7

Wächter, K.-M. (1999). *Die Macht der Ohnmacht: Leben und Politik des Franz Xaver Ritter von Epp (1868–1946)*. Peter Lang.

Wagman, I. (2016). Locating UNESCO in the historical study of communication. In P. Simonson & D. W. Park (Eds.), *The international history of communication study* (pp. 71–89). Routledge.

Wagner, H. (2023). Media studies in Germany in the context of cultural studies and Franco–German cooperation. *History of Media Studies, 3*. https://doi.org/10.32376/d895a0ea.5644812e

Waisbord, S. (2014). United and fragmented: Communication and media studies in Latin America. *Journal of Latin American Communication Research*, 4(1), 55–77. https://journal.pubalaic.org/index.php/jlacr/article/view/61

Wessler, H., & Averbeck-Lietz, S. (2012). Grenzüberschreitende Medienkommunikation: Konturen eines Forschungsfeldes im Prozess der Konsolidierung. In H. Wessler & S. Averbeck-Lietz (Eds.), *Grenzüberschreitende Medienkommunikation* (pp. 5–18). Nomos.

Wessler, H., & Brüggemann, M. (2012). *Transnationale Kommunikation: Eine Einführung*. Springer VS. https://doi.org/10.1007/978-3-531-94190-5

Wiedemann, T. (2012). *Walter Hagemann: Aufstieg und Fall eines politisch ambitionierten Journalisten und Publizistikwissenschaftlers*. Herbert von Halem.

Wiedemann, T. (2017). Surviving in the journalistic field: The Catholic journalist Walter Hagemann's rollercoaster ride during the Third Reich. *Journalism Studies, 18*(8), 943–959. http://dx.doi.org/10.1080/1461670X.2015.1099462

Zillich, A. F., Riesmeyer, C., Magin, M., Müller, K. F., Pfaff-Rüdiger, S., Rothenberger, L., & Sehl, A. (2016). Werte und Normen als Sollensvorstellungen in der Kommunikationswissenschaft: Ein Operationalisierungsvorschlag. *Publizistik, 61*(4), 393–411. https://doi.org/10.1007/s11616-016-0286-4

Zorn, W. (1959). Franz Ritter von Epp. In O. zu Stolberg-Wernigerode (Ed.), *Neue deutsche Biographie* (Volume 4, pp. 547–548). Duncker & Humblot.

Global media and communication ethics

The tension between universalism and cosmopolitanism

Barbara Thomass

Introduction

In a globalized world, communication involving contact between people of different national origins and cultures has become the norm. Media offerings are distributed globally, but at the same time are culturally shaped and received locally. The intertwining local, national, and global conditions of media production, distribution, and reception, sometimes subsumed under the notion of the *glocal*, offer potential for conflict as well as new possibilities for understanding (Thomaß, 2013, pp. 353–354).

The search for norms that enable this understanding of human coexistence is subject to the respective historical, cultural, and social conditions. In an increasingly globalized world, the question arises as to how universal these norms can or must be. In contrast, there is a call for a cosmopolitan turn in communication studies. In this article, the debate about universalist thinking and its significance for media and communication ethics is shown, as well as its aporias and the challenges it faces in the field of media and communication ethics through cosmopolitan approaches.

Christians et al. (2009, pp. 37–39) identified, from the perspective of Western history, four historical phases in which normative foundations of public communication developed that are still relevant today and still dominate the discourse:

- the classical period from 500 B.C. to 1500, in which the main concern of theorists was the veracity of public discourse within a corporatist order;
- the early modern period (c. 1500 to 1800), in which the main concern was the freedom of those participating in the public sphere in a liberal order;
- the period of modern popular democracies (from 1890 to 1970), in which the social responsibility of those participating in public communication was at the center of aspirations;
- the current "postmodern" period (since the 1970s), in which civic participation in the public sphere is the central target value.

It is important to note that these phases describe the developments in the occidental cultural sphere, and they have been described without knowledge of other communicative-ethical developments. Normative communication theory (even if it did not bear this designation at the periods mentioned) thus revolved around the question of how public discourse should be constituted in a given socio-political context (Christians et al., 2009, p. 65).

Now, international and intercultural communication is characterized by the fact that entities enter into communicative exchange with each other, each of which can or does follow different target values of public communication. Given this plurality of worldviews about the ideal communicative order, conflicts that arise at the interfaces of different orders are predictable. A blatant example was the publication of the Muhammad cartoons in 2005 in a Danish daily newspaper under the title "The Face of Muhammad" (Benatar, 2008; Debatin, 2007; Levey & Moodod, 2009). This series of 12 cartoons focused on the Islamic prophet and religious founder Muhammad. The publication triggered demonstrations and violent riots in many countries around the world, particularly in regions with strong Islamic cultures. This led to diplomatic conflicts between the Danish government and Islamic states as well as a global debate about religious freedom, freedom of the press, artistic freedom, and freedom of expression.

Plurality is set as a value in itself in democratic societies; thus, communication studies, having emerged and differentiated in Western pluralistic societies, is imbued with the fundamental values of their social systems. Most scholars and media practitioners in pluralistic and Westernized societies will agree that the following core values are to be protected in regulating the media sector in the public interest: freedom of expression, democracy, protection and promotion of culture, diversity, universal access, and privacy rights (Babe, 1990; Napoli, 2001), and it is obvious that these values are based on a Western perspective, in which this author is firmly rooted, so that their juxtaposition with values from other parts of the world is still an open task.

This normative bias can be made conscious, and the aforementioned prerequisite of any research can be handled transparently, but it can hardly be overcome. How limited such a view is becomes clear whenever one is confronted with the value systems of other societies and communication cultures. Then, the value of plurality becomes a challenge because it includes the acceptance of other values—cultural autonomy, consensus principle, national ethos, priority of religious views, to name but a few—that may conflict with one's one.

Conflict is thus, on the one hand, inscribed in the encounter or confrontation of different communication cultures. A few years ago, for example, the cultural scene in Germany was shaken by a debate that brought to light the contradictory nature of values: At the internationally acclaimed international art exhibition Documenta 15, the work *People's Justice* by the artist group Taring Padi was initially on display, showing, among other things, a soldier with a Star of David and a pig's face wearing a helmet labeled "Mossad," the name of Israel's foreign intelligence service. Because of this imagery, which was widely interpreted as being anti-Semitic, the banner was first covered up and then taken down. Artistic freedom as a universal value collides here with the high sensitivity to anti-Semitism expected in and by Germany. Not only was it seen as a conflict between the artists' responsibility to act ethically and responsibly and the right to freedom of expression but also it was interpreted as a domination of the host country over the Indonesian artists' freedom of expression.

Avoiding conflicts that can become destructive, on the other hand, can therefore be regarded as the lowest common denominator of all participants. Beyond such a pragmatic determination of a universal norm, it is worth examining how the claim to the universality of norms came into the world in the first place and how it can be justified.

Therefore, this chapter examines the philosophical origins and dimensions of universalism and its historical development. It reveals the contradictory implications of the concept and shows how it became a significant influence in philosophy about the state. With this background, we point to the limits of universality in view of context and power and discuss the relation between cosmopolitanism and universalism in media and communication ethics.

What is universalism?

Universalism is derived from the Latin word *universalis*, which means "general." This ideal indicates an assumption and claim that all the diversity of reality as a whole can be traced to a single principle or law of order. From this, it follows that ideas, ideals, rights, and obligations apply in principle to all human beings. Universalism is therefore a perspective that prioritizes the whole of an entity above singularity, and generality above specificity. The concept contends that some principles and norms are valid for all human beings.

In societal applications, universalism is operationalized in rules and practices that have general widespread value. To work properly, these rules must be generally accepted by society and internalized by its members. When effective, universally applicable rules are a proper basis for the pursuit of justice and provide clarity about the conditions and obligations that apply to each and all. The effectiveness of universal rules is ensured by prioritizing fairness and justice over creating good relations through favoritism, and exceptions to the rules are not tolerated, as they would jeopardize the universal principle.

Universalism can be observed in individualistic societies, as is clearly the case in many Western societies. The prevailing emphasis on social justice, maintaining order, and being able to plan in a thoughtful, rational way are based on respect for values that are presumed to be universal and are considered positive and beneficial for the welfare of all. The disadvantages are keyed to degrees of inflexibility and rigidity in patterns of required adherence to values and rules. Defenders of universalism have been accused of ignoring cultural differences and promoting Western values, which indicates an attitude of ethnocentrism (Benhabib, 1994; Donnelly, 2003; Marko, 2012; Namli, 2018; Vattimo, 2007).

This brief overview of universalism hints at a complex development of thinking about "the universal" that spans from ancient Greece to the French Revolution and beyond. The ideal that some fundamental things are universal and should be respected as such is the cornerstone of significant codifications pertaining to the rights and liberties of free peoples. This is evident, for example, in the American Declaration of Independence (Jefferson, 1776), which claims, "all men are created equal." Thus, the context of universalism has been usefully applied from the broadest and most fundamental claims of universal rights to technical systems of communication, transportation, and other matters of social welfare. It is timeless in principle because universalism claims that, once identified, universal values are all-encompassing and always valid.

Universalist thinking entered normative views of media ethics and communication at an early stage. Since universalists postulate universal principles and norms that should apply to all people, in terms of media and communication ethics, this means that certain ethical standards and responsibilities should apply universally to journalists and media organizations. Thus, truthfulness and objectivity in reporting and the unhindered circulation of information have been established as universal values, according to which journalists endeavor to provide and circulate accurate and balanced information, regardless of cultural or national differences. As early as the 1970s, the debate about the so-called "New World Information and Communication Order" showed how much this Western view clashed with the countries of the Global South, which were then referred to as developing countries, because they feared the intolerable cultural influence of international news agencies.

Today, such universal claims are less often accepted as valid and are often challenged. The all-embracing pretension of universalism has been challenged by its conceptual counterpart: particularism—the attitude that underlines the specifics of single cases, the uniqueness of a given culture, or the individuality of a person. The degree to which universalism is universal is arguable. The claim has limits because in application the practice excludes certain groups—typically minorities and the disadvantaged—in various ways. The principle is more transcendent than the practice. As hinted at in the American claim noted above, the universalism of human rights excluded women until they won their rights through struggle. And there are many regions where women have not won their rights to this day.

After this short explanation of the concept of universalism, we will see how the development of universalist thinking became an influential intellectual abstraction about the nature and value of the human being that has inspired political philosophers and the political shaping of societies.

How did universalism emerge and develop?

Universalism emerged not as a fact in the real world but rather as an intellectual reality or construct. The principle of *quod semper, quod ubique, quod omnibus* (Tönnies, 2001)—forever, everywhere, for everybody—postulated that from a distanced viewpoint, all human beings are equal. This thought became a manifestation of material power. However, the aporias inherent in universalism have come to light since the beginnings of the concept of universalism.

The Greek philosophers Plato and Aristotle were early advocates of universalism. They constructed views of an ideal state in which every free man (only) is empowered and appointed to discuss and decide public affairs in open debate. The Greek polis corresponds to this phenomenon, which contains a germ of universalism. However, the Greek philosopher Seneca addressed the tension between those included in the universal and those excluded: "Although everything is permitted against a slave, there is something that is not permitted by the common right of every living being against a human being, because he is the same nature as you" (as cited in Tönnies, 2001, p. 66).

Later, in ancient Rome, there was a fundamental shift in the legal system. The legal rights of the *pater familias*—the head of a greater family—were extended to every Roman citizen, to the effect that every man was at liberty to act on his own rights. This grounds a point that Talcott Parsons (1971) stipulated: "Modern society originated only in a single evolutionary arena, the West, i.e. essentially in Europe, which inherited the legacy of the western half of the Roman Empire" (p. 10). This aligns with the views of Max Weber, who, in his work *The Protestant Ethic and the Spirit of Capitalism* (1904/1930), posited that cultural phenomena occurred on the soil of the Occident, which were of universal significance and validity in a developmental direction.

This strand of philosophic history suggests a noble ideal but also indicates good reason for associating universalism with the critique of imperialism. All too willingly, Western thinking has claimed for too long and too exclusively advances in human progress for itself and has relegated other societies and cultures as backward and deficient by its own self-acclaimed standards.

Tönnies (1995) described the origins of universalism as endeavoring to legitimate the claims of a group (i.e., the particular but not all groups) amid competing claims by other groups in differentiating society. This view was supported by Schweppenhäuser (1998) who claimed that early stages of bourgeois self-knowledge were characterized by the will to assert oneself as a particular social group fighting for the preservation of its prerogatives against the central power of a community and the nobility of a given time.

The idea and ideal of unalienable individual rights emerged in the Enlightenment, which created the basis for the modern notion of universalism that is based on a recognition of the power of the human spirit. The philosophers of Enlightenment aimed to identify general laws for all mankind and to follow them as such.

All of this explains why the universal is not manifest in the material world but is an influential intellectual abstraction about the nature and value of the human spirit. The French Revolution converted these intellectual claims into a basis for real power in the timeless triad of "liberty, equality, and fraternity." The proclamations of a revolutionary bourgeoisie, especially the Declaration of Human and Civil Rights of 1789 and the French Constitution of 1791, transformed the particular bourgeois interest into a universal human interest. The constitutions of modern societies almost universally believe that *citoyens* in democracies have the same freedom rights and the right of self-preservation.

Nonetheless, the foregoing has already noted the persistent problem that universalistic thought, since its inception, has never referred to everyone in practice but always to select groups. Beyond this privileged selection, the same "universal" rights were not accorded to large groups of people: slaves, women, colonized peoples, children, and so forth. Societies have made corrective progress, however, and today, the universalism principle is being extended to the idea of animal protection, the rights of animals, and the rights of nature. The universal is becoming more universal than before.

However, the underlying problem persists. François Noël Babeuf, an agitator during the French Revolution and founder of the Conspiracy of the Equals, was among the first to point out the inherent contradiction of universalism—a philosophic claim of rights for all that in practice excludes many. He took the French Revolution's idea of equality to a radically serious level and, with his notion that all people are factually equal and should enjoy equal rights and conditions in economic terms, created the first attempt to realize the concept of a classless society in political practice (Buonarroti, 1909). Later socialists referred to his ideas.

Karl Marx, author of the seminal *Capital: A Critique of Political Economy* (1867) and the political pamphlet *The Communist Manifesto* (1848), defended the idea of a classless society in which no one should have privileges resulting from possession of the means of production. While he had claimed to turn the philosophical thinking of Hegel upside down—which means to ascribe to it a material base—we can state that Marx has turned universalism upside down, as he spelled out the material prerequisites of equality.

Universalism in the twentieth century

In the twentieth century, individual rights as a universal enfranchisement were codified in the International Declaration of Human Rights (1948). This was the result of a long process in which the idea of human rights gained a symbolic and politically fundamental significance for thinking about how to create a more peaceful world order. The profound crises experienced in the international order during the second half of the twentieth century inspired reflection on how to create a fairer world order.

These crises included the devastation of World War II, independence for former colonies and the often messy aftermath, an awareness of the interdependence of states that spread in the 1970s, and the dissolution of Soviet power with the collapse of socialist states in the 1990s. Also important after 1945 was the question of how to organize the Nuremberg Trials of Nazi war criminals. This revived the debate as to whether there were inalienable rights that would validate the intervention of the international community. The judgments handed down in the Nuremberg Trials were based on the conviction that individuals—regardless of respective national legislation—had a range of universal rights that warranted the conviction of war criminals on this basis (Sands, 2016).

The creation of the United Nations (UN), whose founding documents included the Universal Declaration of Human Rights (1948), was a monumental step forward in the codification of universal rights and the assertion that they applied universally to all individuals on earth. This did not emerge from a vacuum. In the 1920s, international law scholars and internationally oriented lawyers had given thought to the meaning and possibility of international human rights. From 1948 onward, the concept has been a persistent and foundational aspect of the human rights system that grounds the UN. The establishment of the UN was linked to the question of how more effective mechanisms of international security and peacekeeping could be established (Eckel, 2019). This was followed by the confirmation of a human rights regime in the Council of Europe, adopted as a corresponding convention in 1950.

The leaders of the anti-colonial struggle relied on the promise of human rights to legitimate their cause but did not shy away from noting the persistent problem of a hypocritical Western deception, which they had experienced for centuries. Human rights were not universal in the practice of the colonial masters (Nkrumah, 1962). In the United

Nations General Assembly, in which the former colonial states had the same voting rights as all members, past human rights violations were discussed. During the escalations of the Cold War, the culprits of the most violent abuses of human rights were accused. This also contributed to the instrumentalization of human rights discourse for respective international interests (Westad, 2007).

Significant advances are evident in the Declaration on the Grant of Independence to Colonial Countries and Peoples in 1960, the inclusion of the right to self-determination of peoples in UN rules in 1966, and the Convention on the Elimination of All Forms of Racial Discrimination in 1965 (Jensen, 2016). These represented significant milestones on the road to universal rights in the international order (UN, 1994). The attempt of the former colonial states to wrest the commitment to a new world economic order from the industrialized West in the 1970s was also based on the idea of human rights and their universality. Although this did not bring any immediate result, and was hotly contested by the USA in particular, a right to development was proclaimed in 1986 (Office of the High Commissioner for Human Rights [OHCHR], 1986).

The Universal Declaration of Human Rights (1948) included communications rights—framed at that time as information rights—and the debates concerning the interpretation of these rights gave rise to the claim of a New World Information and Communication Order (NWICO) (Servaes, 1984). The debate about the NWICO within the United Nations Educational, Scientific and Cultural Organization (UNESCO) unfolded in the 1970s and 1980s and ultimately led to the USA temporarily withdrawing from UNESCO. This example illustrates the problems of universalist thinking within applied and globalized media and communication ethics (see also Averbeck-Lietz at el. in this book).

When UNESCO was founded in 1946, a commitment to freedom of information was one of its basic principles. The strong position of the media of the former colonial powers was condemned as media imperialism, particularly by the young governments of the post-colonialist states. The aim of the NWICO was to push back the reporting of the Western media in favor of the local and regional media. The preamble to the UNESCO Declaration of 1978 explicitly called for a "New World Information and Communication Order."

The declaration represented a compromise between a commitment to the so-called free flow of information and concessions to the (socialist) countries of the East and the countries from the Non-Aligned Movement, which wanted to prevent the unbridled flow of information from capitalist states. This, in turn, was interpreted by Western states as a license to control journalists and the media. In 1984, the USA withdrew from UNESCO, followed a year later by the UK. In 1989, UNESCO abandoned the term "information order." In view of the extremely unequal communication relations between the North and the South, the universalist claim of unbridled media freedom, which was a market freedom under the given economic conditions, met with massive resistance from the weaker parties, who, however, were unable to prevail. The debate about NWICO thematized UNESCO's claim of a central role in defining culture's place in international relations and the limits of state sovereignty with a view to media and communication (Brendebach, 2021).

In the 1970s, a new and growing awareness of environmental hazards caused by humans as a consequence of industrialization emerged. The importance of resource management and the need for sustainability has grown in the decades since, accompanied by the idea of a universal claim across generations to an intact environment (Kaiser & Meyer, 2016; Macekura, 2015).

The Conference on Security and Cooperation in Europe (CSCE) process in Europe, the activities of opponents of dictatorship in Latin America, and dissidents in Eastern Europe, as well as the growing emphasis on human rights in the Global South, dramatically demonstrate the diversity of national, cultural, and historic contexts in which protagonists of universal rights operate. These also demonstrate the volatility and difficulty of advancing a unified political project. Nevertheless, universal human rights have become a persistent issue and are now a foreign policy reference point for many Western governments (and beyond).

The collapse of the Soviet Union and the end of dictatorships in Argentina, Brazil, Chile, the Philippines, and South Korea, as well as Greece, Portugal, and Spain, and the fall of the apartheid regime in South Africa have given rise to the hope that universal human rights could be enforced worldwide. At the 1993 UN Human Rights Conference, 171 states declared their support for the universal application of human rights (OHCHR, 1993). The question is not in the principle but rather in the practice: How should and would this be implemented, and should international interventions to enforce rights be universally recognized? This is the subject of ongoing debates in the face of diverse conflict situations.

In postmodern thought, universalism is subject to a variety of criticisms. The most far-reaching critique is drawn from the impossibility of achievement, because the world and its inhabitants are too diverse to be subsumed under a universal view. Another powerful critique from a cultural perspective links this to the problem of relativism. Perhaps less is universal than particular, because everything is relative to varying degrees. Moreover, who are the protagonists of Western societies that they should be allowed to proclaim their values as having general validity? European human rights concepts arise from a specifically European civil, liberal, and secular historical project, as indicated earlier in our brief look at Greek and Roman philosophers. Therefore—so the argument goes—they cannot easily (or at all!) be "transferred" to other cultures.

Another serious critique stems from a multiculturalist perspective. As Jürgen Habermas keenly observed, behind universalistic legal claims the particular assertiveness of a particular collective are concealed on a regular basis (Habermas, 1997) Habermas argued that the moral universe extends to all natural persons, however, and in that light has been regarded as perhaps the most famous living "universalist" (Köhl, 2003).

The fact that all existing societies do not grant universal rights to all people does not make the idea obsolete or invalid. The problem in the formulation of human rights, and especially in their establishment as rights for all humans, hinges on factors that prevent their realization, and these are constitutively inscribed (Schweppenhäuser, 1998). Property, security, and freedom are the central criteria of universal human rights but are also equated with individual rights, especially in antagonistic competitive societies that do not question the appropriateness of a capitalist orientation. This creates a contradictory situation: The foundation of human rights—which has a universal claim to validity—and the worldwide establishment of social conditions that would not only permit this formulation, but almost require it, at the same time blocking their realization.

The continuation of freedom and property rights based on universal human rights has enormous implications for the media sector in democratic societies and for democratic rights, which are also based on universalism but are in conflict. We address this next.

Universalism and media in the welfare state

The step from universalism as a philosophical tradition of thought to the universal claim of citizens to certain state services of common interest is enormous. It is generally explained with reference to the construction of a welfare state. The principle of equality, which is so important for democratic states, has never been fully realized, as noted above. This is mainly due to prevailing ownership structures and discrepancies between rich and poor citizens. In an ongoing struggle over conflicting ideas on how to reconcile freedom and equality, modern democracies have increasingly set themselves the task of providing at least equal opportunities for all citizens to live in dignity and have an essential degree of social security.

This is how the modern welfare state was created, with the essential task of subsidizing certain meritorious goods that the market does not or cannot provide due to the need for profitability. Perhaps some degree of supply is provided by the market but not to a sufficient extent in relation to the need for these goods. Such merit goods include education, security (including social security), public infrastructure, and culture. These are services that benefit everyone in society, that is, the public at large. Hence, they must be provided as public services. In the late nineteenth and early twentieth centuries, mediated social communication was defined as a merit good.

The principle that media are merit goods that can and should be made available to every citizen to the same extent and at the same quality is the core value of the universal service obligation. This obligation implies a universal human right and requires societies to provide an adequate infrastructure for their delivery and performance. Everyone should be reached by postal services, everyone should be supplied with electricity, and all people need access to a telephone connection—and today, access to the Internet. People who are not able to access online media suffer a "digital deficit" that puts them, and their life chances, at risk compared with people who have access. This is especially true for the citizen's right to be informed. Golding (2017) argued that despite the abundance of information available online, there is a growing inequality of access to quality information. This deficit of access to high-quality online services can, when it comes to deficits of quality of information, translate into a citizen detriment (Thomass, 2019). The debate about the implementation of universal services has always centered on the scope and quality of the service. Today, for example, the question in modern industrial societies is not only whether Internet access exists but also whether it meets the requirements of broadband connection. The invention of public service broadcasting (PSB) had its origins almost 100 years ago and was established in many countries that were structured by policy and practice to provide universal services, such as social welfare states. It is one important materialization of universalistic thinking in the realm of communication. And vice versa, universalistic thinking has inspired communications studies in many respects.

There is another line of important argumentation that still legitimates universalism in the media that has significant implications today. This is the argument for media responsibility—the social responsibility of media—to strengthen democracy. Universalism is a prerequisite for realizing that mandate as a project of addressing universal rights. From this perspective, citizens' communication and information rights are the focus of discussion: "The logic is simple. Democracy needs citizens who are equally informed; thus, they must be guaranteed equal access to all relevant information" (Nieminen, 2019, p. 58). This makes the case for citizens' communication and information rights explicit. The basic elements are derived from, among others, international treaties and conventions, such as the UN Declarations of Human Rights, the European Convention on Human Rights (ECHR), and the Charter of Fundamental Rights of the European Union (Horowitz & Nieminen, 2016). Nieminen (2019) proposed five areas of communication and information rights, which all refer to the universalist claim of rights: rights to access, availability, critical competence, dialogue, and privacy (p. 58).

Universalism and cosmopolitanism

With a cosmopolitan turn in communication studies, which should strive for the given diversity of communication cultures and the inclusion of glocal perspectives, as stated in the introduction, we hope for and expect enriched communication studies that take note of the diversity of communication conditions on the globe and test its own theoretical arguments and empirical questions against the background of the diversity of these communication conditions. But in view of the high importance of the equality of cultures that cosmopolitanism propagates, it must be asked what to de-Westernize and to become more cosmopolitan means in consequence. How does this relate to universalism?

It is argued in this book to push forward a less hegemonic approach that presents multiple and critical perspectives on what is understood as international and inter- and transcultural in our discipline. The inclusion of a broader range of ideas and comparative research, the bridges to other disciplines, and the sound consideration of context knowledge are the benefits of this approach.

However, the enrichment that cosmopolitan communication studies would experience through deep internationalization comes up against limits when one realizes that a strongly normative component is inherent in communication studies. There are relevant questions concerning international and intercultural communication, such as those pertaining to the concepts of the recipient, objectives of political communication, the role of self-image and practice in journalism, public relations theories, communication policy, and the ways media and communication ethics are dealt with. Nevertheless, normative assumptions and presuppositions will always flow into these research questions, methods, research designs, and interpretations of results, which will ultimately result from values that predominantly go back to the philosophy of the Enlightenment (Karmasin et al., 2013; Zillich et al., 2016). In a volume on the normativity of communication studies, we argued that these assumptions and presuppositions should always be made transparent and explicit as part of the research process (Karmasin et al., 2013, p. 473).

The integration of diverse perspectives into a cosmopolitan-oriented media and communication ethics will have to provide approaches to the questions of how freedom of speech can be balanced with the perseverance of culture and tradition and how self-determination can be organized in relation to the requirements of collective and social cohesion in media production, distribution, and reception. As answers to these questions are far from being easy, the focus will be on who can enforce which values in which contexts, with which arguments, and also which positions of power in the background.

Cosmopolitan communication studies can promote the sensitization for the implicit presuppositions of research of Western/Northern provenance, and thus enrich one's own state of research and open it up for new questions. In particular, the discrepancy between postulated values and real inter- and transcultural communication relations, which are inherited, (re)produce disadvantages, exclude participation, and perpetuate positions of power, result in new research questions that can only be grasped with a cosmopolitan view of these communication relations.

However, it is precisely the normative dimensioning of many research fields and subdisciplines in communication studies that is in tension with the basic understanding of cosmopolitanism. This becomes clear when one considers how a cosmopolitan would argue against universal values and what positions a universalist would take against a cosmopolitan approach.

Some examples of questions within applied ethics may illustrate this. The universalistic demand for unconditional equality of women and all genders is opposed to the defense of the family, which is postulated as a high value in many cultures. Animal rights conflict with dietary habits; a pluralistic conception of religious freedom cannot be reconciled with the culturally justified and defended unity of a nation with one religion. For ethical questions of communication, a comparable tension arises: Who speaks? Who is allowed to speak? How sensitive should the media be to questions of religion? Should the media address the new, the unknown, or the problematic in a society, or should they preserve prevailing social patterns or contribute to social cohesion? The treatment of such questions in research is based on different normative assumptions, which ultimately include values that are culturally conditioned. The tension between cosmopolitanism and universalism is thus rooted in the different assessment of whether universal values should apply or whether all cultural values have an equal existence.

Conclusion

The history of the concept of universalism is an ongoing attempt to determine what is universal for all people and to embed it in norms. Since the Enlightenment, universalism has been fundamental to Western state constitutions, which were later adopted by many countries around the world. Universalist claims to validity, however, have always been criticized because of the presumption of comprehensive applicability, especially when the application has excluded so many who are also entitled to claim the right. As articulated in the Universal Declaration of Human Rights of 1948, universalist ideas have become a model for the international order. An essential component of human rights—the freedom of communication—has achieved decisive importance for the media order of pluralistic states.

Universalism, as a notion of the boundless universality of consensual norms and values, has led to the adoption of the Declaration of Universal Human Rights of 1948. The ideas of the equal rights of all people expressed therein enter into the notions of ideal communication relations that inspire and often guide research in communication studies. These ideas have led to a sustained hegemonic representation of positions in Western/Northern communication research. It is the legitimate interest of researchers from the Global South to overcome this hegemony, and researchers from the North/West to overcome the theoretical and empirical truncations that result from this hegemony. But even as citizens of the world, as cosmopolitans see themself, they must ask themselves which norms and values are guiding, shaping, or even merely accompanying their research. The demand for equality of cultures does not exempt us from taking a stand when universal values are affected.

References

Babe, R. E. (1990). *Telecommunicatins in Canada: Technology, industry and government*. University of Toronto Press.

Benatar, D. (2008). Cartoons and consequences. *Think*, 6(17–18), 53–57. https://doi.org/10.1017/s1477175600002980

Benhabib, S. (1994). In defence of universalism – Yet again! A response to critics of situating the self. *New German Critique*, 62(Spring–Summer), 173–189. https://www.doi.org/10.2307/488515

Brendebach, J. (2021). Contested sovereignties: The case of the "New World Information and Communication Order" at UNESCO in the 1970s. In G. Feindt, B. Gissibl, & J. Paulmann (Eds.), *Cultural sovereignty beyond the modern state: Space, objects, and media* (pp. 106–127). De Gruyter Oldenbourg. https://doi.org/10.1515/9783110679151-006

Buonarroti, P. (1909). *Babeuf und die Verschwörung für die Gleichheit*. Verlag von J. H. W. Dietz Nachf.

Christians, C. G., Glasser, T. L., McQuail, D., Nordenstreng, K., & White, R. A. (2009). *Normative theories of the media: Journalism in democratic societies*. University of Illinois Press. https://www.jstor.org/stable/10.5406/j.ctt1xcjws

Debatin, B. (Ed.). (2007). *Der Karikaturenstreit und die Pressefreiheit: Wert- und Normenkonflikte in der globalen Medienkultur/ The cartoon debate and the freedom of the press: Conflicting norms and values in the global media culture*. Lit.

Donnelly, J. (2003). *Universal human rights in theory and practice* (2nd ed.). Cornell University Press.

Eckel, J. (2019). Menschenrechte und die Gestaltung der internationalen Ordnung im 20. Jahrhundert. In P. Geiss, D. Geppert, & J. Reuschenbach (Eds.), *Eine Werteordnung für die Welt? Universalismus in Geschichte und Gegenwart* (pp. 263–288). Nomos.

Golding, P. (2017). Citizen detriment: Communications, inequality, and social order. *International Journal of Communication*, 11, 4305–4323. https://ijoc.org/index.php/ijoc/article/view/6673

Habermas, J. (1997, February 4). Der interkulturelle Diskurs über Menschenrechte. *Frankfurter Rundschau*.

Horowitz, M. A., & Nieminen, H. (2016). European public service media and communication rights. In G. F. Lowe & N. Yamamoto (Eds.), *Crossing borders and boundaries in public service media* (pp. 95–106). Nordicom.

Jensen, S. L. B. (2016). *The making of international human rights: The 1960s, decolonization, and the reconstruction of global values*. Cambridge University Press. https://doi.org/10.1017/CBO9781316282571

Kaiser, W., & Meyer, J.-H. (Eds.). (2016). *International organizations and environmental protection: Conservation and globalization in the twentieth century*. Berghahn. https://doi.org/10.3167/9781785333620

Karmasin, M., Rath, M., & Thomaß, B. (Eds.). (2013). *Normativität in der Kommunikationswissenschaft*. Springer VS. https://doi.org/10.1007/978-3-531-19015-0_17

Köhl, H. (2003, September 22–26). *Moralischer Universalismus und der Begriff der moralischen Gemeinschaft: Eine Gespenstervertreibung* [Conference presentation]. GAP.5: Fifth International Congress of the Society for Analytical Philosophy, Bielefeld, Germany. http://www.gap5.de/proceedings/pdf/539-546_koehl

Levey, G. B., & Modood, T. (2009). The Muhammad cartoons and multicultural democracies. *Ethnicities*, *9*(3), 427–447. http://dx.doi.org/10.1177/1468796809337427

Macekura, S. J. (2015). *Of limits and growth: The rise of global sustainable development in the twentieth century*. Cambridge University Press. https://doi.org/10.1017/CBO9781139680509

Marko, J. (2012). Ethnopolitics: The challenge for human and minority rights protection. In C. Corradetti (Ed.), *Philosophical dimensions of human rights: Some contemporary views* (pp. 265–291). Springer. https://doi.org/10.1007/978-94-007-2376-4_14

Namli, E. (2018). Critique of human rights universalism. In M. Stenmark, S. Fuller, & U. Zackariasson (Eds.), *Relativism and post-truth in contemporary society* (pp. 123–140). Palgrave Macmillan. https://doi.org/10.1007/978-3-319-96559-8_8

Napoli, P. M. (2001). *Foundations of communications policy: Principles and process in the regulation of electronic media*. Hampton Press.

Nieminen, H. (2019). Inequality, social trust and the media: Towards citizens' communication and information rights. In J. Trappel (Ed.), *Digital media inequalities: Policies against divides, distrust and discrimination* (pp. 43–66). Nordicom.

Nkrumah, K. (1962). *Towards colonial freedom: Africa in the struggle against world imperialism*. Heinemann.

Office of the High Commissioner for Human Rights. (1986). *Declaration on the right to development*. https://www.ohchr.org/EN/ProfessionalInterest/Pages/RightToDevelopment.aspx

Office of the High Commissioner for Human Rights. (1993). *World conference on human rights*. http://www.ohchr.org/EN/ABOUTUS/Pages/ViennaWC.aspx

Parsons, T. (1971). *The system of modern societies*. Prentice Hall.

Sands, P. (2016). *East West street: On the origins of "genocides" and "crimes against humanity"*. Weidenfeld & Nicholson.

Schweppenhäuser, G. (1998). Die Aporie des menschenrechtlichen Universalismus. *Kritische Justiz*, *31*(2), 260–265. https://www.jstor.org/stable/24000636

Servaes, J. E. (1984). *Major events, documents and resolutions concerning the New International Information Order (1945–1984)*. Thammasat University.

Thomaß, B. (2013). Normativität in der internationalen und interkulturellen Kommunikation. In M. Karmasin, M. Rath, & B. Thomaß (Eds.), *Normativität in der Kommunikationswissenschaft* (pp. 353–370). Springer VS. https://doi.org/10.1007/978-3-531-19015-0_17

Thomass, B. (2019). Economic inequality, appraisal of the EU and news media. In J. Trappel (Ed.), *Digital media inequalities: Policies against divides, distrust and discrimination* (pp. 95–112). Nordicom.

Tönnies, S. (1995). *Der westliche Universalismus: Eine Verteidigung klassischer Positionen* (2nd ed.). Westdeutscher. https://doi.org/10.1007/978-3-663-09645-0

Tönnies, S. (2001). *Der westliche Universalismus: Die Denkwelt der Menschenrechte* (3rd rev. ed.). Westdeutscher. https://doi.org/10.1007/978-3-322-80841-7

United Nations. (1994). *The United Nations and apartheid, 1948–1994*. https://unesdoc.unesco.org/ark:/48223/pf0000105439

Vattimo, G. (2007). A prayer for silence. In J. D. Caputo & G. Vattimo, *After the death of God* (J. W. Robbins, Ed.; pp. 89–113). Columbia University Press.

Weber, M. (1930). *The protestant ethic and the spirit of capitalism* (T. Parsons, Trans.). Charles Scribner's Sons, George Allen & Unwin. (Original work published 1904)

Westad, O. A. (2007). *The global cold war: Third world interventions and the making of our times*. Cambridge University Press. https://doi.org/10.1017/CBO9780511817991

Zillich, A. F., Riesmeyer, C., Magin, M., Müller, K. F., Pfaff-Rüdiger, S., Rothenberger, L., & Sehl, A. (2016). Werte und Normen als Sollensvorstellungen in der Kommunikationswissenschaft: Ein Operationalisierungsvorschlag. *Publizistik, 61*(4), 393–411. https://doi.org/10.1007/s11616-016-0286-4

Diversity within the media
Cosmopolitan perspectives on social positions and structures in Germany

Christine Horz-Ishak

Introduction

The aim of this chapter is to develop a conceptual framework and theoretically fruitful perspectives for the study of demographic diversity in the media structures from a cosmopolitan point of view. Inspired by the (self-)critical debate about a cosmopolitan approach in communication studies that is going on for a while now, the guiding question is how far the diversity paradigm may enrich this debate and helps unveil how inequality and social positions are shaped in and by media in respect to a communication studies definition (Badr et al., 2020; Waisbord & Mellado, 2014; Wasserman, 2020).

The accelerated cultural diversification of Western societies gives rise to challenges such as racism and social inequality, particularly for individuals who "embody diversity" (Ahmed & Swan, 2006, p. 96). The eight major diversity categories (ethnicity, gender, age, sexual orientation, (dis)ability, religion, social status, and race) are the starting points that are commonly used for analysis and will be critically assessed later in this chapter.[1] Diversity within the media, or "Media Diversity," refers to the presence (or absence) of pluralistic perspectives, and viewpoints within organizations and media structures, particularly focusing on minority voices (Douglas, 2022). This idea of Media Diversity is of particular relevance to communication studies, given the pivotal role of the media as social constructors and reproducers of social realities (Bayer, 2013).

However, to understand diversity as a structural and global phenomenon, a cosmopolitan approach in cosmopolitan communication is imperative (Badr et al., 2020; Badr & Ganter, 2021; Siapera, 2010). Beck and Sznaider (2006) proposed a lived, reflexive form of cosmopolitanism that "emerges from within" in diverse contexts and manifests

---

1 The concept of "race" is replaced by "migration background" in this context due to the historical connotations with the Nazi era in German-speaking countries (Vinz & Schniederig, 2010). "Ethnicity" is replaced by "migration background" or "history," thus encompassing all individuals vulnerable to racism and discrimination within the German context.

distinctive attributes (p. 9). This internal perspective is the focus of this chapter because before carrying out international comparative studies, an understanding of the contexts of comparison must first be available, because even in contexts we are familiar with, we often do not understand communication phenomena ourselves (Grüne, 2016). This also applies, for instance, to the diasporic media and the communicative realms of minorities. From the viewpoint of a cosmopolitan critique, it is hence necessary to first gain insight into the German context, because even in the West, such as in the USA and Germany, the focus categories of difference vary significantly while speaking of demographic diversity.

While US-American research gather expertise on racist and sexist portrayals of marginalized groups, the approach in German communication studies has been different. It prioritizes framing and content analysis with regard to Muslims and immigrants, as anti-Islamic sentiment and biased media reporting further exacerbate long existing negative public discourse about these groups (Hafez, 2016, 2019; Richter & Paasch-Colberg, 2023; Schiffer, 2005). Furthermore, the German diversity approach is rather top-down and politically instrumentalized. Thus, it is reasonable acknowledging context-specific definitions first, before recognizing diversity as a global phenomenon.

In Germany, approximately 27% of the population has a migration background, 9.4% live with a disability, 7% identify as LGBTQ+, and about 6.6% as Muslims (Pfündel et al., 2021; Statistisches Bundesamt, 2022, 2024; YouGov, 2021). These and other groups at the margins remain vulnerable to exclusion and discrimination. For example, there remains a significant lack of communication research, particularly about classism in German media (Theißl, 2023; Hanitzsch & Rick, 2021). Whereas the underrepresentation, misrepresentation, and invisibility in media content is well documented in national and international communication studies, notable research gaps persist regarding "structural" diversity (Horz et al., 2020; Fengler & Kreutler, 2020; Saner et al., 2024).

A critical diversity approach that is proposed here can be useful to offer crucial insights into the processes that (re-)construct social positions in the media sector and the conditions that influence how content is produced, shared, and consumed (see also Ganter in this book; Napoli, 1999). Secondly, if marginalized groups participate in the media at all, it raises concerns of whether this is an emancipatory project or an "elite capture" where hegemonial groups define whose perspectives should be included in the first place (Bayer, 2013, pp. 27–28.; Táíwò, 2022). Moreover and most importantly, Media Diversity in a broader sense can be linked to the normative idea of a pluralistic media system and liberal public spheres, which ultimately makes it a political issue that needs further investigation (Padovani et al., 2022).

A critical diversity approach, further developed to an analytical tool for cosmopolitan communication research bears the potential to help analyze how social structures shape each other within the media sector, which includes institutional, organizational, and governmental structures. It may shed light on the relationalities that emerge within systems and internal organizational structures, focusing on interdependencies between individuals and the organization rather than formal hierarchies. Network and organizational analysis of diversity-related structures can therefore address power dynamics within legacy media and digital networks between disparate groups. This chapter examines to what extent the diversity paradigm can enhance our understanding of the meaning and the complexity of power structures and structural discrimination in the media sector, as well as media participation, identity formation, networks, and international connectivity of minorities through their own media outlets. Instead of presenting a comprehensive conceptual framework from a Western perspective, the chapter is meant as a starting point for further dialogue with scholars from diverse contexts. After a brief and critical evaluation of the theoretical discourse on the diversity paradigm, the current trajectory of German communication studies follows.

Unboxing the diversity paradigm from a cosmopolitan perspective

A comprehensive theory of diversity that transcends paradigmatic debates does not yet exist, neither in Germany nor internationally, as its meaning shifts depending on the context (Erdur, 2024; Hipfl, 2023; Klein, 2020). A possible baseline to develop a heuristic tool is to explore how diversity has been historically and socially constructed (Vertovec, 2021, p. 1275).

The long-standing diversity discourse has paved the way for its analytical exploration (Nieswand, 2021). The most fundamental definition assumes that differences between individuals are a social and cultural fact, particularly in post-migrant and postcolonial societies. As such, demographic diversity inherently challenges the prevailing models of normality, including those perpetuated in the media associated with the White, male, and Western citizen. An increasing awareness and social learning process about diversity in Germany is currently inspired by more prominent migrant voices and knowledge in public debates. Diversity has thus become an integral part of self-understanding processes within the super-diverse society, despite the accompanying tensions (El-Mafaalani, 2020).

Two key strands of the paradigmatic debate are particularly relevant: normative-democratic approaches and utilitarian approaches. The normative-democratic strand focuses on promoting equality, fairness, and respect. It is rooted in the US-American civil rights movement and resolutions of international organizations. The UN for example, asserts that accessibility to the media is vital for all people (United Nations, 2018). In communication studies, this is reflected in the idea that media access is a crucial resource of "cultural citizenship" (Klaus & Lünenborg, 2004). However, access alone does not ensure trust, as migrant communities' trust in German media has declined over time, due to biased reporting (Sachverständigenrat für Integration und Migration [SVR], 2021). In that sense, fostering Media Diversity is not only an ethical exercise, but also a task for the whole of society. Therefore, it is one that is ultimately political in nature and intended to prevent power dynamics that perpetuate inequality formed by a history of exclusion and discrimination (Decker et al., 2022; Zick et al., 2011). The normative-democratic approach to diversity asserts the emancipatory potential of minority communication phenomena. Diversity is then a "cosmopolitan almost narrative," despite the overall "top-down approach" (Sarikakis, 2014, p. 89). The utilitarian strand, the second of the two, links demographic diversity to economic utility. Within the neoliberal agenda, diversity is legitimized as a resource to employ the uniqueness of one's competencies and value in a market economy (Lentin & Titley, 2008). This perspective originates in the 1980s and 1990s in the USA as a reaction to the influx of women, Black individuals, and immigrants into the labor market (Johnston & Packer, 1987). The new diversity had to be managed to prevent conflicts disrupting the emerging neoliberal economy. This commercialization of empowerment, which is individualized and market-oriented, is not aimed at making structural changes. Instead, the success of minorities is anchored on an individual level without addressing the underlying systemic inequalities (Banet-Weiser, 2015). It obscures the unequal power relations that exist within organizations, hindering the ability to challenge them (Ahmed, 2012). Moreover, the categorizations of difference themselves raise concerns of being excessively hierarchical and implicitly rooted in a White/Western liberal framework that neutralizes "histories of political struggles" (Ahmed, 2012, p. 96). The "photogenic diversity" or "diversity washing" that often comes along with diversity Management serves as a superficial display of difference without substantive change (Nieswand, 2021, p. 8).

As a result, Critical Diversity Studies, which consists of an interdisciplinary merging of feminist theory, postcolonialism, and critical race theory criticizes the utilitarian approach for undermining the emancipatory potential of diversity due to its Janus-faced nature (Bayer, 2013, pp. 36–37). Hence, it is "less about analysis and more about management, less about theory and more about application" (Hardmeier & Vinz, 2007, p. 27). However, research faces the dilemma that it may reinforce stereotypes when employing pre-constructed categories. This can only be resolved through "pragmatic deconstruction" to reveal processes of othering and social positioning (Knapp, 2008, p. 167). Consequently, a "non-positivistic, non-essentialist" definition of the diversity categories is necessary (Zanoni et al., 2010, p. 10). In this view, difference is not an inherent reality, but because of ongoing "interpretative actions," it is redefined through ongoing symbolic interaction – hence communication processes (Fuchs, 2007, p. 17).

The discursive lines of the paradigmatic debates thus address the question of whether or not diversity is an empirical description of the demographic plurality of a society, which simply needs to be managed and made useful for human resources. Or, on the contrary, if it is a critical reading, which produces and represents attributions of difference along which power asymmetries, hegemonic practices, and inequality that can be examined and rectified in a non-essentialist manner and that can also be linked to discover emancipatory communication phenomena. In the media, these debates manifest in unique ways. Media are embedded systems with context-related functions for society. However, even public service media, seen as merit goods, cannot escape the competitive market environment, with diversity management often reduced to pragmatic human resource practices (Thomass & Dupuis, 2016, p. 25). Thereby, diversity can also serve as a "Trojan horse" for transforming power dynamics from within (Jones & Stablein, 2006, p. 145). However, the emancipatory potential of access to and visibility in the media should not be overrated. The assumption that simply recruiting marginalized individuals will automatically lead to the pluralization of perspectives in the public sphere is lacking complexity (Meier et al., 2024). In fact, some evidence suggests that the boundaries between inclusive and exclusive debates on diversity are becoming increasingly blurred, fostered by the emergence of new affective communities that normalize right-wing ideologies (Zahn & Lünenborg, 2024). An inevitably interdisciplinary cosmopolitan approach has to address these blurred lines from a critical perspective.

A cosmopolitan critique of diversity-related communication research in the German context

To get closer to a diversity framework, we need to self-critically address current blind spots (Badr et al., 2020). To do so, we must first acknowledge that there is a robust research output on media representation of marginalized groups, particularly migrants and Muslims within German-speaking communication studies (Hafez & Schmidt, 2020; Zick et al., 2011). Findings indicate that media representations play a pivotal role in the reinforcement of negative images of and violence against immigrants, refugees, and Muslims (Arlt & Wolling, 2016; Beck, 2017; Benček & Strasheim, 2016). Journalistic reporting often victimizes these minorities or denotes them as dangerous intruders and objects of state measures, while rarely portraying them as active agents in their own narratives (Fengler & Kreutler, 2020; Holzberg et al., 2018; Richter & Paasch-Colberg, 2023). The cosmopolitan critique, however, underscores the significant limitation of German communication studies on representation and content analysis with an "ethnic lens" (Badr et al., 2020; Glick Schiller et al., 2006). In this context, the term *Vielfalt* (diversity) is frequently employed to signify demographic variety and is often involved in othering (Larrazet & Rigoni, 2014). The emphasis on specific group identities over others neglects the complexities of intersectionality. Due to the scarcity of demographic data in German legacy media, structural analysis is difficult in the first place. For instance, the number of journalists with a migration background can only be estimated at 1% to 4% of the workforce, with a suspected but unverified increase among trainees (Kraemer et al., 2020). Figures about more powerful positions, such as editors in chief, show that only 6.4% have migration backgrounds in Germany's most widely consumed legacy media. However, the reasons for the existing homogeneity of leadership positions and imbalances in the "workforce" and "programming" need further investigation (Horz et al., 2020).

Most importantly, representation studies rarely draw conclusions about the importance of structural diversity as an influencing factor on observed biased and negative media images. Obviously, communication studies do not always ask the right questions that are necessary to generate their own rationales to give impulses to policies and political strategies. Moreover, findings on structural diversity do only sporadically translate into media practice, such as changes in personnel and recruitment structures in media organizations (Meier et al., 2024; Saner et al., 2024). Only a few international comparative studies about media systems and democratic functions address diversity as one structural item, among others basically in the sense of gender diversity (Padovani et al., 2022; Trappel & Tomaz, 2021). One critical point is the emphasis on politically charged research topics such as media representation of migration, refugees, and Islam. Communication studies in Germany has aligned closely with domestic and foreign policy interests, explicitly or implicitly legitimizing disintegration policies intertwined with neoliberal agendas (Horz, 2016; Richter et al., 2023). One example is that approaches such as the "intercultural integration" of migrants in the media were co-opted into official policies (Geißler, 2010; Presse und Informationsamt der Bundesregierung, 2011, pp. 320–327). A common focus in this respect is the question of whether media consumption aids integration or not and if the pre-constructed "others" need special attention. As a result, to date, "mediated symbolic realms" often reinforce the idea of discrete ethno-media and segregation

(Sarikakis, 2014, p. 90). This perspective overlooks questions about migrants' communicative performance, their participation in cultural production, and the significance of communication spaces. Consequently, diverse media cultures and outlets remain unrecognized as integral to German culture. This essentialist view constructs an "us vs. them" dichotomy, which fails to acknowledge culture as a dynamic process and the interconnectedness of a globalized contexts and lifeworlds (Couldry, 2007, p. 250; Hafez & Grüne, 2021). The biased perspective and compliance with official politics underlines the necessity to revise the existing concepts, epistemologies, and foci of research. One goal of a cosmopolitan approach is to challenge the binary between "native" and "migrant," exploring the "in-between" spaces of post-migrant societies (Foroutan, 2019; Badr et al., 2020, p. 300). Consequently, research with a cosmopolitan perspective needs to challenge and transcend the political instrumentalization of this subject matter, particularly in terms of fostering a more impartial and objective investigation.

Another critical point is a prerequisite of a cosmopolitan perspective, namely the awareness that German communication studies have largely neglected other communities, for example, LGBTQ+ individuals. As noted by Comella and Sender (2013, p. 2564), sexuality research is an emerging subfield within international communication studies. In their book, Dhoest et al. (2017) provided a comprehensive analysis of LGBTQ+ media and culture in Europe, exploring media programming by queer individuals, their creation of outlets and programming, and their participation in counter-publics and online political debates. Notably, the collection lacks contributions from German scholars. In contrast, even undergraduate students in the USA are engaging in intersectional research on Black LGBTQ+ individuals and social media empowerment (Bonjo, 2017). The study of Black, African, and Afrodiasporic people in workforce structures and as media actors in Germany is another field that is still embryonic. Admittedly, a major obstacle is the lack of reliable demographic data on these groups in Germany, underscoring the need for more research. Estimates of the "Afrozensus" suggest that there are around 1.27 million people in these communities, with 71% having been born in Germany (Aikins et al., 2021; Mediendienst Integration, n.d.). Despite being a small minority, research on anti-Black racism and embodied differences in media content and structures could lead to valuable results for research related to other People of Color (PoC). These and more shortcomings in German communication studies raise questions about whether the objectives of funding institutions are to blame, or rather a broader lack of cosmopolitan thinking (Richter et al., 2023). Fortunately, a growing responsiveness to issues of diversity, racism, and discrimination in editorial staff structures and decision-making bodies can be observed; however, this is currently mostly confined to a focus on the meso levels of news media (González Hauck et al., 2024; Grabenheinrich, 2023; Meier et al., 2024).

To sum up, there is a need for more politically and practically impactful communication studies—not instrumental ones—with the aim to better address significant knowledge gaps about structural inequalities and the richness of communication phenomena in superdiverse societies. As discussed above, the current issues can be at least partly attributed to the dominance of hegemonial perspectives in communication studies and their strong migration policy orientation toward the study diversity.

Toward a heuristic framework of diversity for a cosmopolitan communication approach

To achieve a holistic structural analysis of Media Diversity, the proposed analytical tool combines Stephen Vertovec's (2021) social science approach to diversity with Philipp Napoli's (1999) deconstruction of the latter as part of the normative media pluralism model to propose a heuristic matrix for empirical research across various media fields and levels.

Vertovec's (2021) conceptualization of diversity as a "social organization of difference" (p. 1275) is based on the abovementioned normative-democratic strand and critical approaches to the diversity paradigm. He suggests a diversity approach that helps unveil structures of inequality in "distinct times, scales and contexts" (Vertovec, 2021, p. 1273). As such, diversity "is not intended for explicating mechanisms but for interpreting how social structures shape each other" (Vertovec, 2021, pp. 1275–1276). Vertovec referred to diversity as a structural category, as discussed above. He has taken into account three key intertwined dimensions involved in reproducing difference and shaping inequalities: configurations, representations, and encounters of diversity. However, the following overview will mainly emphasize configurations and encounters, as both are still underrepresented topics in communication studies, while being inherently necessary to gain insight into the structural conditions that frame content production and the ways social groups are represented.

Configurations of diversity are the "organizational arrangements of diversity" that shape social positions through political and organizational structures. They describe "both vertical and horizontal inequalities" that intersect with race, class, and gender, evolving over time and encompassing historical, cultural, and economic factors (Vertovec, 2021, p. 1276; Lünenborg et al., 2011). These structural patterns affect consciousness, group interactions, and access to power and privileges, forming a "matrix of domination" (Andersen & Collins, 2018, p. 400). These configurations highlight not just the numbers but the social positions of diverse staff, such as journalists or camerapersons, and the extent to which domination manifests itself, that is, in legacy media (Graf, 2011; Grabenheinrich, 2023). Journalistic innovations like "entrepreneurial journalism," however, emphasize the transformation of existing inequalities, as this type of journalism is potentially less hierarchical due to more flexible structures (Suhr, 2023, pp. 1, 4).

Encounters of diversity refer to dynamic social interactions across social categories, manifesting in both mediated and interpersonal communication (Vertovec, 2021, p. 1282). As Gerhards and Neidhardt (1990) claimed, these micro-publics form the communicative core of the public sphere. By emphasizing these encounters, the focus is on investigating how interactions between media professionals with and without diversity characteristics are structured.

These encounters are often informal and laden with stereotypes about a respective minority, formed over time in (mediated) communicative interactions. Consequently, unequal status among staff members in media organizations may reinforce exclusionary practices, whereas staff with equal statuses are less likely to be excluded (Allport, 1954; Vertovec, 2021, p. 1282). The unequal access to media professions and job allocation, i.e.

due hurdles such as unpaid internships in media organizations, results in an unequal status for minority staff members. Communicator research highlights, for instance, how minority staff in legacy media—often those with migration backgrounds—are sometimes treated as experts on minority issues or, conversely, as tokens or intruders (Graf, 2011). These encounters are both structured and structure hegemonial and discriminatory practices. In contrast, alternative independent media outlets such as channels and hashtags online can challenge dominant discourse and media images by their own perspectives and self-representation, thereby potentially empowering previously silenced voices. Thus, it is crucial to intensify research as to what extent these communication phenomena actually do so and to what degree they facilitate transcultural connections, community, and identity building, as well as the development of critical counter-publics. A cosmopolitan approach to communication studies can thus benefit from the concept of encounters in terms of diversity research by questioning the communicative conditions that foster a convivial workplace or, on the contrary, simply reconfigure existing stereotypes and discrimination (Cottle, 1998). Vertovec's (2021) broader social science concept discussed above has now been merged with Napoli's (1999) media pluralism concept to shift diversity to communication studies and to come closer to an analytical framework for a cosmopolitan communication approach.

Philip M. Napoli (1999) deconstructed the "diversity principle," which he perceived as part of the media pluralism model.[2] Whereas "diversity" denotes the multiplicity of perspectives within a system, "pluralism" serves as the structural framework that facilitates them.

Both normative concepts are reflected in the German broadcasting system: "internal pluralism" refers to the range of perspectives within public service media (PSM), and "external pluralism" pertains to the multiplicity of commercial outlets. This "dual system" reflects the normative concept of pluralism, wherein both social and cultural diversity play a crucial role. Nevertheless, apart from the normative idea, power imbalances and bureaucratic structures within legacy media are key reasons why diverse perspectives are is often silenced (Cottle, 1998). Hence, it is necessary to analyze the paradoxes that dilute the normative ideal. Napoli (1999, p. 11) identified three pillars of media pluralism: source diversity, content diversity, and exposure diversity. Since our focus is on structural aspects of diversity in the media, we concentrate on examining source diversity, although all three elements are certainly interrelated.

Source diversity is divided further into two components: (a) ownership diversity, again subdivided into outlet and programming[3] and (b) workforce diversity. Both the components and the subcomponents will be described below.

---

2 In the English-speaking context, the term diversity was used until the late 1990s as a synonym for what we understand today as media pluralism. Napoli integrated both pluralism and diversity (in the sense that we use it currently and in this chapter) into the term diversity.

3 Napoli confusingly employs the terms "programming" and "content diversity" synonymously, alternating between the two, although the category named "content" is a separate factor of his model (that is not discussed here). For reasons of clarity, the term "programming" is used in this chapter, because it is more focused on the making of content and not the content itself.

Ownership diversity: Outlets and programming

Ownership diversity as one structural aspect of source diversity includes the diversity related to a wide range of media programming and media outlets. A research focus on outlets may include commercial TV channels or PSM that are owned by the general public. Developed in the era of linear media, Napoli's hypothesis faces new challenges with the platformization and proliferation of networked communities on social media, such as YouTube, Twitter, and TikTok. Content creators, for example, are owners and creators of their accounts at the same time. These platforms have created novel avenues for agency, participation, and media production, requiring a rethinking of publicness in a decentralized media landscape fueled by the platform economy (Lünenborg, 2019, p. 319; Schankweiler & Straub, 2023; Walsh, 2024).

Workforce diversity

The context dependency of diversity discussed earlier can be exemplified with source diversity's second structural aspect of workforce diversity in its contextualized setting: for example, whereas the regulatory framework requires broadcasting outlets in the USA to compose their staff according to "the demographics of the surrounding market" (Napoli, 1999, pp. 13–14), the German broadcasting regulation is less systematic. According to Napoli, workforce diversity includes not only editorial staff but also management, administrative personnel, technical professionals, and self-regulatory bodies across various media outlets with the potential to pervade traditional power structures. Workforce can certainly be expanded into digital outlets by investigating content producers and their role in the production process, usually addressed by Production Studies (Mayer et al., 2009).

Hence, this approach can help to promote network analysis instead of communicator surveys that cannot explain the "how" and "why" of certain power structures. The broad focus across all professional levels in the workforce helps to better understand the actor constellations and structural settings at play.

The critical source diversity concept

The critical concept proposed here is based on the synthesis of Vertovec (2021) and Napoli's (1999) approaches, which were discussed above. This synthesis holds promise, as it asks new questions about structural inequalities and, at the same time, emphasizes bottom-up research perspectives. Such an approach integrates the earlier interpretations and unleashes their potential for empirical research on the axes of power as well as bottom-up participation, reflecting the plurality of socially situated knowledge and perspectives within media structures and organizations. Figure 1 explains the synthesized model. The framework is shaped by Napoli's source diversity (1999), which comprises ownership (outlet and programming) and the workforce diversity. Vertovec's (2021) approach of diversity as "the social organization of difference" is what occurs within this framework. Both components and subcomponents shape each other, as indicated in Figure 1.

Figure 1: Overview (source: author's compilation)

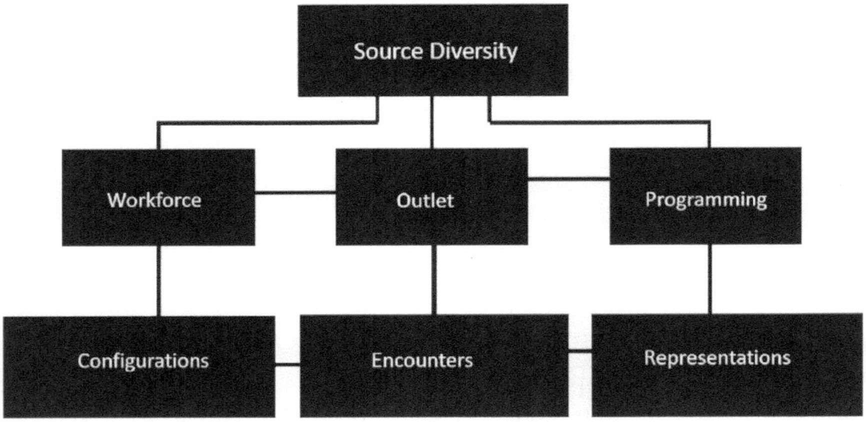

Possible research areas

The approach proposed here can transcend the conventional political-utilitarian approach to this subject matter and, above all, is less biased. Ultimately, it can be incorporated into various research areas. Empirical research of Media Diversity could start with one of Vertovec's (2021) dimensions of "social organization of difference" combined with one of Napoli's structural aspects. Table 1 shows exemplary research areas with an intersectional approach.

This approach allows for systematic examination of how diversity categories intersect. The reciprocal formation processes of Media Diversity can then be analyzed. Due to the limited space of this chapter, it is not possible to exhaustively discuss research areas or offer a detailed delineation of the interrelated set of meanings. To provide a more concise summary, few interwoven aspects of Vertovec's (2021) dimensions are outlined.

Table 1: Diversity matrix as an analytical tool: Examples of possible research areas (source: author's compilation)

	Configurations *Examines how the respective category is structurally organized (e.g., composition and status of the workforce)*	**Encounters** *Investigates how interactions occur (e.g., collaboration between different groups and professions, interaction with stakeholders and audiences)*	**Representation** *Explores how diversity is represented (e.g., representation of minorities in the media)*
Workforce *Analyzes the diversity of professionals in a respective outlet involved in media production*	Composition and status of diverse staff; identity; professional roles (Intersectional and more combinations)	Power asymmetries; discrimination; professional networks (Intersectional and more combinations)	token; diversity expert; equally treated (Intersectional and more combinations)
Outlets *Examines the diversity and structure of media platforms or accounts*	single person or group; community members only (Intersectional and more combinations)	participation in public discourse; counterpublics; emancipatory approach; self-explaining (Intersectional and more combinations)	identity and community representation (Intersectional and more combinations)
Programming *Examines how diversity is reflected in productions and how it is presented*	issue orientation; community orientation (Intersectional and more combinations)	interaction with community; included in programming (Intersectional and more combinations)	self-presentation; voice (Intersectional and more combinations)

According to Napoli (1999) ownership is subdivided into outlet and programming. For practical reasons both specifications replace the overarching term ownership.

Configurations and workforce

A thorough understanding of Media Diversity, and what it actually means, requires expanding the focus beyond newsroom structures. NGOs such as the German "Neue Deutsche Medienmacher*innen" publish science-based assistance to foster diversity at the professional level. Additionally, governance structures need more attention because of their importance in addressing systemic barriers in media institutions (Ganter & Badr, 2022; Neue Deutsche Medienmacher*innen, 2022). Noteworthy, these critical topics have largely been left to non-academics and contracted research to explore.

Gendered power dynamics, exemplified by the #MeToo cases, gender pay gaps, and the lack of support for care responsibilities, particularly in Western media institutions, raises concerns about the role of configurations within the workforce in traditional media outlets, thus demanding more attention from a comparative perspective (Orosz & Löer, 2018). Context-led research on this topic seems rewarding, as some international comparative studies have indicated (Trappel & Tomaz, 2021). However, there remains a clear need to intensify our efforts to analyze the reproductions of power structures within media organizations from a cosmopolitan point of view. Class disparities within the media workforce also remain a blind spot in German research: one survey found that 95% of media trainees in Germany come from academic backgrounds, raising concerns about the misrepresentation of other social classes (Kraemer et al., 2020). This highlights a gap in scholarly research that requires urgent measures, because these and other foci of Media Diversity research reveal the quality of media pluralism and the public sphere (Theißl, 2023). We can then look at how online outlets take up and respond to the issue to learn more about the networked bandwidth of discourse around diversity in the media.

Configurations and ownership

The role of digital outlets in spreading alternative narratives, both inclusive and exclusionary, is also a crucial area of analysis. Influencers such as Madeline Stuart, a model with Down syndrome, exemplify how normative-democratic and utilitarian strands of diversity purposefully intersect on digital outlets for community-building and monetization at the same time, reflecting broader trends in Media Diversity informed by the platform economy (Christensen-Strynø & Eriksen, 2020). Other outlets can follow this trend by propagating exclusionary content. For example, so called Tradwife influencers on TikTok promote ultraconservative gender roles that are also propagated by right-wing ideologies, blurring the lines between family values and political extremism—and at the same time are self-made women, which can generate high income with their accounts. TikTok's recommendation algorithms often contribute to the normalization of such views (Sykes & Hopner, 2024). A study from Media Matters on tradwives revealed that 30.6% of the content on TikTok's "For You" page included this type of material, even within a single afternoon of browsing (Little, 2024).

Issues such as disability, social status, and gender at intersections are crucial for a cosmopolitan understanding of the workforce in contemporary communication phenomena and the evolution of networked publics (boyd, 2011). This also means engaging with emerging fields, such as queer production studies, which explore how "queers produce their own media outlets both within and outside of 'mainstream' cultural industries" (Martin, 2018, p. 7). Queer studies, in particular, offer critical insights into how norms, gendering, and heteronormativity are reproduced or challenged in the media (Drushel, 2017; Ernst, 2024).

Encounters and outlets & programming

Platforms such as YouTube and TikTok allow individuals to act as content creators, influencing the media landscape in ways that transform user behavior and aspirations. This also plays a role in transcultural identity formation and international protest cultures, as digital outlets offer new spaces for encounters as a prerequisite for collective action and representation (Monshipouri & Prompichai, 2018; Schankweiler & Straub, 2023).

Encounters in co-determination bodies of PSM broadcasting councils should also be taken into account more seriously to gain insight into how structural exclusion mechanisms affect marginalized groups like women, disabled people, Sinti and Roma, LGBTQ+, PoC, and others (Neue Deutsche Medienmacher*innen, 2022). Comparative studies show that PSM still struggle to dismantle barriers to minority access, although they are obliged to integrate and reflect on society as a whole (McGonagle, 2014; Horz-Ishak & Thomass, 2021). This paradox persists despite demographic changes and the growing intersectional complexity of social inequalities. Structural barriers preventing those with experiences of racism and discrimination from joining broadcasting bodies therefore require attention and systematic investigations. Network analysis seems to be a suitable method to examine how diversity dimensions interact and intersect in these encounters (McRuer, 2006; Zahn & Lünenborg, 2024).

Further examination of encounters beyond traditional media outlets is essential to be able to respond to current trends. Bottom-up approaches, such as the radical pluralism model, have the potential to enhance our understanding of transcultural communication across diverse contexts. A cosmopolitan approach should investigate how encounters in specific outlets contribute to community building, whether through hegemonic, counter-hegemonic, or emancipatory efforts. Social networks like #BlackLivesMatter can be viewed not only as protest cultures but also as nodes of intersecting communicative encounters that have the potential to evolve into broader social movements involving various actors. This requires a combination of established and innovative methodologies, including ethnographic techniques and network analysis.

Exploring innovative methods from a cosmopolitan point of view

The abovementioned phenomena create a resonance space for empirical studies on Media Diversity with a cosmopolitan perspective. Epistemic approaches, heuristics, and methods need to be decolonized to avoid replicating boundaries in research (Badr & Ganter, 2021). Addressing these issues allows for insights beyond critiques of media representation and is important to advancing cosmopolitan research. Some useful attempts have been made by scholars from the Global South that promote to de-Westernize approaches by including local and regional contexts into theory building (Mano & milton, 2021; Moyo, 2020).

In order to gain a truly cosmopolitan perspective, it is necessary to adopt subject-oriented methods. In contrast to what I would call a "parachute research" approach, which involves the usually brief fieldwork to gather data, "embedded research" is a more comprehensive and immersive methodology. It entails deeper engagement with the subject matter, allowing researchers to gain a nuanced understanding of the lived experiences of their research subjects and partners. Clifford Geertz's (1987) ethnographic method of "dense description" is an appropriate means of approaching the "ethics of listening" (Wasserman, 2013). Standpoint theory, a feminist social theory approach that acknowledges individual's unique perspective that shapes the perception of the world, could integrate advocacy elements into research and reveal socially situated knowledge. For example, feminist scholar Sandra Harding (1994, pp. 284–285) demanded that researchers "reinvent" themselves by involving research subjects in methodology generation, concept development, research design, data collection, and result interpretation. A participatory approach can hence include minority knowledge and demonstrate respect for the reluctance of vulnerable groups to cooperate with powerful groups such as researchers. For instance, interviewing the people involved in diasporas necessitates a considerable investment of time and resources to establish rapport and trust prior to the commencement of research (Horz, 2014). By incorporating interdisciplinary approaches from postcolonial studies or ethnography, cosmopolitan communication scholars can create new analytical frameworks to better understand the meaning of structural diversity in the media and its societal implications, laying the groundwork for internationally comparative studies on Media Diversity (Johanssen & Garrisi, 2020).

References

Ahmed, S. (2012). *On being included: Racism and diversity in institutional life*. Duke University Press. https://doi.org/10.2307/j.ctv1131d2g

Ahmed, S., & Swan, E. (2006). Doing diversity. *Policy Futures in Education*, 4(2), 96–100. https://doi.org/10.2304/pfie.2006.4.2.96

Aikins, M. A., Bremberger, T., Aikins, J. K., Gyamerah, D., & Yıldırım-Caliman, D. (2021). *Afrozensus 2020: Perspektiven, Anti-Schwarze Rassismuserfahrungen und Engagement Schwarzer, afrikanischer und afrodiasporischer Menschen in Deutschland, Berlin*. Each One Teach One; Citizens for Europe. https://afrozensus.de/reports/2020/

Allport, G. W. (1954). *The nature of prejudice*. Addison-Wesley Publishing.

Andersen, M. L., & Collins, P. H. (2018). Why race, class, and gender matter. In D. B. Grusky & J. Hill (Eds.), *Inequality in the 21st century* (pp. 400–401). Routledge.

Arlt, D., & Wolling, J. (2016). The refugees: Threatening or beneficial? Exploring the effects of positive and negative attitudes and communication on hostile media perceptions. *Global Media Journal – German Edition*, 6(1). https://globalmediajournal.de/index.php/gmj/article/view/47

Badr, H., & Ganter, S. A. (2021). Towards cosmopolitan media and communication studies: Bringing diverse epistemic perspectives into the field. *Global Media Journal – German Edition*, 11(1). https://doi.org/10.22032/dbt.49164

Badr, H., Behmer, M., Fengler, S., Fiedler, A., Grüne, A., Hafez, K., Hahn, O., Hamidi, K., Hanitzsch, T., Horz, C., Illg, B., Litvinenko, A., Löffelholz, M., Radue, M., Richter, C., Thomaß, B., & Töpfl, F. (2020). Kosmopolitische Kommunikationswissenschaft: Plädoyer für eine "tiefe Internationalisierung" des Fachs in Deutschland: Ein wissenschaftliches Positionspapier. *Publizistik*, 65(3), 295–303. https://doi.org/10.1007/s11616-020-00576-6

Banet-Weiser, S. (2015). Keynote address: Media, markets, gender: Economies of visibility in a neoliberal moment. *Communication Review*, 18(1), 53–70. https://doi.org/10.1080/10714421.2015.996398

Bayer, J. (2013). *Media diversity in Deutschland: Kulturwissenschaftliche Perspektiven auf journalistische Praxis* [Doctoral dissertation, Ludwig-Maximilians-Universität Munich]. Electronic Theses of LMU Munich. https://doi.org/10.5282/edoc.16086

Beck, M. (2017, September 18). *Securitization of refugees in Europe*. E-International Relations. http://www.e-ir.info/2017/09/18/securitization-of-refugees-in-europe/

Beck, U., & Sznaider, N. (2006). Unpacking cosmopolitanism for the social sciences: A research agenda. *British Journal of Sociology*, *57*(1), 1–23. https://doi.org/10.1111/j.1468-4446.2006.00091.x

Benček, D., & Strasheim, J. (2016). Refugees welcome? A dataset on anti-refugee violence in Germany. *Research and Politics*, *3*(4), 1–11. https://doi.org/10.1177/2053168016679590

Bonjo, E. (2017). LGBTQ People of Color and digital spaces of empowerment. *Sprinkle*, *10*, 13–24.

boyd, D. (2011). Social network sites as networked publics: Affordances, dynamics, and implications. In Z. Papacharissi (Ed.), *A networked self: Identity, community and culture on social network sites* (pp. 39–58). Routledge.

Christensen-Strynø, M. B., & Eriksen, C. B. (2020). Madline Stuart as disability advocate and brand: Exploring the affective economies of social media. In J. Johanssen & D. Garrisi (Eds.), *Disability, media, and representations: Other bodies* (pp. 35–50). Routledge.

Comella, L., & Sender, K. (2013). Doing it: Methodological challenges of communication research on sexuality [Editorial introduction]. *International Journal of Communication*, *7*, 2560–2574. https://ijoc.org/index.php/ijoc/article/view/2507

Cottle, S. (1998). Making ethnic minority programmes inside the BBC: Professional pragmatics and cultural containment. *Media, Culture and Society*, *20*(2), 295–317. https://doi.org/10.1177/016344398020002008

Couldry, N. (2007). Researching media internationalization: Comparative media research as if we really meant it. *Global Media and Communication*, *3*(3), 247–271. https://doi.org/10.1177/1742766507082569

Decker, O., Kiess, J., Heller, A., & Brähler, E. (Eds.). (2022). *Autoritäre Dynamiken in unsicheren Zeiten: Neue Herausforderungen – alte Reaktionen?* Psychosozial. https://doi.org/10.30820/9783837979190

Dhoest, A., Szluc, L., & Eeckhout, B. (Eds.). (2017). *LGBTQs, media and culture in Europe*. Routledge.

Douglas, O. (2022). The media diversity and inclusion paradox: Experiences of black and brown journalists in mainstream British news institutions. *Journalism*, *23*(10), 2096–2113. https://doi.org/10.1177/14648849211001778

Drushel, B. E. (2017). A state of the union: LGBTQ representations and the concept of community. In J. Campbell & T. Carilli (Eds.), *Locating queerness in the media: A new look* (pp. 11–22). Lexington Books.

El-Mafaalani, A. (2020). *Das Integrationsparadox: Warum erfolgreiche Integration zu mehr Konflikten führt*. KiWi-Taschenbuch.

Erdur, O. (2024). Kritik der diversen Vernunft. *Mittelweg 36*, *33*(2), 2–22.

Ernst, W. (2024). Phänomene des Werdens: Intersektionalität, Queer, Postcolonial, Diversity und Disability Studies als Orientierungen für die Medienforschung. In J. Dorer, B. Geiger, B. Hipfl, & V. Ratković (Eds.), *Handbuch Medien und Geschlecht: Perspektiven und Befunde der feministischen Kommunikations- und Medienforschung* (pp. 57–72). https://doi.org/10.1007/978-3-658-20707-6_6

Fengler, S., & Kreutler, M. (2020). *Stumme Migranten, laute Politik, gespaltene Medien: Die Berichterstattung über Flucht und Migration in 17 Ländern* (OBS-Arbeitspapier 39). Otto Brenner Stiftung. https://www.otto-brenner-stiftung.de/fileadmin/user_data/stiftung/02_Wissenschaftsportal/03_Publikationen/AP39_Migration.pdf

Foroutan, N. (2019). *Die postmigrantische Gesellschaft: Ein Versprechen der pluralen Demokratie*. transcript.

Fuchs, M. (2007). Diversität und Differenz: Konzeptionelle Überlegungen. In G. Krell, B. Riedmüller, B. Sieben, & B. Vinz (Eds.), *Diversity Studies: Grundlagen und disziplinäre Ansätze* (pp. 17–34). Campus Verlag.

Ganter, S. A., & Badr, H. (Eds.). (2022). *Media governance: A cosmopolitan critique*. Palgrave Macmillan. https://doi.org/10.1007/978-3-031-05020-6

Geertz, C. (1987). *Dichte Beschreibung: Beiträge zum Verstehen kultureller Systeme*. Suhrkamp.

Geißler, R. (2010). Mediale Integration von ethnischen Minderheiten: Der Beitrag der Massenmedien zur interkulturellen Integration. In Abteilung Wirtschafts- und Sozialpolitik der Friedrich-Ebert-Stiftung (Ed.), *WISO Diskurs: Zur Rolle der Medien in der Einwanderungsgesellschaft* (pp. 8–22). Friedrich-Ebert-Stiftung. https://library.fes.de/pdf-files/wiso/07394-20100820.pdf

Gerhards, J., & Neidhardt, F. (1990). *Strukturen und Funktionen moderner Öffentlichkeit: Fragestellungen und Ansätze*. Wissenschaftszentrum Berlin für Sozialforschung. https://bibliothek.wzb.eu/pdf/1990/iii90-101.pdf

Glick Schiller, N., Çaglar, A., & Guldbrandsen, T. C. (2006). Beyond the ethnic lens: Locality, globality, and born-again incorporation. *American Ethnologist*, *33*(4), 612–633. https://doi.org/10.1525/ae.2006.33.4.612

González Hauck, S., Paasch-Colberg, S., & Pöggel, T. J. (2024). *Zwischen Anerkennung und Abwehr: (De-)Thematisierungen von Rassismus in Medien, Recht und Beratung: Fokusbericht des Nationalen Diskriminierungs- und Rassismusmonitors*. Dezim-Institut. https://www.rassismusmonitor.de/publikationen/zwischen-anerkennung-und-abwehr-de-thematisierungen-von-rassismus-in-den-medien-recht-und-beratung/

Grabenheinrich, M. (2023). *Journalismus und Diversity: Umgang mit kultureller Diversität in der journalistischen Praxis und Konsequenzen für die Aus- und Fortbildung*. Springer VS. https://doi.org/10.1007/978-3-658-39189-8

Graf, H. (2011). Examining ethnicity in German newsrooms. In H. Graf (Ed.), *Diversity in theory and practice: News journalists in Sweden and Germany* (pp. 121–147). Nordicom.

Grüne, A. (2016). *Formatierte Weltkultur? Zur Theorie und Praxis globalen Unterhaltungsfernsehens*. transcript.

Hafez, K. (2016). Compassion Fatigue der Medien? Warum der deutsche "Flüchtlingssommer" so rasch wieder verging. *Global Media Journal – German Edition*, *6*(1). https://globalmediajournal.de/index.php/gmj/article/view/50

Hafez, K. (2019). "Die verhängnisvolle Neigung der Medien…": Plädoyer für einen Humanitären Journalismus. *Communicatio Socialis*, *52*(4), 490–502. https://doi.org/10.5771/0010-3497-2019-4-490

Hafez, K., & Grüne, A. (2021). *Grundlagen der globalen Kommunikation: Medien – Systeme – Lebenswelten*. UVK. https://doi.org/10.36198/9783838555515

Hafez, K., & Schmidt, S. (2020). Rassismus und Repräsentation: Das Islambild deutscher Medien im Nachrichtenjournalismus und im Film. *Bundeszentrale für Politische Bildung*. https://www.bpb.de/lernen/bewegtbild-und-politische-bildung/themen-und-hintergruende/314621/rassismus-und-repraesentation-das-islambild-deutscher-medien-im-nachrichtenjournalismus-und-im-film/

Hanitzsch, T., & Rick, J. (2021). *Prekarisierung im Journalismus*. Ludwig-Maximilians-Universität. https://www.ifkw.uni-muenchen.de/lehrbereiche/hanitzsch/projekte/prekarisierung.pdf

Harding, S. (1994). *Das Geschlecht des Wissens*. Campus.

Hardmeier, S., & Vinz, D. (2007). Diversity und Intersectionality: Eine kritische Würdigung der Ansätze für die Politikwissenschaft. *Femina Politica*, *16*(1), 23–32.

Hipfl, B. (2023). Doing diversity of, in and with media: Challenges and potentials. *Global Media Journal – German Edition*, *13*(1). https://doi.org/10.22032/dbt.57975

Holzberg, B., Kolbe, K., & Zaborowski, R. (2018). Figures of crisis: The delineation of (un)deserving refugees in the German media. *Sociology*, *52*(3), 534–550. https://doi.org/10.1177/0038038518759460

Horz, C. (2014). *Medien – Migration – Partizipation: Eine Studie am Beispiel iranischer Fernsehproduktion im Offenen Kanal*. transcript.

Horz, C. (2016). The politics of diversity and public service media in Germany. In C. Richter, I. Dupuis, & S. Averbeck-Lietz (Eds.), *Diversity in transcultural and international communication* (pp. 35–65). LIT.

Horz, C., Boytchev, H., & Neumüller, M. (2020). Wie divers sind deutsche Medien? In *Viel Wille, kein Weg: Diversity im deutschen Journalismus* (pp. 24–57). Neue Deutsche Medienmacher*innen. https://neuemedienmacher.de/zahlen-fakten/diversitaet-im-journalismus/

Horz-Ishak, C., & Thomass, B. (2021). Germany: Solid journalistic professionalism and strong public service media. In J. Trappel & T. Tomaz (Eds.), *The media for democracy monitor 2021: How leading news media survive digital transformation* (Vol. 1, pp. 197–256). Nordicom.

Johanssen, J., & Garrisi, D. (Eds.). (2020). *Disability, media, and representations: Other bodies*. Routledge. https://doi.org/10.4324/9780429469244

Johnston, W. B., & Packer, A. E. (1987). *Workforce 2000: Work and workers for the twenty-first century*. Hudson Institute. https://eric.ed.gov/?id=ED290887

Jones, D., & Stablein, R. (2006). Diversity as resistance and recuperation: Critical theory, poststructuralist perspectives and workplace diversity. In A. M. Konrad, P. Prasad, & J. K. Pringle (Eds.), *Handbook of workplace diversity* (pp. 145–166). Sage.

Klaus, E., & Lünenborg, M. (2004). Cultural Citizenship: Ein kommunikationswissenschaftliches Konzept zur Bestimmung kultureller Teilhabe in der Mediengesellschaft. *Medien und Kommunikation (M&K)*, *52*(2), 193–213. https://doi.org/10.5771/1615-634x-2004-2-193

Klein, U. (2020). Diversity Studies und Diversitystrategien: Plädoyer für eine Theoretisierung der Praxis und für eine Konzeptualisierung der Theorie. In B. Kortendiek, B. Riegraf, & K. Sabisch (Eds.), *Handbuch interdisziplinäre Geschlechterforschung: Geschlecht und Gesellschaft* (pp. 1–11). Springer VS. https://doi.org/10.1007/978-3-658-12500-4_117-1

Knapp, G. A. (2008). Equality, difference, deconstruction. The benefits of theoretical approaches in women's and gender studies for practice. In G. Krell (Ed.), *Chancengleichheit durch Personalpolitik* (5th ed., pp. 163–172). Wiesbaden.

Kraemer, L., Tautz, D., & Hagemann, N. (2020, November 4). Wie divers ist der ARD-Nachwuchs? *Journalist*. https://www.journalist.de/startseite/detail/article/wie-divers-ist-der-ard-nachwuchs

Larrazet, C., & Rigoni, I. (2014). Media and diversity: A century-long perspective on an enlarged and internationalized field of research. *InMedia*, *5*. https://doi.org/10.4000/inmedia.747

Lentin, A., & Titley, G. (2008). More Benetton than barricades? The politics of diversity in Europe. In G. Titley & A. Lentin (Eds.), *The politics of diversity in Europe* (pp. 9–28). Council of Europe.

Little, O. (2024, May 1). *Tradwife influencers are quietly spreading far-right conspiracy theories*. Media Matters for America. https://www.mediamatters.org/tiktok/study-tradwife-influencers-are-quietly-spreading-far-right-conspiracy-theories

Lünenborg, M. (2019). Affective publics. In J. Slaby & C. von Scheve (Eds.), *Affective societies: Key concepts* (pp. 319–329). Routledge.

Lünenborg, M., Fritsche, K., & Bach, A. (2011). *Migrantinnen in den Medien: Darstellungen in der Presse und ihre Rezeption*. transcript.

Mano, W., & milton, v. c. (2021). *Routledge handbook of African media and communication studies*. Routledge.

Martin, A. L., Jr. (2018). Introduction: What is queer production studies/why is queer production studies? *Journal of Film and Video*, *70*(3–4), 3–7. https://doi.org/10.5406/jfilmvideo.70.3-4.0003

Mayer, V., Banks, M. J., & Caldwell, J. T. (Eds.). (2009). *Production studies: Cultural studies of media industries*. Routledge. https://doi.org/10.4324/9780203879597

McGonagle, T. (2014). Public service media and cultural diversity: European regulatory and governance frameworks. In K. Horsti, G. Hultén, & G. Titley (Eds.), *National conversations: Public service media and cultural diversity in Europe* (pp. 63–82). Intellect.

McRuer, R. (2006). *Crip theory: Cultural signs of queerness and disability*. New York University Press.

Mediendienst Integration. (Ed.). (n.d.). *Zahlen und Fakten: Schwarze Menschen in Deutschland*. https://mediendienst-integration.de/gruppen/schwarze-menschen.html

Meier, K., García-Avilés, J. A., Kaltenbrunner, A., Porlezza, C., Wyss, V., Lugschitz, R., & Klinghardt, K. (Eds.). (2024). *Innovations in journalism: Comparative research in five European countries*. Routledge. https://doi.org/10.4324/9781032630410

Moyo, L. (2020). *The decolonial turn in media studies in Africa and the Global South*. Palgrave Macmillan. https://doi.org/10.1007/978-3-030-52832-4

Monshipouri, M., & Prompichai, T. (2018). Digital activism in perspective: Palestinian resistance via social media. *International Studies Journal*, *14*(4), 42–63. https://www.isjq.ir/article_89793.html?lang=en

Napoli, P. M. (1999). Deconstructing the diversity principle. *Journal of Communication*, *49*(4), 7–34. https://doi.org/10.1111/j.1460-2466.1999.tb02815.x

Neue Deutsche Medienmacher*innen. (Ed.). (2022). *Welche Gesellschaft soll das abbilden? Mangelnde Vielfalt in Rundfunkräten und was dagegen hilft*. https://neuemedienmacher.de/zahlen-fakten/rundfunkraete/

Nieswand, B. (2021). Die Diversität der Diversitätsdiskussion. In A. Röder & D. Zifonun (Eds.), *Handbuch Migrationssoziologie* (pp. 1–26). Springer VS. https://doi.org/10.1007/978-3-658-20773-1_17-1

Orosz, M., & Löer, W. (2018, September 13). *WDR: Neue Details beleuchten Probleme mit der Aufarbeitung von sexuellem Missbrauch*. Correctiv. https://correctiv.org/aktuelles/wdr-metoo/2018/09/13/wdr-neue-details-beleuchten-probleme-mit-der-aufarbeitung-von-sexuellem-missbrauch/

Padovani, C., Belluati, M., Karadimitriou, A., Horz-Ishak, C., & Baroni, A. (2022). Gender inequalities in and through the media: Comparing gender inequalities in the media across countries. In J. Trappel & T. Tomaz (Eds.), *Success and failure in news media performance: Comparative analysis in the Media for Democracy Monitor 2021* (pp. 79–100). Nordicom.

Pfündel, K, Stichs, A., & Tanis, K. (2021). *Muslimisches Leben in Deutschland 2020*. Bundesamt für Migration und Flüchtlinge. https://www.deutsche-islam-konferenz.de/SharedDocs/Anlagen/DE/Publikationen/Studien/mlid-2020-lang.html?nn=598216

Presse und Informationsamt der Bundesregierung. (2011). *Nationaler Aktionsplan Integration: Zusammenhalt stärken – Teilhabe verwirklichen*. https://api.bptk.de/uploads/2012_01_31_Nationaler_Aktionsplan_Integration_e9cdd413a1.pdf

Richter, C., & Paasch-Colberg, S. (2023). Media representations of Islam in Germany: A comparative content analysis of German newspapers over time. *Social Sciences & Humanities Open*, *8*(1), Artikel 100619. https://doi.org/10.1016/j.ssaho.2023.100619

Richter, C., Grüne, A., Hafez, K., Fiedler, A., Behmer, M., Horz-Ishak, C., Badr, H., Litvinenko, A., Hahn, O., Radue, M., Sarısakaloğlu, A., Löffelholz, M., Fengler, S., Illg, B., Hamidi, K., Hanitzsch, T., & Thomaß, B. (2023). Die "tiefe Internationalisierung" der deutschen Kommunikationswissenschaft? Eine Evaluation der Personal- und Forschungsstrukturen sowie der Lehrprogramme deutscher Hochschulen. *Global Media Journal – German Edition*, *13*(1). https://doi.org/10.22032/dbt.57904

Sachverständigenrat für Integration und Migration. (2021). *Auf Empfang gestellt? Aktuelle Befunde zur Mediennutzung und zum Medienvertrauen der Bevölkerung mit Migrationshintergrund*. https://www.svr-migration.de/publikation/medienvertrauen/

Saner, M., Körner, M., & Lugschitz, R. (2024). Diversity and inclusion: "Difference matters." In K. Meier, J. A. García-Avilés, A. Kaltenbrunner, C. Porlezza, V. Wyss, R. Lugschitz, & K. Klinghardt (Eds.), *Innovations in journalism: Comparative research in five European countries* (pp. 126–134). Routledge. http://dx.doi.org/10.4324/9781032630410-16

Sarikakis, K. (2014). Identity and diversity in European media policy: Crisis changes everything(?). In K. Donders, C. Pauwels, & J. Loisen (Eds.), *The Palgrave handbook of European media policy* (pp. 81–98). Palgrave Macmillan. https://doi.org/10.1057/9781137032195_4

Schankweiler, K., & Straub, V. (2023). Bildproteste für die Freiheit im Iran: Die Memefication des Widerstands in den Sozialen Medien. *21: Inquiries Into Art, History, and the Visual*, 4(1), 97–110. https://doi.org/10.11588/XXI.2023.1.93820

Schiffer, S. (2005). *Die Darstellung des Islams in der Presse: Sprache, Bilder, Suggestionen. Eine Auswahl von Techniken und Beispielen*. Ergon Verlag.

Siapera, E. (2010). *Cultural diversity and global media: The mediation of difference*. Wiley-Blackwell.

Statistisches Bundesamt. (2022). *Sozialleistungen: Schwerbehinderte Menschen*. https://www.destatis.de/DE/Themen/Gesellschaft-Umwelt/Gesundheit/Behinderte-Menschen/Publikationen/Downloads-Behinderte-Menschen/schwerbehinderte-2130510219004.pdf?__blob=publicationFile

Statistisches Bundesamt. (2024). *Migration und Integration: Bevölkerung nach Migrationshintergrund und Altersgruppen* [Data set]. https://www.destatis.de/DE/Themen/Gesellschaft-Umwelt/Bevoelkerung/Migration-Integration/Tabellen/migrationshintergrund-alter.html

Suhr, M. (2023). Diversität und (Anti-)Diskriminierung im Kontext journalistischer Neugründungen: Eine Forschungsagenda. *Global Media Journal – German Edition, 13*(1). https://doi.org/10.22032/dbt.57907

Sykes, S., & Hopner, V. (2024). Tradwives: Right-wing social media influencers. *Journal of Contemporary Ethnography, 53*(4), 453–487. https://doi.org/10.1177/08912416241246273

Táíwò, O. O. (2022). *Elite capture: How the powerful took over identity politics (and everything else)*. Pluto Press. https://doi.org/10.2307/j.ctv2g591sq

Theißl, B. (2023). Klassismus als Thema in Medien und Journalismus. *Kulturelle Bildung Online*. https://doi.org/10.25529/8gz7-6404

Thomass, B., & Dupuis, I. (2016). Media implications of diversity: As a reality, a normative concept and as a pragmatic concept for human resources management. In C. Richter, I. Dupuis, & S. Averbeck-Lietz (Eds.), *Diversity in transcultural and international communication* (pp. 15–34). LIT.

Trappel, J., & Tomaz, T. (Eds.). (2021). *The media for democracy monitor 2021: How leading news media survive digital transformation* (Vol. 1 and 2). Nordicom.

United Nations. (2018). *Convention on the rights of persons with disabilities* (CRPD/C/GC/6). https://docstore.ohchr.org/SelfServices/FilesHandler.ashx?enc=6QkG1d%2FPPRiCAqhKb7yhsnbHatvuFkZ%2Bt93Y3D%2Baa2qtJucAYDOCLUtyUf%2BrfiOZckKbzS%2BBsQ%2BHx1IyvGh6ORVZnM4LEiy7ws5V4MM8VC4khDIZJSuxotVqfulsdtPv

Vertovec, S. (2021). The social organization of difference. *Ethnic and Racial Studies, 44*(8), 1273–1295. https://doi.org/10.1080/01419870.2021.1884733

Vinz, D., & Schiederig, K. (2010). Gender und Diversity: Vielfalt verstehen und gestalten. In P. Massing (Ed.), *Gender und Diversity: Vielfalt verstehen und gestalten* (pp. 13–44). Wochenschau Verlag.

Waisbord, S., & Mellado, C. (2014). De-Westernizing communication studies: A reassessment. *Communication Theory, 24*(4), 361–372. https://doi.org/10.1111/comt.12044

Walsh, T. (2024). TikTok as a site of social protest in Iran's Gen-Z uprising. *Discourse & Society, 35*(5), 625–650. https://doi.org/10.1177/09579265241234351

Wasserman, H. (2013). Journalism in a new democracy: The ethics of listening. *Communicatio, 39*(1), 67–84. https://doi.org/10.1080/02500167.2013.772217

Wasserman, H. (2020). Moving from diversity to transformation in communication scholarship. *Annals of the International Communication Association, 44*(1), 1–3. https://doi.org/10.1080/23808985.2019.1706429

YouGov. (2021). *Survey results*. https://www.lsvd.de/media/doc/3168/lgbt_support_survey.pdf

Zahn, P. H., & Lünenborg, M. (2024). *#FrauenSagenNein – Bridging the divide: Analyzing the affective network of gender-critical alliances* (SFB 1171 Working Paper 03/2024). Refubium. http://dx.doi.org/10.17169/refubium-43108

Zanoni, P., Janssens, M., Benschop, Y., & Nkomo, S. (2010). Unpacking diversity, grasping inequality: Rethinking difference through critical perspectives. *Organization, 17*(1), 9–29. https://doi.org/10.1177/1350508409350344

Zick, A., Küpper, B., & Hövermann, A. (2011). *Intolerance, prejudice and discrimination: A European report*. Friedrich-Ebert-Stiftung. https://library.fes.de/pdf-files/do/07908-20110311.pdf

Beyond mediatization

The importance of non-mediated political and social communication in international comparative research

Kai Hafez & Anne Grüne

Introduction

A cosmopolitan orientation of media and communication studies and the internationalization of scholarship in general must go hand in hand with an expansion of interest in communication phenomena beyond a technology-centric logic. The aim of this chapter is to work out a stronger integration of interactive non-mediated communication in international comparative communication research, to discuss its theoretical potential and fields of application, and to strengthen communication-ecological thinking.

Both aspects—communication ecology and international comparisons—are inextricably linked. International communication research will require a comprehensive understanding of political and social communication if it wants to make meaningful comparisons beyond the boundaries of a mere national understanding of science (see also Badr et al., 2020). A selective view of media developments only, as is usually practiced in media systems analyses, will not suffice but must be accompanied by an understanding of political and sociocultural structures and communication practices in other regions of the world. Thus, non-mediated communication should be taken seriously as an independent form of political and social communication.

Conversely, a comprehensive communication ecology approach benefits from international comparisons, as the functionality of non-mediated communication can be sharpened by comparing their political and sociocultural frameworks. Integrating perspectives from the Global "North" and "South," this chapter aims to conceptualize various arenas of non-mediated political and social communication. It discusses the recent developments of parliamentary plenary, committee, and symbolic communication in consolidated democracies, transformation states, and authoritarian countries. It reflects on the perspectives and deficits of internal communication in political parties, and it engages in a comparative analysis of the quality of governmental decision-making communication across different political systems. Moreover, it also outlines the relevance of publicly hidden private communication encounters and today's "salon" communication, as well as the communicative use of the street as part of the overall social transformation process.

These arenas are potential counterbalances to hegemonic political or media discourses. Their practices need to be included in any deeper assessment of media and communication freedom around the world. In some countries, where the mediated public sphere is controlled, civil societies and lifeworlds can still be vibrant and politicized. In more free systems, in turn, discourses can become biased and civil society lethargic. The cosmopolitan approach considered here incorporates this openness to learn from what can be regressive and progressive at the same time in transformations across the globe.

Theoretical introduction

Communication ecology and non-mediated communication

Non-mediated communication is included most systematically in the theorization of the school of media and communication ecology. Michael Giesecke (2007), an important scholar in the German tradition, limited the definition of "media" and media communication to written (print) and/or technical mass media. The basic mode of communication of these media is discursive-monological, and they are thus in opposition to non-mediated communication characterized by face-to-face interaction. A broad definition of "media" based on primary aspects such as language, facial expressions, gestures, or non-linguistic mediators such as money would make it more difficult to differentiate between monologic and dialogic social communication.

The metanorm of communication ecology is not "mediatization" (as will be discussed in detail below), but the reflection of the balance and imbalance between media communication and non-mediated communication in modern society. Giesecke (2007) described communication ecological approaches as the study of the "interdependence of elements, of systems and networks" and media and communicators as "equal elements of communicative networks" (p. 256). He also talks of the coevolution of different forms of communication. Depending on the type of communication ecology theory formation (biometaphorical, technical, systemic, etc.), the relationships between the elements are subject to the different dynamics of balance/imbalance, autonomy/interdependence, etc.

Activating the ecological paradigm is essential if we are to better understand the sustainability of communication processes. For Giesecke (2007), technical media have their own specific functions and were developed in the course of the modernization of industrial society. Modern knowledge societies no longer function purely interactively but require text and storage media. However, this equally important interaction is suspended for the time of media consumption. In Giesecke's (2007) view, we now find ourselves in a media-centered imbalance at the end of the industrial age (p. 210). Like capitalism, the entire modern age is oriented toward the accumulation and consumption of information. Non-mediated communication is therefore seen by Giesecke as a "constantly necessary counterweight" (p. 211) to an overly strong media imprint; it should be relearned again when individual medial, written, and image-based and when social information processing is necessary. From this perspective, the question arises as to why non-mediated communication has not been included to a greater extent in the development of theories in communication and media studies.

Non-mediated communication is either placed solely and autonomously in the focus of microanalyses in communication studies, for example, in sociological communication theories, the study of interpersonal communication, or intercultural communication research, or it is examined as a subordinate process of media communication, primarily in reception and opinion research. Although in media ecology theorization, face-to-face communication can itself be regarded as an environment (Strate, 2017, p. 16), the question of interaction often remains limited to how humans interact with different technical and symbolic media (e.g., Postman, 2000; Scolari, 2012).

The same applies to the related medium theory. Non-mediated contexts, for example, in traditional oral cultures, are taken into account here, but ultimately the social function of media is again prioritized, and a relatively linear leading media imprint from traditional to postmodern societies is assumed (e.g., Meyrowitz, 2018). Joshua Meyrowitz (2018) has self-critically mentioned that too little emphasis has been placed on the co-existence of forms of communication, given that "people in literature societies continue to speak, and those in electronic cultures still read, write, and use print" (p. 636). Thus, by focusing on a supposedly leading media culture, other simultaneous developments in communication cultures are likely to be neglected.

Related approaches to non-mediated communication in international comparisons

A coherent communication ecology theory, which goes beyond a fundamental paradigmatic debate and allows for an operationalization that can be applied to concrete phenomena of social and political communication, does not yet exist. Moreover, a communicative "ecosystem" is sometimes only understood terminologically as a digital extension of communication networks (Yousuf, 2018), which leaves out non-mediated communication.

Comparative media-ethnological studies on communication culture, in turn, pay attention to non-mediated communication but often have strongly essentialist features (e.g., "whining" is described as a German way of speaking; see Winchatz, 2017) and are therefore, in our view, not considered capable of being theorized in the sense of social science. The same applies to parts of intercultural communication, where behavioral patterns and non-verbal forms of expression are sometimes interpreted as characteristics of macrosystems (nation, culture), whereas heterogeneous socialization and cultural learning effects tend to be ignored (see Hafez & Grüne, 2022, pp. 111–112).

Several communication theories at least reveal similarities to a communication ecology approach that includes the various forms of communication equally. For example, Jürgen Habermas's (1995) theory of communicative action can be interpreted as communication ecology insofar as "society" is conceptualized here in a duality of system and lifeworld and the juxtaposition of "strategic" and "communicative action." Accordingly, the lifeworld can be thought of as a primarily non-mediated sphere of interaction, as it is based on everyday social experiences. Moreover, Habermas's (1962/1992) critique of the "colonization of the lifeworld" in his work on the structural change of the public sphere outlines a critique of the imbalance between the social levels of communication.

Jürgen Gerhards and Friedhelm Neidhardt (1990) also conceptualized public communication on three levels, whereby, in addition to the media, theme-centered public assemblies (*Versammlungsöffentlichkeiten*) and encounter publics (lifeworld encounters) play a role and thus include non-mediated communication. In his concept of the "networked society," Richard Münch (1995) also assumed that communication must take place at all levels (in different publics and also non-mediated) in order to bring the basic communicative logics (e.g., theatrical/conflictual vs. rational) to bear and thus enable a link back to the lifeworld (pp. 104–107).

In addition to public sphere theory, approaches to political communication consider the relevance of interpersonal communication. The role of social interaction has already been emphasized in the idea of the "two-/multi-step flow of communication" in research on opinion leaders (e.g., Katz & Lazarsfeld, 1955; Merton, 1968) and has also been recently discussed in the context of the communication mediation model (Shah et al., 2017). However, the actual influence of the different levels of communication in political opinion making and civic action has been contested (e.g., Chaffee, 1986), and we still lack a comprehensive theoretical integration.

Cultural studies-inspired audience research also assigns an important role to the non-mediated communication processes that follow the reception of media in shaping social discourse (media appropriation). However, even in canonical texts, non-mediated communication remains diffuse, less studied, and not conceptually refined (Hall, 1980). Something similar could be said about the fields of development communication and urban communication.

In non-Western communication studies, a critical examination of linear media effects and a discussion of "alternative" or "indigenous" communication have been found in Latin American debates since the 1960s. This tradition is also linked to the philosophy of Buen Vivir and highlights the important role of dialogue, listening, and participation in social communication processes (Calderón et al., 2018; Huesca & Dervin, 1994). The Latin American school of "mediaciones" (e.g., Martín-Barbero, see below) in particular turned away from media centrism and examined popular communication practices of the population at large (Saucedo Añez, 2019, pp. 5–6). In this view, media representations do not displace other forms of communication but are closely linked to non-mediated communication as social communication. Only where they displace indigenous, non-literate cultures, for example, are modern media understood as invasive. A new communicative hybridity (in line with communication ecology) is then called for to rebalance them.

We can find similarities to African debates here. According to Ngũgĩ (1986), every language conveys a specific cultural sign system and thus ensures its survival. For example, the much-discussed African philosophy of Ubuntu is stored in African proverbs (Sesanti, 2022; see also Radue et al. on "prefix journalism" in this book). Language and thus orality (especially in the case of high illiteracy) becomes the core of community, nation-building, and social development, according to Ngũgĩ (1986), who also noted that although writing is not rejected, it tends to be seen as a colonialist impulse that separates people from oral traditions (p. 17). Alternative forms of communication, such as so-called "oramedia" (e.g., talking drums) in African societies, are also interpreted in this sense as interactive and independent techniques whose community-building function is especially emphasized in contexts in which access to modern media is exclusive or fragile due to inequality (e.g., Ugboajah, 1982; Wilson, 2008). The "exaggeration of writing" (Bergermann, 2012, p. 268) at the cost of oral communication becomes a widespread topos of postcolonial media theory.

Postcolonialism and anti-colonialism are often closely linked to communication ecology thinking, even if not explicitly. Nevertheless, it is lamented that an independent communication theory is still in its infancy and often exhausted in diffuse demands for de-Westernization (Mano & milton, 2021a, p. 7)—which, however, makes little sense if one also considers the communication ecology tradition as part of the Western sciences.

Cultural recentering in the tradition of Ngũgĩ (1986) also includes a repositioning of communication and media strategies, which comprise digital media in opposition to mass media (Frassinelli, 2021, p. 51). Critics of Ngũgĩ have pointed out that culture should not only be understood as verbal language but also as observational and imitative communication (Issa, 2022). Non-mediated communication, then, is to be understood as a complex act that includes observation as well as interaction (Hafez & Grüne, 2022, pp. 8–17).

Lastly, contributions on the "Arab street" from the last few decades are also interesting for a communication ecology approach. A change in the concept of the "street" is recognizable, whereby the theoretical debate has reacted to political changes instead of theoretically integrating them into an overall concept. Originally, the term "street" was equated with public opinion in authoritarian Arab states (Pollock, 1993). Since the 2003 Iraq War, transnational Arab television (e.g., Al-Jazeera) has been included in the analyses, as the informal political meetings of Arab anti-war protests resonated with the new media (Lynch, 2003, p. 69; see also Zayani, 2008). During the Arab Spring following the year 2010, a further shift of the concept toward non-mediated communication became apparent: "virtual and public spaces came into a mutual synergy and produced a formidable potential for mobilizing a broad variety of actors" (Salvatore, 2013, p. 225).

Thus far, non-mediated communication is already part of theory formation in Western and non-Western research traditions. However, this comes along with a general lack in conceptualizing the interplay between different forms of communication, and comprehensive attempts to more precisely define the balance between the various modes of communication in society are rare.

One constructive example is the media dependency theory (Ball-Rockeach, 1985; Ball-Rokeach & DeFleur, 1976). Here, the authors tried to integrate both the interdependencies of the media with other social systems (politics, economy) and their reciprocal relationship with interpersonal and individual logics on the part of the media audience into one model. This approach has also been extended to new media environments, such as the Internet. It has been shown that interpersonal communication in networks of acquaintances and relatives can influence mediatized communication on Internet platforms and must therefore be regarded as an influential component of the overall communication environment (Ball-Rokeach et al., 2001; Matei & Ball-Rokeach, 2003). However, the relevance of these communication levels for cultural and political developments (democratization processes, values, ideologies, etc.) still represents a major research deficit.

A critique of the mediatization approach

In some respects, the natural antagonist of communication ecology is mediatization research. In contrast to communication ecology, mediatization research does not focus on the entirety of communication processes, only on media communication, which creates the impression of techno-capitalist determinism. While non-Western criticism of public sphere theory remains largely within the framework of mediatization thinking in its critique of Western philosophy, rationality, and the social categories of citizenship (e.g., Gunaratne, 2006; Kang, 2021), there are numerous critics of the mediatization approach worldwide. Jesús Martín-Barbero (1993) essentially denied that a social and political movement is shaped by media logic. In his view, it is not so much that the media shape society as that they bring the long-term cultural and structural driving forces of society to bear. Mikhail Buralkin and Svetlana Chernenkaya (2020) also argued that mediatization theory does not recognize the social dynamics of society. According to the authors, media are less subject to their own functional logic than to the logic of other social systems. In a similar vein, Marko Ampuja (2014, pp. 353–369) denied that media logic is a central social meta process, as it is too radically conceived and overlooks social continuities—above all, the role of capitalism. However, he also accused critics of a strong media impact of not having worked out their alternative concepts.

Functions of non-mediated communication

To understand the significance of communication ecology, it is important to describe the functions of non-mediated communication and its interactions with media communication in more detail. We therefore propose eight balance functions below, which can be divided into a narrower political dimension and a broader sociocultural dimension. This list, of course, is not exhaustive:

1. *Inclusion–exclusion balance:* Media discourses are necessarily limited in their participatory function. Mediatization is often more exclusion than inclusion. Non-mediated communication, in contrast, offers extended opportunities for participation for those who would otherwise not be heard.
2. *Consensus–difference balance:* Non-mediated communication offers social compensation for discursive deformations, especially in the mass media. If media are too consensus-oriented, it offers an arena for differentiation; if media are too polarized, new social consensuses can be created in the realm of non-mediated communication. Non-mediated communication is thus part of the social pluralism of opinions and knowledge.
3. *Transparency–opacity balance:* Authors such as Gerhards and Neidhardt (1990) and Jay Blumler (2002) have emphasized the necessity of permeability and connection between the social micro, meso, and macro levels of the public sphere, as otherwise the abbreviating media logic endangers social stability. Non-mediated communication, however, tends to move to invisible backstages, which enables necessary routine communication (e.g., of the political system) but at the same time makes transparency and linkage of the levels of media communication and non-mediated com-

munication more difficult. Communication ecology asks for the rules of a "healthy" rebalancing.

4. *Strategy–dialogue balance:* While media communication often reinforces discursive-monological and even persuasive communication, non-mediated communication can promote a dialogical, understanding-oriented exchange. Interactive and dialogic communication are not the same thing, but non-mediated communication can help stabilize the formation of opinion in a society.
5. *Community–society balance:* Communication ecology is not about a romantic relapse into the days of small interactive communities, but it is about strengthening the sense of community in the larger social context. If non-mediated communication is neglected, a loss of trust in the media threatens to escalate into a loss of trust in society. If, in turn, non-mediated communication is practiced too excessively, tribalism and polarization may be the result.
6. *Diversity–sustainability balance:* The communication-ecological society generates fewer infodemics but ideally finds a balance between information generation and knowledge production made possible by non-mediated communication.
7. *Innovation–reproduction balance*: Societies must both ensure their stability and enable necessary development, reform, or even transformation. In this interplay, a balance between different forms of communication is crucial. While media communication relates to a hegemonically formed archive of social knowledge, non-mediated communication offers alternative storage patterns through oral traditions, reconfigurations of oblivion, and knowledge production through dialogue (Flusser, 2003).
8. *Aesthetic/ritual–utilization balance:* Media communication is always subject to certain logics of utilization and exploitation (information, entertainment, education, profit, etc.) due to its integration into systemic production contexts. In contrast, aspects of non-mediated communication are more independent if we consider that the production of society itself is its purpose. Its social benefits can range from ritualized community communication to aesthetic commentary on the world. These forms of communication can create a residue of unused special knowledge, which can be called up in crises.

Stages and arenas of non-mediated communication

Erwing Goffman (1959) introduced the concept of "stages" to communication science. We also use the term "stage" in the context of communication ecology. Stages in our approach can be characterized as public as well as non-public and medial as well as non-medial, and on each stage, we find several arenas of communication, as the following figure (see next page) illustrates.

While the mediatization thesis assumes a shift of social and political communication to the frontstage of the media, communication ecology focuses its attention on: 1) the nonpublic backstage of the political system (e.g., political decision-making communication) and everyday social encounters (e.g., private households) and 2) public or semipublic ("admission-controlled") and more or less organized sidestages of assembly communication.

At its core, the concept of communication ecology assumes that all stages must be used effectively and connected with each other, on the one hand, through personal networks (e.g., opinion leadership or the multi-step flow of communication), and on the other hand, through technical and spatial conditions (e.g., digitalization, street communication). The following contribution on comparative non-mediated communication focuses in particular on the sidestage, but includes interactions with the frontstage (media) and the backstage.

Figure 1: Communication stages and arenas of social and political communication (source: author's compilation)

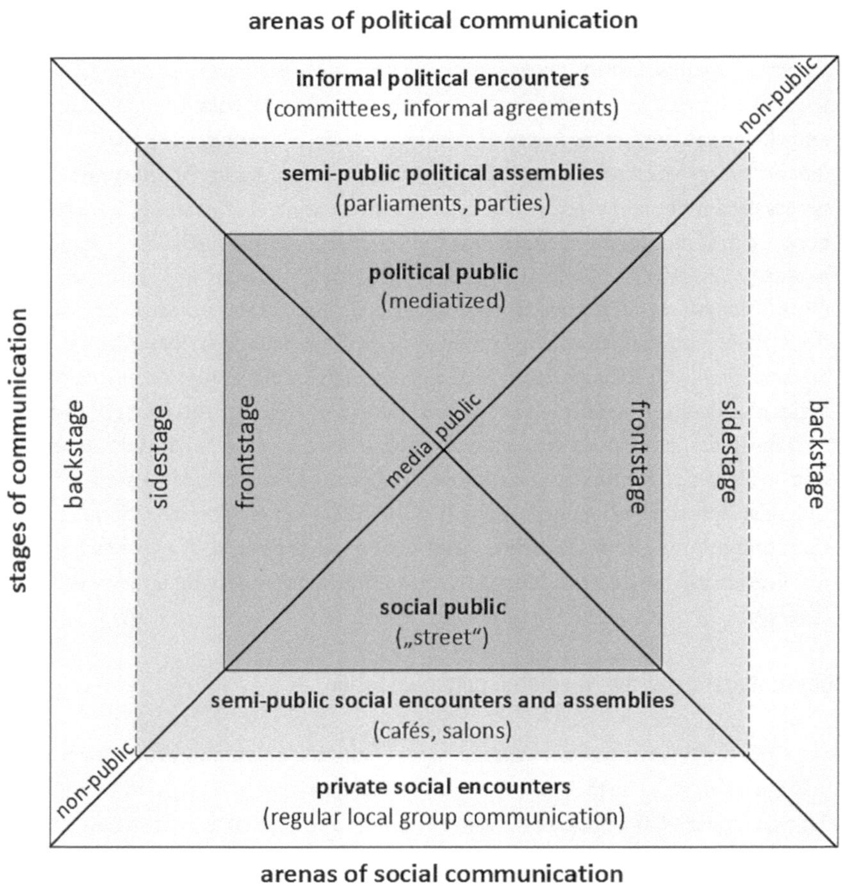

Non-mediated communication in international comparative research

The analysis of many arenas of non-mediated communication lacks country-specific and international comparative research worthy of the name. Handbooks on political communication as well as on communication studies in Africa, for example, are, with a few exceptions, such as PR and propaganda research, very media-centered; comparative approaches are rarely found there (e.g., Mano & milton, 2021b; Olukotun & Omotoso, 2017). Political science has a more comparative orientation but only marginally deals with communication issues (e.g., Boix & Stokes, 2007). The same applies to sociology, where comparative analyses can be found, for example, on family and friendship structures or urbanization developments, but with minor interest in communication.

This chapter certainly cannot fully compensate for such deficits, but it is intended to provide examples of the theoretical benefits that international comparative research in the field of non-mediated communication can generate. It must be made clear that we can neither cover all continents and countries nor all problem areas but can only show a certain panopticon of the academic potential of current and future research.

We devote ourselves to the following arenas in the two main areas of political and social communication:

- Political communication:
  - Parliamentary communication
  - Party communication
  - (Government) decision-making communication
- Social communication:
  - Regular local group communication
  - Salon communication
  - Street communication

We address some of the central issues of non-mediated communication that take place both on the sidestages of semi-public assemblies and encounter communication (parliaments, parties, salons, streets) and on the backstages ([government] decision-making communication, private regulars' meetings).

Our theoretical orientation is largely based on the functional catalog of communication ecology developed above. Other conceivable fields of research such as political opinion leaders, political advertising, political campaigns, or socialization are omitted due to space constraints.

Arenas of political communication

Parliamentary communication

Parliamentary communication takes place in various ways and in diverse locations. In the plenary chamber, discursive and strategic communication is predominantly cultivated, often with theatrical accompaniment; interactive communicative action generally takes place on the backstage in committees and informal circles, where even political opponents can negotiate compromises in dialogue (Patzelt, 1998). To understand parliamentary communication, it is therefore essential to research not only visible plenary communication but also invisible committee communication. Unfortunately, this rarely happens because these areas are not easily accessible to most researchers.

The increased public presence of parliaments in the contemporary media age has ironically led to an ever greater shift in real decision-making communication from parliaments to committees and backstage; the plenary sessions resemble pseudo-dialogical productions (Burkhardt, 2016). These productions are only part of non-mediated communication to a limited extent; they are more of an intersection of media communication and non-mediated communication. However, the actual communicative substance of many parliaments as problem solvers and legislators is found on the backstage of parliaments, where interpersonal communication is predominant.

Most of today's studies of parliamentary communication are analyses of speeches, rhetoric, and discourse in plenary sessions (e.g., Proksch & Slapin, 2014). In terms of communication ecology, they are interesting insofar as they represent an arena for the public that provides deeper insights into the actions and thoughts of elites than media analyses are generally able to do. In addition, a comparison of different countries would provide interesting insights into whether, for example, "unparliamentary expressions," as Amakali (2016) determined for Namibia, have become a global standard. Such analyses are important for an understanding of political cultures worldwide, but, as already noted, they say little about the effectiveness of the institution of parliament, as this often takes place in inaccessible committees and informal circles and needs more research for clarification.

Symbolic non-verbal communication in parliaments has also been studied in numerous countries. It has been shown that the non-verbal level is a central universal code of political communication (Abélès, 2006; Pietsukh, 2021). The importance of rituals and symbols in political territorial demarcation and in a battle of opinions, which can (but do not necessarily have to) represent social struggles, should not be underestimated. Non-mediated communication can only be captured here anthropologically and through observational research, which to date has largely been available for individual countries only and not for global comparison.

The number of existing studies is significantly larger at the intersection between non-mediated communication and media communication. European studies have shown that parliamentary debates are strongly influenced by the media agenda (Vliegenthart et al., 2016). While this seems to confirm the mediatization thesis, it is completely unclear whether such effects really play a role worldwide, that is, also in so-called "flawed" democracies, where oligarchs or even authoritarian elites more strongly shape parliaments. In this context, an integrated analysis of another intersection between non-mediated communication and media communication, that of the digital (interactive?) exchange between citizens and parliaments, so-called e-parliamentarianism, would also be valuable. Parliaments are not only related to the media environment, but they also act on the sidestage of direct contact with citizens and can open up in this direction precisely through digital means. The interactive quality of parliamentary websites is relevant here. Ideally, it would be possible for the strong media influence of the plenary debates to be reversed into a strong citizen influence of the members of parliaments, which could in principle be noticeable in decisive committees. However, such complex communication flows have not yet been recorded by science. The existing website analyses merely show that consolidated democracies can lag just as far behind in terms of interactive design as transition states and autocracies, for example, in the Arab world and Africa (e.g., Arrif, 2020; Oni et al., 2016), and that some young democracies are clearly ahead in this respect (Joshi & Rosenfield, 2013). A global parliamentary report has also shown that a country such as Uganda, which has received little attention from international research, now holds citizen hearings in parliament, that is, it incorporates elements of oral, non-mediated communication (Power, 2012, p. 32). Without overemphasizing the compensatory function of parliamentary communication in (pseudo-)parliaments, it is nevertheless possible that in some societies with severe restrictions on freedom, political non-mediated communication takes on a kind of valve function and compensates for deficits in public participation in media. To sum up, more research is needed.

Overall, even this brief overview shows that numerous phenomena can and should be examined in terms of communication ecology. The internal communicative logic of parliaments to mediate between plenary and committee communication is just as important here as the external logic of engaging on different levels of encounter, assembly, and media publics. In particular, these interactions, interdependencies, balances, and imbalances between non-mediated communication and media communication play a role in the communicative understanding of politics and have hardly been researched to date. Only by taking a communication-ecological view of parliaments can their roles in the functional areas of participation, transparency, innovation, and promoting transformation (see the section entitled Functions of non-mediated communication) really be assessed.

Party communication

Political parties have multiple functions in political systems. They organize the formation of political opinion, and the personnel for government offices are recruited from their midst. They are important links between the state and civil society and are therefore central arenas of the social public sphere. Internal party communication serves to inform the party members, develop new policies, coordinate activities, and mobilize emotions (Osei-Kufuor, 2016, p. 11). In addition to written texts such as party manifestos and advertising, oral negotiations in all party branches—from districts to the national leadership —are of central importance. In terms of communication ecology, non-mediated communication on the sidestage of parties is therefore important for balancing media communication and for social functions, such as inclusion, participation, compensation, etc.

Intuitively, the type of political system seems to be decisive for assessing the global conditions of party communication. Accordingly, intraparty communication should flourish in consolidated democracies, while it would be much more limited in defective democracies, hybrid systems, and authoritarian states. This is supported, for example, by the fact that parties are very unpopular in most Arab states (Hegasy, 2000) and have remained of little relevance in the Arab Spring of 2010. The strong role of Islamists in Egypt, for example, and the imbalances in the revolutionary process that led to the overthrow of President Mursi were the fault not only of the military but also of communication problems in political movements and parties that were unable to consolidate their own policies and bridge the strong rivalry between them (Hafez, 2017; Mustafa, 2021). Parties neither developed clear committee structures to enable internal participation nor did they generally succeed in professional external communication via press releases or social media marketing (Kassem, 2016). Such studies cannot be generalized without larger comparative studies that have not yet been conducted. However, many authors have come to similar conclusions for sub-Saharan African countries such as Ghana, Nigeria, or Kenya, namely that the parties here generally do not communicate inclusively and are often enough no more than mobilization tools in the hands of oligarchies. Therefore, disregard for internal party communication leads to a loss of trust (Kipronoh et al., 2017; Osei-Kufuor, 2016). In Nigeria, party leaders are not elected at elective congresses but are preselected by so-called "god fathers" (Simbine, n.d., p. 9). Processes of argumentative exchange in the plenary and of dialogue within and between groups and factions are thus blocked, and opinion forming is shifted from the formal to the informal level, which makes parties dysfunctional if such developments become too strong.

All of this is reminiscent of Huntington's (1968) thesis of the lack of institutionalization of developing societies. Overall, therefore, party communication around the world appears to be in an even worse state than communication in relatively state-related institutions, such as parliaments, where at least in some countries positive approaches can be seen. A conclusive judgment is not yet possible, especially not for Latin America, large parts of Asia, or Oceania, which are not discussed here.

It is important to note that the contrasts between consolidated democracies and other political systems need not always be absolute. Thomas Meyer and Lew Hinchman's (2002) criticism of what they called "mediocracy" bemoaned the decline of internal party democracy in Germany in favor of media tactics exploited by small leadership levels. According to the authors, party conferences are staged and "Americanized" for the media and are thus increasingly losing genuine participation, even in established democracies. Other analyses come to a mixed conclusion: the Obama election campaigns in the USA in particular were evidence of lively parties, although media pressure was also evident (Cohen et al., 2008). However, none of the studies to date have analyzed the entire communication flow from the grassroots to the party leadership from a communication science perspective. The internal democracy of parties is considered to be functioning when party elections work; however, the communication culture that precedes these elections is not examined. In reality, even in the "Global North," party leaders often have (too) strong an influence on the party agenda and personnel development; grassroots democracy functions in acclamation mode. The fact that young people in particular tend less toward parties and more toward political movements could indicate that even in established democracies, internal party communication is often not optimal and that the hoped for participatory and compensatory function of non-mediated communication is not realized effectively enough. It is therefore unfair to point the finger at Asia, Africa, and Latin America, because the actual deficits of party communication tend toward global universality (see also Thomass in this book).

The standard argument for the lack of success of many parties in young democracies is their frequent character as clientelist parties that gather around a leading figure, often an oligarch (Hagopian, 2007). Accordingly, parties in transition societies would not be able to function as mass parties beyond social cleavages and moderate internal social conflicts. From this perspective, hopeful experiments in democracy, such as in Tunisia, fail due to a lack of internal electoral procedures and dictatorial top-down mechanisms in political parties (Brody-Barre, 2013, p. 222). This politological and sociological argument is certainly valid. However, the appeal of an additional communication ecology analysis lies in the fact that it can show that this is not only about power structures (e.g., oligarchs) and institutionalization deficits (clientelism) in developing states, but that a lack of internal party communication also contributes to the decline of parties in the center of consolidated democracy in North America, Europe, and parts of Asia. There is also a growing trend toward the development of populist one-person parties (from Trump in the USA to Sahra Wagenknecht in Germany), which enable new feelings of community and turn parties back into spaces of experience—albeit with a tendency toward neo-authoritarianism. Loss of trust in established mass parties is therefore not only a question of power and social structure but also a factor that depends on the communication cultures of non-mediated communication, the balancing of which is extremely important for the stabilization of democracy worldwide.

(Government) decision-making communication

Political decision-making by governments is a complex process with numerous communication science facets, which are, however, hardly acknowledged by the discipline of media and communication studies. Instead, they are left to political science, which still has a limited understanding of communication as a resource. Although the influence of media coverage on government action has often been discussed—in the field of foreign policy, for example, in the debate on the so-called "CNN effect" (Robinson, 2002)—these are more external influences. Yet, the internal aspects of the decision-making process play a major role in the so-called "policy cycle" of decisions. For example, political decisions go through the following stages: problem definition (agenda setting), policy formulation, implementation, evaluation, revised problem definition, etc. (Knill & Tosun, 2011). In addition to agenda setting and framing itself, the influence of actor networks, group communication, and information processing is decisive. Once again, one could ideally assume that the freer and more democratic a political system is, the more qualified the individual action steps are: democracies would therefore have to make more rational decisions than authoritarian systems, which are characterized by numerous limitations and increments (censorship, limited actor access, limited scientific rigor, etc.).

However, we now know that democratic governments also proceed very differently—often depending on the leadership style of the top decision-makers (e.g., Donald Trump's uninformed populism), so that rationality can be limited across systems (referred to as bounded rationality; Knill & Tosun, 2011, p. 376). At the same time, it can be assumed that processes of non-mediated communication also take place in hybrid and authoritarian political systems, which may not be transparently visible, but may nevertheless exist and are important for a comparative assessment of existing communication ecologies worldwide. China's NGOs, for example, are not as publicly active as they are in direct communication contact with Chinese government agencies. Even when they operate as digital activists, their activities are often based on multimodal communication, including non-mediated interactions and networks. Moreover, online–offline relationships are explicitly described by some authors as the basis of complex "media ecologies" (Cheong & Gong, 2010; Cheong & Yang, 2017, p. 8).

Even clientelism, which is generally considered to be pronounced in authoritarian systems, is a form of communication. The specific form of (neo-)patrimonialism, for example, is explicitly based on personal networks that must not be patriarchal—that is, characterized by ethnic or tribal affiliation. At the same time, even where the theory of patrimonialism is comparative and examines authoritarian or hybrid states, such as Iraq and Pakistan, it has so far paid too little attention to the communicative competencies of consultative authoritarianism, focusing instead, for example, on the personality traits of the top decision-maker (Kinne, 2005). A more detailed examination of a) the networks and b) the modes of interaction within these networks (for example, how openly do they speak?) would be highly relevant (even if they are unfortunately hardly accessible methodologically). Communicative access to the authoritarian decision-maker is certainly extremely limited, but it is by no means certain that authoritarian communication flows must always and everywhere be inferior to democratic procedures. On the contrary, a policy logic can gain even more direct validity, while interests perceived as irrelevant, which must be represented in a democracy, can be more easily ignored. At the same time, however, this suppression also quickly leads to undercomplexity and development deficits in politics and society.

Whether there are national policy styles that can be distinguished, for example, as "liberal-pluralist" vs. "statist" (Knill & Tosun, 2011, p. 383), remains a matter of debate from a communication science perspective. After all, it cannot be ruled out that compensatory elements of communication ecology can also occur in certain "soft" forms of authoritarianism. While "transparency" is certainly no longer a function, authoritarian systems can certainly open up communication channels in the area of non-mediated communication that provide them with important information and allow subjects a hidden form of participation. This can be done without, at the same time, generating undesirable legitimization through publicity, which would undermine the character of the regime and generate domino effects up to and including revolutionary upheavals because more and more people appear as claimants. Perhaps understanding non-mediated communication better is the only way to explain the functionality of certain, often long-lasting, authoritarian systems. Of course, this is not a justification of such systems, but merely a proposal for better analysis in the sense of a more targeted criticism of them.

Arenas of social communication

Regular local group communication

The idea of the "regulars' table" (similar to the German *Stammtischkommunikation* or "kitchen talks") is intended to metaphorically describe the private backstages of social communication, whereby the focus is not on arbitrary forms of private interaction but rather on the aspects of everyday communication in which political and social topics are negotiated. In this context, the appropriation of media and their content takes a prominent place. However, from a communication ecology perspective, non-mediated forms of communication are also relevant, in particular, the rather informal private communication in small groups and communities that refer to social discourses or even prepare them. While interpersonal follow-up communication is the subject of international comparative media appropriation and reception research, there has been limited comparative research on non-mediated everyday communication.

Initially, media domestication research at least contextualized the use of media within the realm of everyday private encounters. Research has indicated how social structures are reflected in the domestic appropriation of media, such as gender and power relations or concepts of privacy (e.g., Berker et al., 2006; Silverstone & Hirsch, 1992). Non-mediated communication always remains a part of this social context in everyday life and thus provides explanatory potential for private media practices. For example, the appropriation of media by women in patriarchal societies can only be interpreted against the background of how patriarchal structures translate into social interactions. However, if non-mediated family or couple communication has already emancipated itself from patriarchal social structures, the private realm stands in stark contrast to visible public or media culture. Consequently, women's emancipative media practices cannot easily be attributed solely to the potential of technical media, as it might be merely part of previous social transformation in the everyday social world.

The consideration of non-mediated communication can thus differentiate the relationship between social micro- and macrostructures. This is particularly helpful in international comparisons, where media effects, for example, in democratic transformations, can tend to be overestimated. This was obvious in the aftermath of the Arab Spring, when academic debates highlighted the role of Internet platform communication ("Facebook revolution") in the dynamic of the revolution. Nevertheless, everyday communication about dissatisfactions with politics in interpersonal networks and on the streets, as well as protest actions of civil society, have played at least as decisive of a role as new media tools (Anas, 2016; Badr, 2015; Hafez, 2015).

This also applies to regressive development trends in consolidated democracies, where social media communication shows destructive effects, although media communication can by no means be the only explanation for anti-democratic behavior. Social cohesion is a comprehensive process that involves multiple layers of communication behavior (e.g., Bürgel et al., 2019). Two things become clear here. On the one hand, political contexts affect the function of socially mediated communication. One would hardly want to interpret privatized hate speech and conspiracy narratives on social media as a resistant and emancipative media practice of democratic citizens, while loud criticism of the political system is certainly understood as such in authoritarian contexts. On the other hand, the question arises as to whether social innovations or regressions do not have a more sustainable precursor in non-mediated communication cultures than new technologies suggest. These are embedded in a wide variety of communication patterns and are shaped by them. Critics of the mediatization thesis, such as Martín-Barbero (1993), Ampuja (2014), and Buralkin and Chernenkaja (2020), have suggested precisely these connections (see the section A critique of the mediatization approach). In the current situation where various democracies are under threat, we can ask whether the non-mediated communication cultures of right-wing extremist groups in nonpublic encounters were overlooked for too long before they were able to develop their potential on social media.

Even more clearly than the complex social contexts of media practices, media follow-up conversations show their potential for social analysis by targeting the interpretative resources within and between societies. Particularly in the context of global media entertainment, a series of international comparative studies have uncovered the patterns of interpretation contained in private lifeworlds (e.g., Gillespie, 1995; Grüne, 2016; Liebes & Katz, 1990/1994; Wagner & Kraidy, 2023). These not only reveal global similarities and differences but document everyday interactions about values, imagined identifications, and image constructions, which can be in harmony with media discourses or in discrepancy with them (in the sense of our function of the consensus–difference balance, see the section Functions of non-mediated communication). However, in this field of research, media content remains the central reference even for interactive non-mediated interpersonal and group-related negotiation.

Purely non-mediated communication on backstages is central to strengthening community and cohesion on its own. Angela Keppler's (2008) work on "table talks" continues to be an important reference for how communication can take on a social orientation function in private groups, such as the family. Even without the simultaneously shared orientation toward media, this lifeworld microcosm at the dining table becomes an important space in which the small group reassures itself in solidarity and practices social roles and positioning. In accordance with the basic sociological relevance of everyday dialogue in private encounters for mutual understanding and social cohesion, these can also initiate irritations with common senses and hence, become the nucleus for social change.

For international comparative communication studies, these encounters have important research potential. For here lies the yet undeciphered repertoire of invisible and silent opinions, the analysis of which could enrich our understanding of social transformation. This is particularly evident in authoritarian systems, where public criticism is prevented, but where it can be assumed that hidden dialogues in the lifeworld still contain a heterogeneity of political knowledge (e.g., Wolfe, 2018). Revolutions in recent decades suggest that lifeworld transformations must have preceded the mobilization of civil society. Without at least a subtle practice of critical readings, the approval and participation of mass protests can hardly be explained. However, the potential of this pre-political arena of communication for understanding change has not yet been sufficiently realized in research. Although the relevance of kitchen talks is apparent, we know far too little about the dinner table conversations in the context of today's crises—ranging from opinions in Russian homes in the current state of the war in Ukraine and state propaganda, patterns of hidden criticism about the Chinese government to the non-mediated conversations of right-wing extremist groups in Europe, or the many young peer groups trying to challenge climate politics. A consistent inclusion of non-mediated communication could break up some essentialist equations of political regime types and the actual sensitivities within societies.

This is not to deny the difficulty of methodically recording private communication, which certainly contributes to the fact that this relatively inaccessible area of social communication ecology is given less attention. At the same time, however, enormous efforts are being made in the research of digital media (e.g., big data tools and digital humanities), which in turn invalidates the argument made above and makes the development of methods itself an accomplice to the gap in recognizing the backstages in communication ecology.

Salon communication

We define salon communication as the semipublic and access-restricted space between the mere private spheres as described above and the visible public spheres, thus alluding to the sidestages of social communication. In salons, communication is already linked to regulated organization and role allocation and thus serves to illustrate the transition from pre-civil private communication to active civil society action. The simple encounters of the lifeworld today include far more salons than was the case with Habermas's (1962/1992) bourgeois public sphere. Clubs, social movements, religious communities, especially in their local manifestations, assemblies, and fan communities represent multi-layered theme-centered social institutions in which non-mediated communication rituals structure social order. Here, we can once again focus on the function of dialogical opinion forming, social knowledge production, and innovation.

As early as 1980, the MacBride Report emphasized the importance of interpersonal communication over mass media in the context of democratization tendencies worldwide, as it has a special function for the exchange of information and coordination in communities (MacBride Commission, 2004, especially part II; Rawan, 2001). In this respect, the argument has not lost much of its relevance when one considers the numerous worldwide contexts in which media communication remains unfree. It can be assumed that under these conditions, social negotiation spaces either adapt to systemic pressure or retreat into salons in which understanding-oriented communication is possible without the interference of power structures.

It is, then, worth looking at opinion leader research, which is part of a certain communication ecology tradition, as it examines the reproduction/innovation patterns of social interaction. In the evaluation of social issues, local opinion leaders also have important functions beyond media communication. Especially in supposedly traditional societies, opinion leaders have been studied, for example, in Iran, Pakistan, or Afghanistan, where they are crucial for the dissemination of information (Rawan, 2001, 2002). In the Islamic world, the importance of mosques and markets has been well documented, as they play an important role in the oral transmission of information. Especially before the Iranian Revolution of 1979, these encounters provided a non-mediated communication system that was crucial for the mobilization of people and the black market of information and propaganda (Rawan, 2001; Sreberny-Mohammadi, 1990). The role of traditional opinion leaders also reemerged recently, not least in the context of the global COVID-19 pandemic, and it has been studied in various contexts such as Indonesia or Nigeria (e.g., Agwu et al., 2020; Limilia et al., 2022).

We must assume that even in seemingly modern industrial societies, the conversations about and orientation toward opinion leaders still play an important role. Here, too, new communities have hierarchical communication rituals. From sports clubs to religious communities, from cultural projects to cafés—all these institutions of a broader cultural public sphere offer non-mediated arenas in which social developments of all kinds can be negotiated.

Even now, we find a research impulse in revolutionary contexts that debates the role of the media. This could be observed in the Arab Spring, in protests in Turkey, and also in right-wing populist movements, whose potential for manipulation through social media is often the focus of debate. However, again, it is notable that the question is rarely asked as to whether the revolutions were perhaps heard in the café conversations long beforehand. In any case, the marginalized theory of social change emphasizes that change is a continuum rather than a crisis phenomenon (Tjaden, 1972, pp. 122–124). There are many indications of the non-simultaneity of media and social discourses that can be helpful in explaining scenarios of social change. If mass media, as well as social media, are subject to disruptions, the question must be asked as to how this is reacted to in non-mediated communication contexts.

Street communication

While we have outlined the communication logics of social groups and communities in their sidestages and backstages in previous observations, the consideration of street communication now focuses less on concrete human relationships and more on the public space of everyday life. Here, people have the opportunity to observe society and interact with "strangers," which is, in line with Habermas (1995), an important communicative foundation for the development of knowledge about society, the coordination of actions in everyday life, and social cohesion.

In this respect, the use and interaction of public spaces can teach us much about the mechanisms of exclusion and inclusion. Neighborhood networks develop their communal cohesion in the space between the aforementioned street and the salons, for which they need common communication rituals and a continuous exchange. Studies have discussed the generation-specific exclusion of young people from public spaces, although for this group, direct communication with strangers and "loitering" are important experiences for learning sociality (Lieberg, 1995). However, it is not only for young people but also for all groups in society that places where people come together become relevant; the example of the information function of markets has already been mentioned.

In addition to everyday unplanned encounters, there is also organized, non-mediated communication on the street, for example, at protests. Interactive communication can even be demonstrated in international protest movements by so-called "movement brokers," who bring the movement together and keep it together (Hafez & Grüne, 2022, pp. 139–144). In local contexts, a variety of communication modes are used, ranging from the choreography of people, tents, and barriers to symbolic articulations that can be expressed in poetry, disguises, body painting, and music. In this respect, it is not enough to discuss mediated networks of protests, as social forms of protest are also connected in other ways, and their performative aesthetics have a formative effect in public space. Pnina Werbner et al. (2014) expressed this concisely: "It is not simply that social networks have spread transnationally even as they 'aggregated' massive numbers of individuals from diverse backgrounds within physical national spaces, using modern means of communication; it is that non-verbal images, music, and bodily gestures too have traveled across borders and been incorporated into local vernaculars" (p. 14).

Marjorie Mayo (2005) also drew attention to the importance of local and international encounters between members of different social movements, not with reference to the aesthetic forms of articulation but to the people-to-people exchange of social movements. Using the example of meetings of NGO participants in India, South Africa, and the USA, she argued that the direct exchange of experiences is an important key to new knowledge as well as the emancipative development of a local expert identity among the participants. This confirms the understanding and knowledge promoting character of non-mediated communication in both international and local contact between social movements.

Even if we count artistic forms of articulation such as street art, theater performances, or ritual performances as media communication in a broader media-ecological sense, their mode of communication comes closer to the characteristics of non-mediated communication: here, orality, and direct physical contact are foregrounded. These artistic forms of expressing opinions are only loosely subject to the principles of media discourse production. In this respect, these alternative media conversations also provide a basis for social communication. The existence of street art and graffiti, for example, can be a further indicator of social resistance or a politicized culture. In their everyday presence, it is precisely not a direct utilization in social dynamics that unfolds but possibly an emotional marginal commentary or symbolic reminder. In sum, there is communication whose social benefit sometimes only becomes apparent in retrospect, which can remain hidden. However, if the multiple transformations in a polycentric world need to be analyzed from a communication perspective, we should no longer rely on mediated dynamics alone.

References

Abélès, M. (2006). Parliament, politics and ritual. In E. Crewe & M. G. Müller (Eds.), *Rituals in parliament: Political, anthropological and historical perspectives on Europe and the United States* (pp. 19–40). Peter Lang.

Agwu, P., Ugwu, C. M., & Eke, C. F. (2020). Community engagement in COVID-19 responses: Evidence from qualitative interface with community opinion leaders in Enugu, Nigeria. *ASEAN Journal of Community Engagement*, 4(2), 416–434. https://doi.org/10.7454/ajce.v4i2.1115

Amakali, J. M. L. (2016). Persuasive speech acts in the Namibian National Assembly. *Journal of Advances in Linguistics*, 7(2), 1205–1217. https://doi.org/10.24297/jal.v7i2.5156

Ampuja, M. (2014). *Theorizing globalization: A critique of the mediatization of social theory*. Haymarket.

Ampuja, M., Koivisto, J., & Väliverronen, E. (2014). Strong and weak forms of mediatization theory: A critical review. *Nordicom Review*, 35(S1), 111–124. https://doi.org/10.2478/nor-2014-0107

Anas, O. (2016). Arab social media: From revolutionary euphoria to cyber realism. In B. Gunter, M. Elareshi, & K. Al-Jaber (Eds.), *Social media in the Arab World: Communication and public opinion in the Gulf States* (pp. 45–67). I.B. Tauris. https://doi.org/10.5040/9781350988057.ch-003

Arrif, M. (2020). Measuring the availability of MP's accountability and interactivity on Arab parliamentary websites using content analysis. *International Journal of Management Technology*, 7(1), 1–19. https://ssrn.com/abstract=3519169

Badr, H. (2015). Limitations of the social media euphoria in communication studies. *Égypte Monde Arabe*, 12, 177–193. https://doi.org/10.4000/ema.3451

Badr, H., Behmer, M., Fengler, S., Fiedler, A., Grüne, A., Hafez, K., Hahn, O., Hamidi, K., Hanitzsch, T., Horz, C., Illg, B., Litvinenko, A., Löffelholz, M., Radue, M., Richter, C., Thomaß, B., & Töpfl, F. (2020). Kosmopolitische Kommunikationswissenschaft:

Plädoyer für eine "tiefe Internationalisierung" des Fachs in Deutschland. *Publizistik*, *65*(3), 295–303. https://doi.org/10.1007/s11616-020-00576-6

Ball-Rokeach, S. J. (1985). The origins of individual media-system dependency: A sociological framework. *Communication Research*, *12*(4), 485–510. https://doi.org/10.1177/0093650850120040003

Ball-Rokeach, S. J., & DeFleur, M. L. (1976). A dependency model of mass-media effects. *Communication Research*, *3*(1), 3–21. https://doi.org/10.1177/009365027600300101

Ball-Rokeach, S. J., Kim, Y.-C., & Matei, S. (2001). Storytelling neighborhood: Paths to belonging in diverse urban environments. *Communication Research*, *28*(4), 392–428. https://doi.org/10.1177/009365001028004003

Bergermann, U. (2012). Postkoloniale Medienwissenschaft: Mobilität und Alterität von Ab/Bildung. In J. Reuter & A. Karentzos (Eds.), *Schlüsselwerke der Postcolonial Studies* (pp. 267–281). VS Verlag für Sozialwissenschaften. https://doi.org/10.1007/978-3-531-93453-2_20

Berker, T., Hartmann, M., Punie, Y., & Ward, K. (Eds.). (2006). *Domestication of media and technology*. Open University Press.

Blumler, J. G. (2002). Wandel des Mediensystems und sozialer Wandel: Auf dem Weg zu einem Forschungsprogramm. In H. Haas & O. Jarren (Eds.), *Mediensysteme im Wandel: Struktur, Organisation und Funktion der Massenmedien* (pp. 170–188). Braumüller Wilhelm.

Boix, C., & Stokes, S. C. (Eds.). (2007). *The Oxford handbook of comparative politics*. Oxford University Press. https://doi.org/10.1093/oxfordhb/9780199566020.001.0001

Brody-Barre, A. G. (2013). The impact of political parties and coalition building on Tunisia's democratic future. *Journal of North African Studies*, *18*(2), 211–230. https://doi.org/10.1080/13629387.2012.742365

Buralkin, M. Y., & Chernenkaya, S. (2020). The problem of the scientific status theory of mediatization. In O. D. Shipunova, V. N. Volkova, A. Nordmann, & L. Moccozet (Eds.), *Communicative strategies of information society* (pp. 272–280). European. https://doi.org/10.15405/epsbs.2020.03.02.32

Bürgel, C., Buttgereit, L., Helsper, S., Hoffmann, L., Horn, M., Jochims, N., Nissen, R., & Roßdeutscher, J. (2019). *Medialer Wandel und gesellschaftlicher Zusammenhalt*. Frank & Timme.

Burkhardt, A. (2016). German parliamentary discourse since 1848 from a linguistic point of view. In P. Ihalainen, C. Ilie, & K. Palonen (Eds.), *Parliament and parliamentarism: A comparative history of a European concept* (pp. 176–191). Berghahn. https://doi.org/10.2307/j.ctvgs0b7n.16

Calderón, C. A., Barranquero, A., & Tanco, E. G. (2018). From media to *Buen Vivir*: Latin American approaches to indigenous communication. *Communication Theory*, *28*(2), 180–201. https://doi.org/10.1093/ct/qty004

Chaffee, S. H. (1986). Mass media and interpersonal channels: Competitive, convergent, or complementary? In G. Gumpert & R. Cathcart (Eds.), *Inter/media: Interpersonal communication in a media world* (3rd ed., pp. 62–80). Oxford University Press.

Cheong, P. H., & Gong, J. (2010). Cyber vigilantism, transmedia collective intelligence, and civic participation. *Chinese Journal of Communication*, *3*(4), 471–487. https://doi.org/10.1080/17544750.2010.516580

Cheong, P. H., & Yang, A. (2017). Chinese non-governmental organizations, media and culture: Communication perspectives, practices, and provocations. *Chinese Journal of Communication*, *10*(1), 1–11. https://doi.org/10.1080/17544750.2017.1275084

Cohen, M., Karol, D., Noel, H., & Zaller, J. (2008). Political parties in rough weather. *Forum*, *5*(4), Article 3. https://doi.org/10.2202/1540-8884.1214

Flusser, V. (2003). *Kommunikologie*. Fischer.

Frassinelli, P. P. (2021). Franz Fanon, Ngũgĩ wa Thiong'o, and African media and communication studies. In W. Mano & v. c. milton (Eds.), *Routledge handbook of African media and communication studies* (pp. 43–53). Routledge.

Gerhards, J., & Neidhardt, F. (1990). *Strukturen und Funktionen moderner Öffentlichkeit: Fragestellungen und Ansätze*. Wissenschaftszentrum Berlin für Sozialforschung. https://bibliothek.wzb.eu/pdf/1990/iii90-101.pdf

Giesecke, M. (2007). *Die Entdeckung der kommunikativen Welt: Studien zur kulturvergleichenden Mediengeschichte*. Suhrkamp.

Gillespie, M. (1995). *Television, ethnicity and cultural change*. Routledge.

Goffman, E. (1959). *The presentation of self in everyday life*. Anchor.

Grüne, A. (2016). *Formatierte Weltkultur? Zur Theorie und Praxis globalen Unterhaltungsfernsehens*. transcript.

Gunaratne, S. A. (2006). Public sphere and communicative rationality: Interrogating Habermas's Eurocentrism. *Journalism & Communication Monographs*, *8*(2), 93–156. https://doi.org/10.1177/152263790600800201

Habermas, J. (1992). *The structural transformation of the public sphere: An inquiry into a category of bourgeois society* (T. Burger & F. Lawrence, Trans.). Polity Press. (Original work published 1962)

Habermas, J. (1995). *Theorie des kommunikativen Handelns. Band 1: Handlungsrationalität und gesellschaftliche Rationalisierung/ Band 2: Zur Kritik der funktionalistischen Vernunft*. Suhrkamp.

Hafez, K. (2015). Radically polarized publics and the demise of media freedom in Egypt. *Égypte Monde Arabe*, *12*, 37–49. https://doi.org/10.4000/ema.3397

Hafez, K. (2017). Zivilgesellschaft am Scheideweg der Demokratie: Anerkennungskämpfe in der ägyptischen Öffentlichkeit. In T. Demmelhuber, A. T. Paul, & M. Reinkowski (Eds.), *Arabellion: Vom Aufbruch zum Zerfall einer Region?* (*Leviathan* special vol. 31, pp. 279–300). Nomos. https://doi.org/10.5771/9783845276038

Hafez, K., & Grüne, A. (2022). *Foundations of global communication: A conceptual handbook*. Routledge. https://doi.org/10.4324/9781003255239

Hagopian, F. (2007). Parties and voters in emerging democracies. In B. Carles & S. C. Stokes (Eds.), *The Oxford handbook of comparative politics* (pp. 582–603). Oxford University Press. https://doi.org/10.1093/oxfordhb/9780199566020.003.0024

Hall, S. (1980). Encoding/decoding. In S. Hall, D. Hobson, A. Lowe, & P. Willis (Eds.), *Culture, media, language: Working papers in cultural studies, 1972–79* (pp. 128–138). Routledge.

Hegasy, S. (2000). They dare to speak out: Changes in the political culture of Egypt, Morocco and the Arab world. In K. Hafez (Ed.), *The Islamic world and the West: An introduction to political cultures and international relations* (pp. 146–160). Brill. https://doi.org/10.1163/9789047400370_013

Huesca, R., & Dervin, B. (1994). Theory and practice in Latin American alternative communication research. *Journal of Communication*, *44*(4), 53–73. https://doi.org/10.1111/j.1460-2466.1994.tb00699.x

Huntington, S. P. (1968). *Political order in changing societies*. Yale University Press.

Issa, Z. (2022, July 13). *Language as culture: A critique of Ngugi wa Thion'go*. Medium. https://zinahissa.medium.com/language-as-culture-a-critique-of-ngugi-wa-thiongo-969ed50508ab

Joshi, D., & Rosenfield, E. (2013). MP transparency, communication links and social media: A comparative assessment of 184 parliamentary websites. *Journal of Legislative Studies*, *19*(4), 526–545. https://doi.org/10.1080/13572334.2013.811940

Kang, J. (2021). Old and new questions for the public sphere: Historicizing its theoretical relevance in post-Cold War South Korea. *Media, Culture and Society*, *43*(1), 158–170. https://doi.org/10.1177/0163443720939480

Kassem, T. (2016). The role of political parties in Egypt's political transition: A mirage. *International Journal of Political Science, Law and International Relations*, *6*(5), 11–36.

Katz, E., & Lazarsfeld, P. F. (1955). *Personal influence: The part played by people in the flow of mass communication*. Transaction.

Keppler, A. (2008). *Tischgespräche: Über Formen kommunikativer Vergemeinschaftung am Beispiel der Konversation in Familien* (3rd ed.). Suhrkamp.

Kinne, B. J. (2005). Decision making in autocratic regimes: A poliheuristic perspective. *International Studies Perspectives*, *6*(1), 114–128. https://www.jstor.org/stable/44218356

Kipronoh, L. J., Kihara, P., & Mbebe, J. (2017). Factors affecting strategy implementation in political parties in Kenya: A case of selected political parties in Nairobi County. *International Academic Journal of Human Resource and Business Administration*, *2*(3), 535–555. https://www.iajournals.org/articles/iajhrba_v2_i3_535_555.pdf

Knill, C., & Tosun, J. (2011). Policy-making. In D. Caramani (Ed.), *Comparative politics* (pp. 373–388). Oxford University Press.

Lieberg, M. (1995). Teenagers and public space. *Communication Research*, *22*(6), 720–744. https://doi.org/10.1177/009365095022006008

Liebes, T., & Katz, E. (1994). *The export of meaning: Cross-cultural readings of Dallas*. Polity Press. (Original work published 1990)

Limilia, P., Indriani, S. S., & Alam, P. W. (2022). Opinion leaders and health information: A systematic literature review in Indonesia. *Proceeding of International Conference on Communication Science*, *2*(1), 737–744. https://doi.org/10.29303/iccsproceeding.v2i1.110

Lynch, M. (2003). Beyond the Arab street: Iraq and the Arab public sphere. *Politics and Society*, *31*(1), 55–91. https://doi.org/10.1177/0032329202250170

MacBride Commission. (2004). *Many voices, one world: Towards a new, more just, and more efficient world information and communication order* (2nd ed.). Rowman & Littlefield.

Mano, W., & milton, v. c. (2021a). Decoloniality and the push for African media and communication studies: An introduction. In W. Mano & V. C. milton (Eds.), *Routledge handbook of African media and communication studies* (pp. 1–18). Routledge.

Mano, W., & milton, v. c. (Eds.). (2021b). *Routledge handbook of African media and communication studies*. Routledge.

Martín-Barbero, J. (1993). *Communication, culture and hegemony: From the media to mediations*. Sage.
Matei, S., & Ball-Rokeach, S. J. (2003). The Internet in the communication infrastructure of urban residential communities: Macro- or mesolinkage? *Journal of Communication, 53*(4), 642–657. https://doi.org/10.1111/j.1460-2466.2003.tb02915.x
Mayo, M. (2005). *Global citizen: Social movements and the challenge of globalization*. Zed Books.
Merton, R. K. (1968). *Social theory and social structure*. Free Press.
Meyer, T., & Hinchman, L. (2002). *Media democracy: How the media colonize politics*. Polity Press.
Meyrowitz, J. (2018). Medium theory and cultural transformation. In L. Grindstaff, M.-C. M. Lo, & J. R. Hall (Eds.), *Routledge handbook of cultural sociology* (2nd ed., pp. 629–638). Routledge.
Münch, R. (1995). *Dynamik der Kommunikationsgesellschaft*. Suhrkamp.
Mustafa, I. (2021). *Revolution und defekte Transformation in Ägypten: Säkulare Parteien und soziale Bewegungen im "Arabischen Frühling"*. transcript.
Ngũgĩ, w. T. (1986). *Decolonizing the mind: The politics of language in African literature*. James Currey.
Olukotun, A., & Omotoso, S. A. (Eds.). (2017). *Political communication in Africa*. Springer. https://doi.org/10.1007/978-3-319-48631-4
Oni, A. A., Oni, S., & Ibietan, J. (2016). ICT and democratic parliament in Africa: State of the matter. *Journal of Governance and Development, 12*(1), 71–85. https://e-journal.uum.edu.my/index.php/jgd/article/view/13409
Osei-Kufuor, P. (2016). *A comparative study on internal political party communication policies in Ghana*. United Nations Development Programme. https://info.undp.org/docs/pdc/Documents/GHA/Best%20Practice%20Study%20on%20Internal%20Commucation%20Policies%20for%20Political%20Parties.pdf
Patzelt, W. J. (1998). Parlamentskommunikation. In O. Jarren, U. Sarcinelli, & U. Saxer (Eds.), *Politische Kommunikation in der demokratischen Gesellschaft: Ein Handbuch mit Lexikonteil* (pp. 431–441). VS Verlag für Sozialwissenschaften. https://doi.org/10.1007/978-3-322-80348-1_32
Pietsukh, O. (2021). Nonverbal communication means in the UK parliamentary debates. *Wisdom, 17*(1), 80–94. https://doi.org/10.24234/wisdom.v17i1.461
Pollock, D. (1993). *The 'Arab street'? Public opinion in the Arab world*. The Washington Institute for Near East Policy. https://www.washingtoninstitute.org/policy-analysis/arab-street-public-opinion-arab-world
Postman, N. (2000, June 16–17). *The humanism of media ecology* [Keynote]. Inaugural Media Ecology Association Convention, New York, NY, United States. https://www.media-ecology.net/publications/MEA_proceedings/v1/humanism_of_media_ecology.html
Power, G. (2012). *Global parliamentary report: The changing nature of parliamentary representation*. Inter-Parliamentary Union; United Nations Development Programme. http://archive.ipu.org/pdf/publications/gpr2012-full-e.pdf
Proksch, S.-O., & Slapin, J. B. (2014). *The politics of parliamentary debate: Parties, rebels, and representation*. Cambridge University Press. https://doi.org/10.1017/CBO9781139680752

Rawan, S. M. (2001). Interaction between traditional communication and modern media: Implications for social change in Iran and Pakistan. In K. Hafez (Ed.), *Mass media, politics, and society in the Middle East* (pp. 175–198). Hampton.

Rawan, S. M. (2002). Modern mass media and traditional communication in Afghanistan. *Political Communication, 19*(2), 155–170. https://doi.org/10.1080/10584600252907425

Robinson, P. (2002). *The CNN effect: The myth of news media, foreign policy and intervention*. Routledge.

Salvatore, A. (2013). New media, the 'Arab Spring,' and the metamorphosis of the public sphere: Beyond Western assumptions on collective agency and democratic politics. *Constellations, 20*(2), 217–228. https://doi.org/10.1111/cons.12033

Saucedo Añez, P. C. (2019). Die lateinamerikanische Medien- und Kommunikationswissenschaft zwischen ideologischer Prägung und begrenzter Internationalisierung: Zeit für Selbstkritik. *Global Media Journal – German Edition, 9*(2). https://doi.org/10.22032/dbt.40625

Scolari, C. A. (2012). Media ecology: Exploring the metaphor to expand the theory. *Communication Theory, 22*(2), 204–225. https://doi.org/10.1111/j.1468-2885.2012.01404.x

Sesanti, S. (2022). Humane communication in African language: African philosophical perspectives. In Y. Miike & J. Yin (Eds.), *The handbook of global interventions in communication theory* (pp. 122–135). Routledge.

Shah, D. V., McLeod, D. M., Rojas, H., Cho, J., Wagner, M. W., & Friedland, L. A. (2017). Revising the communication mediation model for a new political communication ecology. *Human Communication Research, 43*(4), 491–504. https://doi.org/10.1111/hcre.12115

Silverstone, R., & Hirsch, E. (Eds.). (1992). *Consuming technologies. Media and information in domestic space*. Routledge.

Simbine, A. T. (n.d.). *Political parties and internal party democracy in Nigeria*. Centre for Democracy and Development.

Sreberny-Mohammadi, A. (1990). Small media for a big revolution: Iran. *International Journal of Politics, Culture, and Society, 3*(3), 341–371. https://doi.org/10.1007/BF01384966

Strate, L. (2017). *Media ecology: An approach to understanding the human condition*. Peter Lang. https://doi.org/10.3726/978-1-4331-4005-1

Tjaden, K. H. (1972). *Soziales System und sozialer Wandel*. Ferdinand Enke Verlag.

Ugboajah, F. (1982). "Oramedia" or traditional media as effective communication options for rural development in Africa. *Communicatio Socialis, 15*(3), 211–221. https://doi.org/10.5771/0010-3497-1982-3-211

Vliegenthart, R., Walgrave, S., Baumgartner, F. R., Bevan, S., Breunig, C., Brouard, S., Bonafont, L. C., Grossman, E., Jennings, W., Mortensen, P. B., Palau, A. M., Sciarini, P., & Tresch, A. (2016). Do the media set the parliamentary agenda? A comparative study in seven countries. *European Journal of Political Research, 55*(2), 283–301. https://doi.org/10.1111/1475-6765.12134

Wagner, M. C., & Kraidy, M. M. (2023). Watching Turkish television dramas in Argentina: Entangled proximities and resigned agency in global media flows. *Journal of Communication, 73*(4), 304–315. https://doi.org/10.1093/joc/jqad001

Werbner, P., Webb, M., & Spellman-Poots, K. (Eds.). (2014). *The political aesthetics of global protest: The Arab Spring and beyond*. Edinburgh University Press.
Wilson, D. (2008). Research on traditional communication in Africa: The development and future directions. *African Communication Research*, 1(1), 47–59.
Winchatz, M. R. (2017). Jammern [Whining] as a German way of speaking. In D. Carbaugh (Ed.), *The handbook of communication in cross-cultural perspective* (pp. 65–75). Routledge.
Wolfe, A. W. (2018). Dialogue and deliberation as agonistic resistance: Designing interactional processes to reconstitute collective identities. *Journal of Public Deliberation*, 14(2), Article 5. https://doi.org/10.16997/jdd.307
Yousuf, M. (2018). Media engagement in networked environments: An ecological perspective. In K. A. Johnston & M. Taylor (Eds.), *The handbook of communication engagement* (pp. 253–268). Wiley Blackwell. https://doi.org/10.1002/9781119167600.ch17
Zayani, M. (2008). Courting and containing the Arab street: Arab public opinion, the Middle East and U.S. public diplomacy. *Arab Studies Quarterly*, 30(2), 45–64. https://www.jstor.org/stable/41858543

A context-led approach to media systems research

Melanie Radue, Johanna Mack & Carola Richter

Introduction

Why is the printed press booming in India but declining in the USA? Why are Swiss citizens voting to keep public service broadcasting fees, while France is abolishing them? Why are press councils a tool of media accountability in one country and a tool of government control in another? Why are the media systems that developed in the countries of the former Soviet Union so different?

It is comparative media systems research that tries to answer these questions. By looking at the structural conditions for as well as the political and economic influences on media production, media ownership, and media use, media system research aims to explain commonalities and differences regarding the development of media in its respective contextual constraints (Esser, 2014, p. 15). It is thus an important field in communication studies, enabling us to "process world knowledge" (Kleinsteuber, 2004, p. 83) by pointing us to a multitude of sometimes contradictory global developments, to test theories regarding their applicability beyond a single case, and to learn from other examples to understand shortcomings and predict possible developments in one's "own" system (p. 68). Yet, current media system research has a strong focus on Europe and North America, that is, a Western bias both in its geographical scope as well as with regard to the models, indicators, and categories predominantly developed in and for Western contexts. This has resulted in media system typologies that reflect a Western normative ideal of how media should function, and what the media–politics relations should look like. For decades, Siebert et al.'s *Four Theories of the Press* (1956), which favored a "social responsibility" or a "libertarian" model of media systems against an "authoritarian" or "soviet-totalitarian" model, has been used to classify media systems around the world in a simplified and often pejorative way, indicating that some countries are "not there yet" in their development toward the preferred liberal-market democracy. This typology was influential until Hallin and Mancini's *Comparing Media Systems* was published in 2004. Their comparison of 18 European and North American countries indicates three possible models of media–politics relations, including the "liberal," the "democratic-corporatist," and the "polarized-pluralist" models. In the follow-up research, many authors have tried to locate countries in the Global South within these models, often concluding that they need to be described as somehow polarized-pluralist—making this model a "catch-all category" (Voltmer, 2011, p. 225). This has produced a media system research tradition which squeezes media systems all around the world into predefined pigeonholes. While such an approach enables us to differentiate the world into, for example, free and unfree, liberal and illiberal, or democratic and authoritarian systems, and thus evaluate developments from a normatively loaded Western point of view, it prevents us from learning more about the specific characteristics of media systems beyond these narrowly defined categories.

Altogether, we are faced with challenges resulting from Eurocentric work traditions, processes, and structures in (communication) studies as a whole. For media systems research, this specifically means that the unadjusted application of predominantly Eurocentric normative theories and models closes our eyes to contextual norms, values, theories, and knowledge as soon as we consider communication contexts that lie outside the much-researched West. This results in blind spots regarding communication phenomena in some regions of the world, an implicit Eurocentric universalism, and a normative hegemonic practice that perpetuates othering and the maintenance of colonial power imbalances in knowledge production and diffusion. Even though increasing globalization has facilitated access to various contexts, the economic, political, and cultural importance of different world regions has increased, and the boundaries between "Western" and "Eastern" (academic) paradigms are becoming increasingly blurred (Kuo & Chew, 2009, pp. 423–427)—this is hardly reflected in current media systems research. It seems as if this long-standing research tradition has made us ask the wrong questions. Media system research should help us to "process world knowledge" and to gain an understanding of how media structures actually shape media practices and how, in turn, practices and policies shape media structures. Building on previous attempts to de-Westernize media system research (see e.g., Curran & Park, 2000; Stremlau, 2013; Voltmer, 2011), this chapter will shed light on the shortcomings of the dominant lines of current media system research and propose instead a context-led approach to media system analysis including a relational research framework.

In the next section, we will first problematize these shortcomings, contrasting them with context sensitive and relational approaches. In the subsequent section, we will then offer some ideas on how to translate these abstract approaches into more concrete research designs and methodologies by building on a review of relevant studies in this field. Our aim is to provide a perspective for a cosmopolitan version of media system research.

Challenges to achieving a cosmopolitan media system research

In the following, we identify four crucial challenges that in our view need to be reflected upon in order to move toward a more cosmopolitan media systems research.

First, we must reflect on how the systems approach is typically used. Mainly emerging from normative theories in political science, the concept of systems is widely used, but often left without a clear definition in media system research (Mancini, 2020, p. 5762). For this reason, we want to raise awareness of the arbitrariness of the use of the term "system" (Rantanen, 2013, p. 257) and emphasize the notion of "context" that should guide media system research. This might also include a reconsidering of the boundaries of the concept of media systems, which in most academic literature refers to professional mass media. We do not argue to dispense with the use of the term "media system" research, but to emphasize its various connotations. In the following, we refer to media systems as a rather general term, as Mancini (2020, p. 5764) does in his critical assessment of the concept of media system: "As an abstract idea, the definition itself of a media system and its borders may vary in relation to the scholar's investigation goals." This illustrates the necessity to always (re)define what is understood as a "media system" for our specific research questions.

A system is mostly understood as a combination of elements with complex structures and logics (Thomaß, 2013, p. 15). Those structures and logics are shaped by their relationships with one another. This becomes clear in Hallin's (2016) definition of a systems approach:

> [A] basic element of the "systems perspective" is the idea that the elements of a system are defined by their relationships with one another, and therefore cannot be understood without reference to the whole pattern of relationships. (p. 2)

In this regard, a system's perspective highlights relations, but strangely enough research often does not put them into focus. So far, the media system analysis literature has tended to focus on characteristics of categories and factors rather than their relations.

At this juncture, a relational research perspective offers a starting point for overcoming the isolated analysis of components and structural conditions and determinations. As all conditions in a media context are determined by social relationships, a relational perspective does not solely study structures and actors themselves, which is characteristic of most "systems"-oriented research (Emirbayer, 1997, p. 282; Häußling, 2010, p. 70). The relational paradigm of network theory shifts the perception of media contexts along fixed categories and dimensions to the perception of social realities stemming from connectivity, processes, and relations.

The appeal of the network concept as an approach for researching media systems lies in the fact that it is situated between the micro and macro levels, that it encompasses dynamics, and that, as a relational approach, it provides a genuine sociological impetus for tracing mechanisms of social integration and the conditions and results of transformation processes (see also Emirbayer, 1997; Emirbayer & Goodwin, 1994). Network analysis usually takes place at several levels, from the micro level of individual relationships to the macro level of network structures in entire societies. There is also an emphasis on structural analysis to identify patterns, centralities, subgroups, and other structural characteristics. Altogether, a network-focused analysis comprises several key concepts and principles, such as (power) relationships as central elements for understanding social structures and actions (Löwenstein, 2017), contextuality as cultural, historical, and structural dimensions (Emirbayer & Goodwin, 1994), dynamics (Hollstein, 2006), and social capital. Thus far, network analysis has mainly been used for investigating interpersonal and organizational communication. It is frequently applied to questions that deal with social formations in and around the Internet and examine organizational structures. However, some studies and approaches emphasize the value of applying network analysis on a macro level to investigate the network relations of media systems (e.g., Castells, 2001, 2007, 2011; Radue, 2022; Rainie & Wellman, 2012; van Dijk, 2021). Still, questions of power relations in media systems are often inadequately addressed. To overcome this problem, decolonial, feminist, or intersectional approaches as specific lenses of relational research frameworks might help to investigate the social realities of media systems by researching rights, resources, and representation. The feminist movement has brought about sophisticated debates about questions of voice and representation over the last decades (Harvey, 2020; Kurasawa, 2004, p. 241). Such perspectives acknowledge the key role of the media in circulating ideologies and its impact on (id)entities. They also help establish a focus on communication power structures in media system research.

In sum, a relational research framework enables a critical perspective by analyzing power relations, social inequality, and structural barriers that are embedded in relationships and contexts. Integrating this lens into approaches for researching media systems can be achieved by putting an emphasis on the relations between the different actors and institutions that are typically regarded in media systems models. Understanding a relational perspective as context-led research thus enables us to approach communication contexts more openly and thus grasp the details of what is relevant on the ground.

Second, the categories used are often normatively loaded with assumptions of what media structures should look like according to a Western perspective. For example, and with regard to analyses of media markets—a key category in media systems research—academics in the tradition of the critical political economy of communication (e.g., Wasko et al., 2011) have already identified that the functioning of media markets is undermined by neoliberal presumptions. However, the reality of market concentration in many Western countries has shown that the sole competition of entities does not guarantee the best functioning of the media (Knoche, 2013). Public service broadcasting is an important counterweight because it "addresses audiences as public citizens, not individual consumers" (Michalis, 2024, p. 131). Yet, with regard to media systems beyond Europe and North America, the neoliberal preference for a privatization of media outlets and infrastructure as being key for independent media has remained dominant in media systems research, even though some authors have viewed it very critically, as these measures have mainly led to pseudo-liberalization and crony media businesses loyal to regime politics (Della Ratta et al., 2015; Richter & Gräf, 2015).

Thus, media system research that takes Western-centric models and concepts as a starting point would find many aspects of media systems in the Global South deficient by default, positioning itself as a norm. In the same vein that Powers and Vera-Zambrano (2018) "argue for paying closer attention to the epistemologies that shape the very ways comparative research on journalism and political communication is produced" (p. 144), we see contextualization as an alternative approach to account for the limitations that universalism brings about.

For us, therefore, cosmopolitan media systems research should be shaped by a relational approach that is informed by the respective local context. When we apply such a context-led research lens when doing media system research, we bring into focus the quality of relations between societal subsystems and the formal and informal structures and processes that shape them. We therewith can also overcome the predominance of normative Western assumptions on what media structures should look like and rather focus on an analysis of how they actually are. Various examples from different parts of the world illustrate how such a relational and context-led approach would allow us to better understand media systems that are shaped by war and conflict, which Western normative assumptions and categories are of little value in explaining (Radue, 2022).

For example, in conflict settings, traditional media channels might be captured by conflicting parties, so that citizens mistrusting these sources turn to informal channels or interpersonal communication. Stremlau (2013) has explained this dynamic with reference to Somaliland, showing how such examples can challenge dominant understandings of media systems (p. 289).

Regarding contexts where information flows do not mainly pass through broadcasting, print, and journalistic online media, scholars have argued that we should think of communication as a transaction happening in a context shaped by different (f)actors and media and put greater emphasis on the specific societal, technological, and economic factors, but also on the nature of media themselves (see also Hafez & Grüne in this book). This is not only of interest in contexts with weak mass media but also regarding the Internet, where strict differentiations between senders and receivers, public, and private communication tend to fade (Ruotsalainen & Heinonen, 2015).

Similarly, research on media systems in Southeast Asia has highlighted the necessity of deeply looking into the contexts to understand locally relevant categories of media systems. In Myanmar, for example, where over 100 different languages are spoken, this diversity is not reflected in the media landscape. Instead, the policy of teaching only Burmese in schools is also mirrored in the media landscape, where we mainly find Burmese publications. Criminalization of unauthorized publication in ethnic languages stifles cultural expression in Myanmar. A context-led approach can reveal "knowledge necessary to understand underlying concepts, or rather their relationships and how they interact, which in turn provide vital insight on the formation of media environments" (Radue, 2022, pp. 172–173). For example, "neither does the existence of a multilingual society alone necessitate the emergence of a multilingual media market, nor is the mere existence of a multilingual media market an indicator for a free and plural media system in a multilingual society" (Radue, 2022, p. 173). This becomes clear only when we delve into the context of social structures.

Such examples illustrate that a research lens that applies certain Western normative assumptions and does not embed media structures into their context can overlook and thus miss the point of how communication works on the ground. To overcome this problem, Wasserman (2013) argued to bring in "'listening' as an ethical value" (p. 77). Basically, we need to listen to the context itself with its own rules, norms, and values, which can speak to us as researchers. Decolonizing media system research is thus the acknowledgment of the importance of context and the respect for the particularity of contextual diversity.

Third, current media system research often fails to de-Westernize its concepts and lenses also because it rarely analyzes media systems beyond the West. On the one hand, there are many geographical blind spots in media systems research beyond Europe and North America. Even if countries of the Global South are included, it is mostly the big and powerful countries that are in focus, such as Brazil and Mexico in Latin America or

South Africa on the African continent, but most of the rest remain blind spots. On the other hand, if media systems from the Global South are analyzed, this is often done using established Western-centric models and categories (e.g., Hallin & Mancini, 2011) that often lead again to a reproduction of common knowledge instead of allowing us to learn about different concepts and phenomena (Badr et al., 2020). A typical reason for leaving out non-European or South American countries in (large N-) comparative studies is the unavailability or unreliability of statistical data such as media outreach and circulation, number of journalists and media outlets, or media ownership data[1]. As the data for testing traditional concepts are missing and the specific contexts are not acknowledged, many regions in the world remain underexplored and blind spots for media system research.

Fourth, another critical aspect is the unit of comparison. Most media system research takes the nation-state as the unit of comparison. However, some authors claim that in an ever more globalized and networked world—with international news agencies and broadcasting companies such as Al Jazeera that connect several countries (see Kraidy, 2011) and especially with the Internet and social media—such boundaries have become obsolete. With a view to platformization and the rise of social media, search engines, and news aggregators that are not (yet) widely regulated by national governments (Rochefort, 2020), one could argue that at least some parts of media systems are not contingent within national boundaries. The massive scale of international migration and refugee movements adds to this. Instead, overlapping scapes may form (Appadurai, 1995/2010) which connect people based on shared language, physical space, economic and social context, and media. In cases such as Afghanistan, Myanmar, Iran, or India with millions of people living abroad and diasporic and exile media at work, how can the boundaries of a media system be defined? The Kurdish media, for example, function as a transnational network that defies Western ideas of journalism not only through its distinctly activist nature but also by going beyond geographically limited spaces (Schamberger, 2021). However, Flew and Waisbord (2015) stressed that despite such transnational tendencies, the nation-state remains one of the most useful "boxes" as it shapes regulations, institutions, and practices most. Yet, depending on the aim of a study, it might be valuable to experiment with, for example, studying the context of global platforms or the media that connect a specific community, such as diasporic groups sharing a language and cultural-historical characteristics. On a more abstract level, Hafez (2007) has suggested that—without neglecting the national borders as determining characteristics of media systems—we should look at "system connectivity" (p. 9) established through, for example, transnational satellite TV, the Internet or international journalism as well as at strategic media influences from abroad, such as public diplomacy and propaganda that aim for a "system change" (p. 13) or transnational media policies and other forms of "system interdependence" (p. 21) that transcend national borders. Again, context-led research and a relational perspective matter here to adequately grasp what shapes a media system.

In this section, we have argued for applying a context-led research perspective in media systems research. This demands in-depth case studies and a relational under-

1 For example, the Media Ownership Monitor (MOM) Latin America, https://latin-america.mom-gmr.org/en/.

standing of structures, including acknowledging power asymmetries between actors and structure as well as taking into consideration transnational relations of some parts of the system. In addition, the application of Western normative concepts to evaluate media structures and system performance needs to be critically reviewed and avoided. The next section will explain how to operationalize this in media system research.

A cosmopolitan approach to media system research

Above, we identified the crucial shortcomings of current media system research. To overcome these shortcomings, we will now offer recommendations on how to cosmopolitanize our practices in media system research. We outline approaches that will facilitate context-led research, thereby enabling us to tackle widespread Eurocentrism in media systems analysis and shed light on (geographical) blind spots in media system research.

Reflecting researchers and media systems' positionality

Datta (2018) argued that "decolonization is an on-going process of becoming, unlearning, and relearning regarding who we are as a researcher and educator" (p. 2). Thus, a first step in doing meaningful media system research is to reflect on one's own positionality. As outlined above, media system research often builds on assumptions from Western societies and democratic political structures, emphasizing "Western values," such as individuality, equality, free markets, or secularism as normative and universal standards (Pokhrel, 2011, p. 321; see also Volk in this book). To acknowledge and consciously identify this internalization of Eurocentrism in most research is a first step. When, for example, only 8% of the world's population is living under conditions of functional democratic structures (Economist Intelligence Unit [EIU], 2022), how can we so often take these conditions as a starting point for our analysis and as granted universal norms? "Unlearning" Eurocentrism as part of context-led research would be a second step to take. We should rethink which assumptions we consider as the base for our studies of media systems and their social, political, technological, economic, and cultural foundations. To get closer to our research contexts, fieldwork, experiences, and international research collaborations are crucial here, including an openness to the meaning of special circumstances. "Relearning" then means (re)thinking the margins or rather thinking from the margins. What can we actually learn from the experiences and structures of seemingly peripheral media systems? This process of relearning also implies the inclusion of those who live and research under enforced, disempowered, and marginalized conditions and trying to exchange with them at eye level.

However, not only must the researchers' positionality be reflected but also the positionality of the media systems under investigation must be considered. In many places in the world, especially in formerly colonized and fragile contexts, external actors have shaped the appearance of media systems. Media development actors—whether on the UN, bilateral, NGO, or grassroots level—impact media landscapes through funding, advocacy, training/capacity building, etc. While they very practically build parts of media systems, they also transport their ideas of norms, roles, and values (Fraser, 2021).

Again, a relational approach can be useful to make these processes visible, taking into account the power hierarchies between the involved actors and between the conceptualizations of media systems that come into play in these cases. The results of such processes may be euphemistically described as hybridity, but this often conceals the fact that there are parallel structures of media systems as they "should be" and as they "are" on the ground. In Afghanistan, the challenges of imposing a system conceived through normative ideas onto a complex and conflict-ridden context have become obvious: despite large-scale international investments of international actors trying to establish a new media system after 2001, at least since the Taliban took over in 2021, media freedom in the Western sense is essentially nonexistent, according to Relly and Zanger (2017). However, they noted, such a system change may happen "in countries with heavy foreign intervention, where imported journalism values are layered upon previous and continued institutional arrangements and where violence and instability continue unabated" (p. 1233). Context-led research can show, for example, how regulations that would support media are adopted but are then misused by authorities to exclude or control journalists. On a smaller scale, this happens if regulations meant to support media are abused for government control, or if precarious working conditions force journalists to accept brown envelopes or per diems, although it goes against their values (Sampaio-Dias, 2019). If international media development actors, for example, due to a lack of baseline studies or less close interaction with partners, do not realize these dynamics, it may lead to a "mimicry" system that combines certain elements that are not what their label claims (Fengler, 2022).

Allowing for a deep (historical and transnational) contextualization

A central approach that pays respect not only to Indigenous knowledge but also to our approach to context-led research is a deep contextualization of media and communication phenomena. Contextualization means to incorporate historical developments and related path dependencies, transnational entanglements, and power relations in society (Giraudy, 2015; Huang, 2003; Matar, 2012; Milton, 2001; Roudakova, 2011; Sparks, 2008; Stremlau, 2013; Voltmer, 2008, 2011). As laid out in Stremlau's (2013) "diagnostic approach," the histories and concepts of the entities we are studying are essential for a contextual analysis, and an understanding of the margins and contingency of media systems is needed in respect to their formal and informal structures and functional logic. As such, citizen media, social media posts, music, interpersonal communication, or public debates may be included in our analyses, too, depending on the context.

Dina Matar (2012) provided us with an example of how to integrate a historical perspective: "in studying the 'here' and 'now', we need to take into contexts both material and immaterial (discursive) 'genealogies'—the particular histories of nation-states, religion(s), capitalist class formations, national, regional and international politics as well as cultural and discursive formations" (pp. 78–79). Bayart (2009) called for a more contextual analysis in which a political entity is understood through the history and concepts of governmentality that its people have invented, rather than through the lens of foreign models of government or media systems that have been designed elsewhere.

Deep contextualization is in line with what the concept of listening (Wasserman, 2013) brings with it. Datta (2018) referred to the practice of decolonization as "research with Indigenous communities that places Indigenous voices and epistemologies in the center of the research process" (p. 2). Kurasawa (2004) emphasized the "recognition of global cultural pluralism" (p. 235) through a practice of cosmopolitanism from below: "upon an ethos of cultural openness that actively seeks out and tries to understand and appreciate ways of thinking and acting found in different societies, as well as listening to the voices of those who are not often heard in the elite cosmopolitan discourse" (p. 240).

In terms of contextual relearning, an emphasis on inductive approaches is key. For example, an inductive comparative study (Radue, 2022) using a relational research framework compared media control processes in Malaysia, Myanmar, and Thailand and concentrated on the connotative context. It found that not only should the relationship between media and politics be relevant for analysis but also that "society-media relationships provide a broad spectrum of analytical starting points" (p. 172). This study inductively detected connotative context factors to understand media control in these three countries. In those cases, according to Radue, the relationships of journalists and media producers with specific elites, such as the military and monarchy in Thailand, Buddhist nationalist elites and the Tadmadaw (Burmese armed forces) in Myanmar, and economic elites like government-linked companies (GLCs) and religious and ethnic elites in Malaysia, provide new insights into the media contexts while the categories from traditional media system analysis fail to do so.

One approach to gaining explanatory weight by analyzing media systems is the inclusion of (transnational) path dependencies. Rodny-Gumede (2020) suggested that postcolonial societies have, in the case of the same colonizer, also developed out of a similar logic, yet undergone different transformations based on historical or cultural factors. Here, historic approaches may be useful (Bastiansen, 2008), which should again be context sensitive and include aspects of the communication system before and beyond the colonizers' communication practices and infrastructure to achieve a full picture. As Serwornoo (2021) has shown, this path dependency extends to information flows within these media systems: he found that the media reporting in Ghana about neighboring African countries carries the same problems as European reporting about Africa—because the Ghanaian media mostly relies on external actors such as international news agencies and the BBC for such information. Researchers have also found shared experiences based on historic, cultural, geographical, and technological factors such as leapfrogging (see e.g., Tereshchuk, 2018). These include the experiences of colonization and state building in multiethnic societies after decolonization, often influenced by external actors. When Ngomba (2012) asked if the "Chinese model" had a place in Africa's media systems, this alluded to the increasing number of actors with growing influence on different sectors in the African continent, including the media.

Relearning and emphasizing context also means identifying possible new categories for media system research. For example, Yin (2008) suggested that "alternative concepts or dichotomies can be explored in building new models, such as an observer-interventionist/activist dichotomy, or commercial–ideological dichotomy" (p. 55). Norris (2009), on the other hand, wrote, "[t]he search for typological schema and categorical classifications of 'media systems' or 'political communication' systems should perhaps be abandoned" (p. 340). Other researchers have suggested the inclusion of new contextual categories, such as cultural values (e.g., Mellado & Lagos, 2013; Willnat & Aw, 2009). In Asian contexts, concepts such as "face," "loyalty," "social harmony," "individualism," and "collectivism" have an impact on all social and political interactions, hence opening up the potential for a variety of new research (Willnatt & Aw, 2009). For the comparison of media freedom in Thailand, Myanmar and Malaysia, Radue (2022) inductively found:

> factors that are crucial to understanding and explaining the formation of the three media contexts; . . . [for example], (de)centrality, social stratifications/polarizations, supremacy of (religious) norms and values, persisting conflicts, national security, seniority, culture of (dis)agreement, multi-ethnicity, multilingual societies, mechanisms of intimidation, rule of law vs. rule by law, media trust, racism. (p. 75)

Also focusing on press freedom, de Albuquerque (2019) drew similar conclusions regarding the often used Western concepts for media analysis and suggested "a postcolonial approach to media/politics relations" (p. 915). Looking at the Brazilian media landscape, he dissected Hallin and Mancini's models (2004) and critically noted that with regard to European case studies, they are often treated as "models from" in terms of being pure research tools. However, when applied to case studies, they are treated as "models to," that is, as "normative parameters" (de Albuquerque, 2011, p. 72). Of course, transfers of categories and approaches need to be carefully proven and justified but are necessary for any serious comparative work, as Kubik (2015) discussed concerning the combination of "context sensitivity with generalizing ambition[s]" (p. 362).

Applying context sensitive methodology

A challenge when researching media systems is the operationalization of rather abstract theoretical ideas, such as "system," into a functional empirical research design. While trying to move away from Eurocentric concepts, the aim to avoid the overuse of existing models and resulting inaccuracies should also be reflected in methodology. Above, we identified a general lack of reliable and statistically relevant data, for example, for media usage figures or media infrastructure, as a common problem in non-Western countries. Thus, more in-depth research would be necessary. Qualitative mapping (Marx, 2023) can provide a basic understanding of the existing elements and actors of a media landscape, their interrelations and roles. This can tie in with relational and network approaches, especially if it is implemented in a participative way, with journalists, experts, or media users sharing their everyday experience rather than the researcher applying a predefined lens and asking just for the expected outcomes. Participatory methods can be useful to ensure that aspects important in the respective context are not tainted or overlooked. These methods must be context sensitive to reach media practitioners and users, including disadvantaged groups. Schönbächler (2023, pp. 141–144) described an example that she applied in Burkina Faso: When the COVID-19 pandemic hit and the security situation became too tense for her to visit and interview female journalists in local radio stations across the country, she used instant messaging techniques to conduct a diary study. Either using their own recording device, a device sent by the researcher, or sending WhatsApp audio messages, female journalists produced regular audio diary records, providing not only a unique insight into their daily work, but also their professional and personal concerns (Schönbächler, 2023, pp. 143–144). As well as enabling research in a context marked by security and health concerns, the study also served as an example of a feminist research perspective because it considered the realities of women journalists' work life and allowed them to express these in a self-determined way.

While context sensitive describing and mapping of media systems should be the first step to allow for a more Indigenous knowledge production to surface, most media system research is interested in comparisons and drawing conclusions on the causes and outcomes of certain processes. Previously, Siebert et al. (1956) stated that beyond describing and comparing media systems, it is important to ask "Why is the press as it is?" (p. 1). Peruško et al. (2020) stressed that different methods must be used depending on whether the research is supposed to be descriptive or explanatory (p. 33). To analyze the different paths of development in post-socialist media systems, they worked with fuzzy set qualitative comparative analysis (fsQCA), which can show causal mechanisms in samples with a medium or large number of cases (p. 33). Comparing these provided insights into the nature of media systems and understanding them as dynamic networks with a process-like character.

Another way of drawing conclusions beyond Western models has been done by Richter and Kozman in their comparative study, published in their text entitled *Arab Media Systems* (2021). Based on descriptions of 18 Arab media systems along rather typical dimensions, such as political system and legal framework, media ownership patterns, and technology and infrastructure, they identified important themes and compared the different states' performances regarding these themes on a spectrum instead of building a clear-cut typology of media systems. For example, they related the particular media ownership patterns, which included media being in the hands of the security apparatus, as well as being assets of crony businessmen or instruments of militias, to their effects on public opinion. The spectrum spanned from creating a confrontational opinion climate in society on one end of the spectrum to a loyalist opinion climate toward the ruling regime on the other end of the spectrum. To answer the question "Why is the press as it is?" the authors not only included the written laws and regulations as a basis for their analysis but also the actual practices of implementing them and the states' approach to media. Such a context-led approach, according to the author, highlights that it makes a difference for the structural build-up of a media system, as to whether or not the regime envisages the media to mobilize for certain causes or to educate the people or to symbolize a wished-for era of modernization.

Conclusion

We have shown why enhancing media system research through context-led research allows us to better understand how media structures shape societies in all parts of the world. While some work has been done in recent years to diversify the research on media systems and include new dynamics, some regions of the world remain "blind spots" that are barely even considered. Consequently, the existing concepts cannot grasp the realities in these contexts, and the over-application of dominant approaches based on Eurocentric case studies leads to distorted or biased views on media systems. This has effects on knowledge production and understandings of norms but also on practical issues, such as media assistance. To achieve an actual cosmopolitan approach to media system analysis, we have therefore stressed the need for and importance of inductive approaches and context-led research. Here, a research focus on historical path dependencies, power imbalances, and relations offers starting points to overcome the shortcomings of established media system research. Instead of Western norm-driven deficit analyses, we can gain new insights by taking findings from seemingly peripheral contexts and concepts seriously.

Beyond voicing our criticism of current media system research, we have provided some hints on how to translate our criticism into more cosmopolitan research practices. As a first step, we need to reflect our positionality and reconfigure it through processes of "unlearning" and "relearning" (Datta, 2018). Understanding the "positionality" of media contexts and acknowledging path dependencies and impacts from different (f)actors in various world regions is essential when we want to go beyond trying to find a one-fits-all typology for media system analyses. Second, by means of deep (historical and transnational) contextualization and inductive approaches, we can pay respect to Indigenous knowledge, voices, and epistemologies. This also means, as a third step, revising our methodologies by means of context sensitive research, relational analysis, qualitative mapping, and participatory methods.

References

Appadurai, A. (2010). *Modernity at large: Cultural dimensions of globalization*. University of Minnesota Press. (Original work published 1995)

Badr, H., Behmer, M., Fengler, S., Fiedler, A., Grüne, A., Hafez, K., Hahn, O., Hamidi, K., Hanitzsch, T., Horz, C., Illg, B., Litvinenko, A., Löffelholz, M., Radue, M., Richter, C., Thomaß, B., & Töpfl, F. (2020). Kosmopolitische Kommunikationswissenschaft: Plädoyer für eine "tiefe Internationalisierung" des Fachs in Deutschland: Ein wissenschaftliches Positionspapier. *Publizistik, 65*(3), 295–303. https://doi.org/10.1007/s11616-020-00576-6

Bastiansen, H. G. (2008). Media history and the study of media systems. *Media History, 14*(1), 95–112. https://doi.org/10.1080/13688800701880432

Bayart, J.-F. (2009). *The state in Africa: The politics of the belly.* Polity Press.

Castells, M. (2001). Bausteine einer Theorie der Netzwerkgesellschaft. *Berliner Journal für Soziologie, 11*(4), 423–439.

Castells, M. (2007). Communication, power and counter-power in the network society. *International Journal of Communication, 1*, 238–266. https://ijoc.org/index.php/ijoc/article/view/46

Castells, M. (2011). *Communication power*. Oxford University Press.

Curran, J., & Park, M.-J. (Eds). (2000). *De-Westernizing media studies*. Routledge.

Datta, R. (2018). Decolonizing both researcher and research and its effectiveness in Indigenous research. *Research Ethics, 14*(2), 1–24. https://doi.org/10.1177/1747016117733296

de Albuquerque, A. (2011). On models and margins. In D. C. Hallin & P. Mancini (Eds.), *Comparing media systems beyond the Western world* (pp. 72–95). Cambridge University Press. https://doi.org/10.1017/CBO9781139005098.006

de Albuquerque, A. (2019). Protecting democracy or conspiring against it? Media and politics in Latin America: A glimpse from Brazil. *Journalism, 20*(7), 906–923. https://doi.org/10.1177/1464884917738376

Della Ratta, D., Sakr, N., & Skovgaard-Petersen, J. (Eds.). (2015). *Arab media moguls*. I.B. Tauris.

Economist Intelligence Unit. (2022). *Democracy index* [Data set]. Our world in data. https://ourworldindata.org/grapher/democracy-index-eiu

Emirbayer, M. (1997). Manifesto for a relational sociology. *American Journal of Sociology, 103*(2), 281–317. https://doi.org/10.1086/231209O

Emirbayer, M., & Goodwin, J. (1994). Network analysis, culture, and the problem of agency. *American Journal of Sociology, 99*(6), 1411–1454. https://doi.org/10.1086/230450

Esser, F. (2014). Methodological challenges in comparative communication research: Advancing cross-national research in times of globalization. In M. J. Canel & K. Voltmer (Eds.), *Comparing political communication in time and space* (pp. 15–30). Palgrave Macmillan. https://doi.org/10.1057/9781137366474_2

Fengler, S. (2022). A comparative analysis of media accountability across the globe: Models, frameworks, perspectives. In S. Fengler, T. Eberwein, & M. Karmasin (Eds.), *The global handbook of media accountability* (pp. 549–602). Routledge.

Flew, T., & Waisbord, S. (2015). The ongoing significance of national media systems in the context of media globalization. *Media, Culture & Society, 37*(4), 620–636. https://doi.org/10.1177/0163443714566903

Fraser, N. (2021). Neue Überlegungen zur Transnationalisierung der Öffentlichkeit. In M. Seeliger & S. Sevignani (Eds.), *Ein neuer Strukturwandel der Öffentlichkeit?* (pp. 139–159). Nomos. https://doi.org/10.5771/9783748912187-137

Giraudy, A. (2015). *Democrats and autocrats: Pathways of subnational undemocratic regime continuity within democratic countries*. Oxford University Press. https://doi.org/10.1093/acprof:oso/9780198706861.001.0001

Hafez, K. (2007). *The myth of media globalization*. Polity Press.

Hallin, D. C. (2016). Typology of media systems. In E. Hannah (Ed.), *Oxford research encyclopedia of politics*. https://doi.org/10.1093/acrefore/9780190228637.013.205

Hallin, D. C., & Mancini, P. (2004). *Comparing media systems: Three models of media and politics*. Cambridge University Press. https://doi.org/10.1017/CBO9780511790867

Hallin, D. C., & Mancini, P. (2011). *Comparing media systems beyond the Western world*. Cambridge University Press. https://doi.org/10.1017/CBO9781139005098

Harvey, A. (2020). *Feminist media studies*. Polity Press.

Häußling, R. (2010). Relationale Soziologie. In C. Stegbauer & R. Häußling (Eds.), *Handbuch Netzwerkforschung* (pp. 63–87). VS Verlag für Sozialwissenschaften. https://doi.org/10.1007/978-3-531-92575-2_7

Hollstein, B. (2006). Qualitative Methoden und Netzwerkanalyse – ein Widerspruch? In B. Hollstein & F. Straus (Eds.), *Qualitative Netzwerkanalyse: Konzepte, Methoden, Anwendungen* (pp. 11–35). VS Verlag für Sozialwissenschaften. https://doi.org/10.1007/978-3-531-90074-2_1

Huang, C. (2003). Transitional media vs. normative theories: Schramm, Altschull, and China. *Journal of Communication, 53*(3), 444–459. https://doi.org/10.1111/j.1460-2466.2003.tb02601.x

Kleinsteuber, H. J. (2004). Comparing mass communication systems: Media formats, media contents, and media processes. In F. Esser & B. Pfetsch (Eds.), *Comparing political communication: Theories, cases and challenges* (pp. 64–86). Cambridge University Press. https://doi.org/10.1017/CBO9780511606991.005

Knoche, M. (2013). Medienkonzentration. In B. Thomaß (Ed.), *Mediensysteme im internationalen Vergleich* (2nd ed., pp. 135–160). UVK.

Kraidy, M. M. (2011). The rise of transnational media systems: Implications of Pan-Arab media for comparative research. In D. C. Hallin & P. Mancini (Eds.), *Comparing media systems beyond the Western world* (pp. 177–200). Cambridge University Press. https://doi.org/10.1017/CBO9781139005098.011

Kubik, J. (2015). Between contextualization and comparison: A thorny relationship between East European studies and disciplinary "mainstreams." *East European Politics and Societies and Cultures, 29*(2), 352–365. https://doi.org/10.1177/0888325414556128

Kuo, E. C. Y., & Chew, H. E. (2009). Beyond ethnocentrism in communication theory: Towards a culture-centric approach. *Asian Journal of Communication, 19*(4), 422–437. https://doi.org/10.1080/01292980903293361

Kurasawa, F. (2004). A cosmopolitanism from below: Alternative globalization and the creation of a solidarity without bounds. *European Journal of Sociology / Archives Européennes de Sociologie / Europäisches Archiv Für Soziologie, 45*(2), 233–255. http://www.jstor.org/stable/23999133

Löwenstein, H. (2017). Pragmatistische-relationale Entwicklungslinien: Eine Einleitung und Hinführung. In H. Löwenstein & M. Emirbayer (Eds.), *Netzwerke, Kultur und Agency: Problemlösungen in relationaler Methodologie und Sozialtheorie.* (pp. 9–27). Beltz.

Mancini, P. (2020). Comparing media systems and the digital age. *International Journal of Communication, 14*, 5761–5774. https://ijoc.org/index.php/ijoc/article/view/14553

Marx, S. (2023). Mapping as critical qualitative research methodology. *International Journal of Research & Method in Education, 46*(3), 285–299. https://doi.org/10.1080/1743727X.2022.2110231

Matar, D. (2012). Contextualising the media and the uprisings: A return to history. *Middle East Journal of Culture and Communication, 5*(1), 75–79. https://doi.org/10.1163/187398612X624391

Mellado, C., & Lagos, C. (2013). Redefining comparative analyses of media systems from the perspective of new democracies. *Communication & Society, 26*(4), 1–24. https://doi.org/10.15581/003.26.36058

Michalis, M. (2024). Whither public service media governance: Looking back, looking ahead. In C. Padovani, V. Wavre, A. Hintz, G. Goggin, & P. Iosifidis (Eds.), *Global communication governance at the crossroads* (pp. 129–144). Palgrave Macmillan. https://doi.org/10.1007/978-3-031-29616-1_8

Milton, A. K. (2001). Bound but not gagged: Media reform in democratic transitions. *Comparative Political Studies*, *34*(5), 493–526. https://doi.org/10.1177/001041400103400 5002

Ngomba, T. (2012). Differing paradigms of media systems development in contemporary Africa: Does the 'Chinese model' have a place? *Journal of Asian and African Studies*, *47*(1), 52–67. https://doi.org/10.1177/0021909611417679

Norris, P. (2009). Comparative political communications: Common frameworks or Babelian confusion? *Government and Opposition*, *44*(3), 321–340. https://doi.org/10.1111/j. 1477-7053.2009.01290.x

Peruško, Z., Vozab, D., & Čuvalo, A. (2020). *Comparing post-socialist media systems: The case of Southeast Europe*. Routledge. https://doi.org/10.4324/9780367226787

Pokhrel, A. K. (2011). Eurocentrism. In D. K. Chatterjee (Ed.), *Encyclopedia of global justice* (pp. 321–325). Springer. https://doi.org/10.1007/978-1-4020-9160-5_25

Powers, M., & Vera-Zambrano, S. (2018). The universal and the contextual of media systems: Research design, epistemology, and the production of comparative knowledge. *International Journal of Press/Politics*, *23*(2), 143–160. https://doi.org/10.1177/1940161218 771899

Radue, M. (2022). Comparing impacts on media freedom in Southeast Asia: Connotative context factors in Malaysia, Myanmar and Thailand. *Global Media and Communication*, *18*(2), 157–179. https://doi.org/10.1177/17427665221097852

Rainie, L., & Wellman, B. (2012). *Networked: The new social operating system*. MIT Press. htt ps://doi.org/10.7551/mitpress/8358.001.0001

Rantanen, T. (2013). A critique of the systems approaches in comparative media research: A Central and Eastern European perspective. *Global Media and Communication*, *9*(3), 257–277. https://doi.org/10.1177/1742766513504175

Relly, J. E., & Zanger, M. (2017). The enigma of news media development with multi-pronged 'capture': The Afghanistan case. *Journalism*, *18*(10), 1233–1255. https://doi.o rg/10.1177/1464884916670933

Richter, C., & Gräf, B. (2015). The political economy of media: An introduction. In N.-C. Schneider & C. Richter (Eds.), *New media configurations and socio-cultural dynamics in Asia and the Arab World* (pp. 25–36). Nomos.

Richter, C., & Kozman, C. (Eds.). (2021). *Arab media systems*. Open Book Publishers. http s://doi.org/10.11647/OBP.0238

Rochefort, A. (2020). Regulating social media platforms: A comparative policy analysis. *Communication Law and Policy*, *25*(2), 225–260. https://doi.org/10.1080/10811680.2020 .1735194

Rodny-Gumede, Y. (2020). Expanding comparative media systems analysis from transitional to postcolonial societies. *International Communication Gazette*, *82*(7), 611–627. ht tps://doi.org/10.1177/1748048519897515

Roudakova, N. (2011). Comparing processes: Media, "transitions," and historical change. In D. C. Hallin & P. Mancini (Eds.), *Comparing media systems beyond the Western world* (pp. 246–277). Cambridge University Press. https://doi.org/10.1017/CBO97811390050 98.014

Ruotsalainen, J., & Heinonen, S. (2015). Media ecology and the future ecosystemic society. *European Journal of Futures Research*, *3*(1). https://doi.org/10.1007/s40309-015-0068-7

Sampaio-Dias, S. (2019). Per diem payments as a form of censorship and control: The case of Guinea-Bissau's journalism. *Journalism Studies*, *20*(16), 2349–2365. https://doi.org/10.1080/1461670X.2019.1593883

Schamberger, K. (2021). *Vom System zum Netzwerk: Medien, Politik und Journalismus in Kurdistan*. Westend. https://doi.org/10.53291/9783949925030

Schönbächler, V. (2023). Instant messaging meets diary studies: Employing WhatsApp in audio diary research with female journalists in Burkina Faso. *ZQF – Zeitschrift Für Qualitative Forschung*, *24*(1), 139–155. https://doi.org/10.3224/zqf.v24i1.11

Serwornoo, M. Y. W. (2021). The coverage of Africa in Ghanaian newspapers: The dominant Western voice in the continent's coverage. *Journalism*, *22*(12), 3013–3030. https://doi.org/10.1177/1464884919887311

Siebert, F. S., Peterson, T., & Schramm, W. (1956). *Four theories of the press: The authoritarian, libertarian, social responsibility, and Soviet communist concepts of what the press should be and do*. University of Illinois Press. https://doi.org/10.5406/j.ctv1nhr0v

Sparks, C. (2008). Media systems in transition: Poland, Russia, China. *Chinese Journal of Communication*, *1*(1), 7–24. https://doi.org/10.1080/17544750701861871

Stremlau, N. (2013). Towards a diagnostic approach to media in fragile states: Examples from the Somali territories. *Media, War & Conflict*, *6*(3), 279–293. https://doi.org/10.1177/1750635213491175

Tereshchuk, V. (2018). African regional media system in post-bipolar era. *Torun International Studies*, *1*(11), 55–66. https://doi.org/10.12775/TIS.2018.005

Thomaß, B. (2013). *Mediensysteme im internationalen Vergleich* (2nd ed.). UVK.

van Dijk, J. (2021). *The network society* (4th ed.). Sage.

Voltmer, K. (2008). Comparing media systems in new democracies: East meets South meets West. *Central European Journal of Communication*, *1*, 23–40.

Voltmer, K. (2011). How far can media systems travel? Applying Hallin and Mancini's comparative framework outside the Western world. In D. C. Hallin & P. Mancini (Eds.), *Comparing media systems beyond the Western world* (pp. 224–245). Cambridge University Press. https://doi.org/10.1017/CBO9781139005098

Wasko, J., Murdock, G., & Sousa, H. (2011). *The handbook of political economy of communications*. Wiley Blackwell. https://doi.org/10.1002/9781444395402

Wasserman, H. (2013). Journalism in a new democracy: The ethics of listening. *Communicatio*, *39*(1), 67–84. https://doi.org/10.1080/02500167.2013.772217

Willnat, L., & Aw, A. (2009). The big unknown: Conclusions about political communication in Asia. In L. Willnat & A. Aw (Eds.), *Political communication in Asia* (pp. 216–228). Routledge.

Yin, J. (2008). Beyond the four theories of the press: A new model for the Asian & the world press. *Journalism & Communication Monographs*, *10*(1), 3–62. https://doi.org/10.1177/152263790801000101

From media to AI governance studies

Decentering established patterns through cosmopolitan critique

Sarah Anne Ganter

Introduction

Media governance is a heterogeneous field comprising political sciences, communication, media economics, sociology, and, increasingly, technological design and programming. As a field of study, governance in media and communication studies also spans a variety of communicative means and types of media, according to their developments over time. Therefore, today, governance in media and communication studies offers an opportunity to be heterogeneous and multifaceted. It subsumes a variety of terms, ranging from media and communication, Internet, and social media platform governance to governance by things and to governance by algorithms and artificial intelligence (AI). Consistent with its conceptual nature, media governance scholars have emphasized and argued for broad conceptualizations of media governance and have pointed to the advantages of dynamic theorizing (Puppis, 2010). This conceptual openness of the field has invited an early critique of media governance as a buzzword, an empty signifier, and an ideologically coined term with little analytical value and as being ambiguous in its use (Ginosar, 2013; Karppinen & Moe, 2013). Together with evolving technological change, the critique of the field has contributed to self-reflections, and calls to reinvigorate the field through theoretical contributions, thus making these more visible to the wider field of media and communication studies (Braman, 2004; Just & Puppis, 2018; Picard, 2016). However, throughout the various self-reflections, the gaping blind spot has been the lack of opening the field up to decolonial, de-Westernizing, intersectional, and cosmopolitan perspectives.

In this chapter, I reflect on the cosmopolitan critique established recently in media governance studies and discuss how it can benefit future work in media governance. This specific area of study has been largely defined by scholars from/in Western contexts. However, it is one of the first areas of study where cosmopolitan critique (Ganter & Badr, 2022) has been proactively proposed as a methodology pushing for epistemological transformation. The cosmopolitan critique in media governance studies aims to include broader conceptual, theoretical, and empirical perspectives in the spirit of speaking with voices from different contexts rather than speaking about their realities. It is a starting point to actively create a governance community conversing beyond the West and, as I argue, a necessity for the field to develop from an emerging area of study to one that fulfills its potential and establishes itself as a main pillar in media and communication studies. Most importantly, the cosmopolitan critique focuses on dialogue and encourages multidimensional media governance perspectives that go beyond applying dominant terms and concepts in non-dominant contexts.

This chapter develops over four sections. First, I review some of the main characteristics of the field, and particularly its development as it has started to flourish more recently due to the rise of platforms and related regulatory challenges and the needed reflections on shifting media ecologies. Second, I discuss cosmopolitan perspectives established in the field by pointing to exemplary work from the literature, and this will help to establish how cosmopolitan critique furthers epistemological transformation. Third, using concrete examples, I explain how cosmopolitan critique benefits media governance studies. Lastly, I discuss the experience of working toward establishing a cosmopolitan critique and argue that we need to work across scholarly, institutional, and pedagogical realms in the field to trigger sustained epistemological transformations through actions on all levels.

Characteristics of Western media governance studies

Deriving from media and communication policy and law research, media governance is not a synonym, but an analytical perspective breaking down media policy processes, values, outcomes, and related power constellations. Braman (2004) describes the policy field as latent and constantly challenging understandings of what belongs to the field; this continues to be true as new technologies constantly create new topics that can be subsumed under media governance. Freedman (2008) refers to media policy as an "umbrella term to describe the whole range of discourses and methods used to shape the behaviour of specific media" (p. 15). Media policies are the result of complex negotiation processes in which political economic and cultural values and objectives are established that broadly shape the context for the production, consumption, influence, and sharing of contents. Mansell and Raboy (2011) describe media and communication policy as a field that "refers to all efforts to influence media and communication systems, including those by the state, industry, and civil society" (p. 13). Consequently, governance considers informal and formal processes and practices that determine the framework in which traditional and new media perform (Hamelink & Nordenstreng, 2007; Kleinsteuber & Nehls, 2011). It is critical to generate knowledge around these frameworks as they also shape the decision-making and behavior of users, media organizations, and various platform and third-party companies that today play an important role in governance aspects, such as data management, content moderation, and copyright. Scholars have attributed multiple applications to media governance as a field of study; however, many scholars have contributed to the field without labeling it as such (e.g., see van Eeten & Mueller, 2013). Media governance is applied as a normative concept, descriptive term, and analytical approach (Donges, 2007). Early on, these three facets of media governance were often intertwined, as academics only slowly revisited, complemented, and amended the many political definitions of governance (e.g., see Raboy, 2002; van Eeten & Mueller, 2013; Woods, 1999). Scholars using media governance as an analytical heuristic for theorizations of policy change and continuity have developed complex and broad understandings that are skeptical about terms like "good governance," which the World Trade Organization (WTO) used early on (Woods, 1999) in their documents and which promoted the idea that governance is necessarily something ideal or good (see Ganter, 2016). As a result, the conceptual and theoretical value of media governance as a perspective has been increasingly foregrounded. In short, assuming this critical perspective, the scholarly field has responded to questions addressing the materiality and discursivity of media and communication policy processes, which transpire through the negotiation of norms and values as well as the (lack of) interactions, their implementation, and the often uneven impacts they have across society.

With that, the field has departed from what van Eeten and Mueller (2013) described as following a misleading focus on centralized institutions and has shifted more into the less convenient but more "disjointed, messy and globally distributed processes that together produce governance" (p. 729). In line with this is the amplification from studying established governance practices, such as self-regulation or co-regulation (Puppis, 2010), toward studying innovations that establish new practices in policymaking (Mansell, 2012). One example here is work looking at policy-hacking, a countering strategy referring to citizens who collectively practice law writing (Hintz, 2016). Another amplification of the narrow perspective on governance as practice is represented by works analyzing protests and civil campaigns accompanying policymaking. Reitman (2016) analyzed the case of the anti-SOPA (Stop Online Piracy Act) protests in the United States as an example of civil society groups contesting elite-focused media governance processes. The study identified several successful strategies of the protests, such as their decentralized nature, speaking up fast and often, powerful visual imagery, engaging with Internet communities, and the crossing of political lines. Löblich (2016) studied the aspect of dissent in civil society groups, using the case of net neutrality debates in the USA, and found that the involvement of civil society groups in media governance debates is often multi-faceted and not unilateral. Other scholars have analyzed cases of limited access to power within policy processes. Kim (2018), for example, finds in the case of media governance in South Korea that media governance processes are steered by powerful media companies and are less accessible for other stakeholders. This is in line with the work of scholars looking at the role of discourses in influencing media policy debates and processes. Ali and Puppis (2018) point to the way powerful media companies can shape media governance discourses through the active use of their agenda-setting power. Ganter and Löblich (2021) argue, based on their discursive media institutionalist framework, that depending on context, media governance discourses can emerge within a broader spectrum of actors which are not homogeneous, and their internal multi-faceted discourses manifest internal and external negotiation of multi-facetted values and norms. Padovani (2018) notes that societal power imbalances and inequities also manifest in media governance processes, practices, and discourses, as well as in unequally distributed consequences of media governance measures. Finally, research increasingly includes citizens' perceptions of policy projects and regulatory measures as a type of policy evaluation. Strycharz et al. (2020), for example, find in a survey study with 1,288 respondents from the Netherlands that citizens were highly aware of the new European Commission's General Data Protection Regulation (GDPR) but perceived very little impact of it on their individual rights.

Consequently, the field of media governance today plays an important role in better understanding conditions under which content is being produced, shared, and consumed—what type of content is visible to whom, where, when, at what cost, and why. However, because of that, it is also more necessary than ever to consider the political, economic, and cultural contexts in which media governance is being studied, defined, and theorized—and to reconsider the generalizability of the field, given that it is originally centered in hegemonic hubs. Therefore, the integration of work from various backgrounds and with countering perspectives has become even more necessary. If we look at the development of the field, we can note how imbalances in perspectives have been established since its early beginnings and how important it is to acknowledge the blind spots of the field and, consequently, to take concrete action to foster reinvigoration of the field through openness, inclusivity, and dialogue.

Two decades of centering the field in hegemonic hubs

Three geographical power hubs have been at the forefront of media governance: (1) the US-American tradition, which largely falls under the Internet governance label, (2) the global governance and policy tradition, and (3) the European tradition, which largely falls under the media and communication governance label. These different governance schools emerged almost simultaneously and cemented structures and research perspectives that have been described as myopic (van Eeten & Mueller, 2013), Western and hegemonic (Alhassan & Chakravartty, 2011; Ganter & Badr, 2022), exclusive, largely White, and patriarchic (Padovani, 2018) and which have only started to evolve from these early attributes.

In the development of the field, the emergence of the Internet contributed largely to the governance turn in media and communication studies (Kleinwächter, 2000; Mueller, 2002). Braman (2013) noted how negotiating standards for what would become the Internet since the late 60s took place in the USA through side notes and informal commenting. In her analysis of the Request for Comments (RFC), addressing points of early Internet design, Braman (2013) showed that these early processes constituted "sociotechnical governance of and by the Internet" (p. 79), which were critical in forming agreement and conflict around standards, values, and their implementations and determining what the Internet should constitute and for whom. The interdependence of the geopolitics of technological design and the emergence of the field is an important factor that has led to geographical imbalances and blind spots shaping the field until today. For a considerable time, technological developments have been studied as originating in the USA, and the idea of the US-American, White, male genius designer of networks, devices, platforms, applications, and standards, values, and practices has influenced media governance studies, as well. In the contrary, work addressing early network development, new technology innovations, and digital platform development generated from outside of the Western/Northern geographical spectrum have by far not played the same role in past media governance work. Even though some publications on developments beyond the USA existed in English early on, the myth of the USA as the sole and central force in creating and structuring digital infrastructures has persisted (e.g., see the critique by Hong, 2022). Examples of such work are contributions to understanding earlier developments such as the Internet revolution in Japan (Coates & Holroyd, 2003), the planning and establishment of governance mechanism for the Internet in India (Shah et al., 2022), or elaborate histories on the development of the Internet infrastructures in Central America (Siles, 2020), Africa (Tankard, 1998; Wasserman, 2017), or the Arab world (Warf & Vincent, 2007).

Regardless, the geopolitical dominance in network development and technological innovation was maintained by a plethora of scholars from and working in the USA, who often chose a national focus and founded a strong US-American school of governance in media and communication studies, which proved to be decisive in the emergence and constitution of the field. This development was also upheld through the field's flagship journals *Telecommunications Policy* (founded in 1976) and *Communication Law and Policy* (founded in 1996) and through emerging strategic centers for policy and governance research.[1] Almost simultaneously, some scholars also started to address questions of national sovereignty (Braman, 2006; Price, 2002; Raboy, 2002) and studied global governance initiatives (Mueller et al., 2007; Siochrú et al., 2002). However, as van Eeten and Mueller (2013) outlined, this work has been criticized for being "myopic" (p. 728) and lacking considerations of power geographies in international entities such as the Internet Corporation for Assigned Names and Numbers (ICANN), the Internet Governance Forum (IGF), Internet Assigned Numbers Authority (IANA), and Internet Engineering task force (IETF) as well as ongoing processes such as the World Summit on the Information Society (WSIS). The latter, however, differed in that the summit aimed at least, on paper, for more heterogeneity and diverse geopolitical representation and contributed significantly to establishing global governance perspectives in media and communication policy studies (Mansell & Raboy, 2011). This body of work was at the start hopeful of multi-stakeholderism and enthusiastic about new forms of discussing frameworks shaping communication and media realities globally; however, soon, the WSIS would become an example of flaws of multi-stakeholder dialogue and Western dominance in setting standards (Musiani, 2013). At nearly the same time, a third power hub emerged that would ultimately combine media, communication, and Internet governance under the label of "media governance" and emphasize theoretical and conceptual developments. This hub was the Germanic school of governance that had its epicenter in Switzerland, predominantly at the University of Zurich, and it fostered neo-institutional perspectives as theoretical foundations in media governance (Donges, 2007; Puppis, 2010). A volume edited in German by Patrick Donges (2007) bundled works of a large group of representatives of media governance scholars from Austria, Germany, and Switzerland. The contributions reflect the early conceptual, theoretical, and normative discussions within Germanic media governance.

---

1 Such as the Internet Governance Project at the School of Public Policy at the Georgia Institute of Technology, or the Berkman Klein Center for the Internet & Society at Harvard University.

Alongside the German-centric developments, media governance as a field also mushroomed across European universities in the Netherlands, Belgium, and the United Kingdom, where new university chairs and research centers labelled as policy and governance have opened since the 2010s. This European section of media governance literature from the start has included a much broader range of topics such as questions around media diversity (Helberger, 2011), regulation of public service media, audiovisual media (Donders et al., 2014; Michalis, 2010), and innovation (Mansell, 2012) as well as specific questions addressing spam regulation (Just et al., 2007) and digital rights (Padovani et al., 2010)—which confirms what Kleinsteuber and Nehls (2011) have summarized as a convergent nature of the field that includes new and old media.

The rise of digital platforms after 2015 led to new labels, such as social media governance (Flew, 2015), governance by things (Schulz & Dankert, 2016), governance of/by platforms (Gillespie, 2017), governance by algorithms (Just & Latzer, 2017), platform governance (Gorwa, 2019), algorithmic governance (Katzenbach & Ulbricht, 2019), and AI governance (Floridi, 2021; Hassan, 2023; Roberts et al., 2021).[2] During this time, voices were already calling for a decentering of media and communication studies (e.g., Waisbord & Mellado, 2014), and suggestions to decolonialize media and communication policy and law as a field were being expressed (Alhassan & Chakravarty, 2011). Regardless, these new labels were proposed from a Western-centered perspective (e.g., Hassan, 2023) and upheld previously created silos, while also reproducing "canonical silences" by ensuring "the continued circulation of influential texts" (Willems, 2023, p. 17) and self-referentiality (Connell, 2007). However, this was also the time when critique regarding the lack of intersectional perspectives (Bannerman, 2020; Padovani, 2018; Smith & Craig, 2023) and the constructive use of these in the field emerged together with first accounts of media governance from the Global South.

Examples of media governance research from the Global South

The three dominant hubs in media governance studies developed simultaneously, but largely in parallel, a phenomenon that Waisbord (2019) described as "siloization" (p. 40). At the same time, media governance as an approach had not spread widely in non-Western contexts (Ganter, 2016; Ganter & Badr, 2022). A review of literature from 2002 through 2019 showed that 45.75% of scholars working with media governance terminologies worked in Europe, 30.72% in North America, 7.19% in Oceania, and only 16.34% were from universities in the Global South (Ganter & Badr, 2022, p. 4). The contributions addressing contexts from the Global South, if published in English, were mainly published in area studies and not in subject-specific journals (e.g., see Yang & Mueller, 2014, on Internet governance in China). Work from or on the Global South addressing questions inherent to media governance studies have rarely used the label of media governance. Creating connections with more frequency through labelling has only started in the past 10 years. One example of this development is the Media and Governance in Latin

2 Some scholars use the terms AI governance and algorithmic governance interchangeably by labelling work on algorithmic governance as AI governance (e.g., see Floridi, 2021; Hassan, 2023).

America conference, which has been taking place at universities around Europe since 2014. Other examples for the uptake of governance research can be found in several centers and institutes of Internet governance located in countries of the Global South.[3] Consequently, this developing stream is creating an important connection between media governance as a field of study and scholarly work from/on countries beyond the central hubs established in previous decades. Bhuiyan (2014) provides a historical analysis of power structures in Internet governance, emphasizing the hegemonic position of the USA and providing the alternative suggestions and critique voiced by stakeholders from the Global South. In his analysis, he emphasized the influence of the state from the perspective of radical justice and proposed multilateralism as an immanent governance process solution to hegemonic structures. In his comprehensive history of media governance in Korea from 1980–2017, Kim (2018) emphasizes the idea that analyzing media governance means analyzing the interactions between government, media market, and civil society. He draws from the media governance framework by Puppis (2010), stipulating that the market had gained more influence and power in Korea over time. Opperman (2018), in his edited volume on Internet governance in the Global South, displays the work of scholars from different countries and times who had explored the meaning and thematical developments and discussions around Internet Governance in the Global South. These studies have shown how scholars have situated themselves proactively in the field, by approaching it through labelling their work accordingly and adding to the existing media governance literature, through applying the terminology for too often invisible contexts.

In addition to these more application-oriented uses of media governance, related critique of its applicability and new conceptual phrasing in relation to the centered media governance literature have only emerged more directly in the past few years (e.g., Akpojivi, 2022; Asthana, 2022; de Albuquerque & de Matos, 2022; Hassan, 2023; Pies, 2022; Raghunath, 2022; Segura & Linares, 2022). This development has been manifested particularly in more visible studies on AI governance in China (Roberts et al., 2021) and in Hassan's (2023) emphasis on a racial and colonial understanding of AI governance. These recent works from scholars working in or on contexts based in the Global South reflect a more direct, dynamic, and confident pushback against and critique of scholarly conceptualizations from the hegemonic centers. Therefore, as voices from the Global South have become more prominent across media governance studies, the question is how to circumvent the reproduction of silos and instigate a lasting epistemic transformation. However, as calls for methodological, theoretical, and academic cosmopolitanism have been voiced more frequently, what has been missing is the application and transfer into an enduring academic ethos. Scholars from the field have started late to instigate this transition, but they have been comparatively early in modeling concrete suggestions for enacting epistemic transformation as an ongoing process. In the following, I will point in more detail to some examples of cosmopolitan critique voiced in the field that has raised

3 Examples include the Internet Governance Institute in Nepal, the Research Center for Internet Development and Governance at Tsinghua University, the African School on Internet Governance, and the Asia Pacific School on Internet Governance.

important questions and enables epistemic wonderings for all media governance scholars.

Cosmopolitanism in media governance studies: Starting with critique, aiming for transformation

Establishing cosmopolitan critique

At the beginning of cosmopolitan perspectives stands the recognition that media governance has not been used frequently as a concept outside of the contexts mentioned above. Scholars from the Global South have questioned governance as being part of a dominant narrative itself (Jose, 2007), claimed the marginalization of countries in global governance processes (Bhuiyan, 2014), found it to be an antiquated concept (Camou, 1995/2020), and came to note that the approach has not been used much so far in their contexts (Allam, 2022; de Albuquerque & de Matos, 2022). In Latin America, for example, media policy and industry studies are a very prominent and strong field that is part of the larger critical media and communication studies movement across the continent (Bolaño, 2020). However, its representatives have for a long time made a point of developing their work independently from other schools (de Albuquerque & de Oliveira, 2021; Paulino & Kaplún, 2020)—while also being largely invisible in Westernized contexts (Ganter & Ortega, 2019).

These circumstances alone can be considered an unvoiced critique that raises questions around the limitations of media governance in the ways in which it has been conceptualized and studied. In this context, it is not possible to just call for and implement academic cosmopolitanism; it is necessary to request and listen to the critique raised as a starting point for dialogue, which then considers this critique. Cosmopolitan critique includes, but goes beyond, the argument for dialogue, recognition, and respect across contexts and cultural spaces and seeks to ingrain those values into academic processes through a *cosmopolitan iteration*. Based on what Seyla Benhabib (2006) called "democratic iteration" (p. 16), cosmopolitan iteration is a proposal for a countering methodology which proceeds through a series of questions to trigger what I call *epistemic wonderings*. The questions leading through the iteration inquire about motivations behind academic work and publishing processes, reasons for involvement of some and exclusion of others, and related power structures. The questions also ask how to ensure that we do not reproduce abyssal thinking and instead circumvent the recreation of closed contact zones, that we identify, think about, and include othered perspectives, ideas, and concepts, and that all cosmopolitan work will create "cosmopolitan contact zones" (de Sousa Santos, 2005, p. 17) that are generative and constructive (Ganter & Badr, 2022). As such, a cosmopolitan critique offers starting points for the cosmopolitan iteration as a reciprocal constructive approach to media governance studies, which is dynamic, self-reflexive, inclusive, and empathetic but not free from disagreement. A cosmopolitan iteration comprises recognition of differences, inclusive differentiation, creation of enabling generative cosmopolitan contact zones, and intercultural translation.[4] All of this shows that the implementation of academic cosmopolitanism through critique and iteration requires time and resources to support the researchers, journal editors, students, and reviewers' ability to establish, maintain, and engage in cosmopolitan (net)work(s). Cosmopolitan critique needs to be voiced, listened to, considered, and implemented on academic, institutional, and pedagogical levels. Implementing academic cosmopolitanism requires recognizing one's own limitations, established belief systems, approaches, and practices to be able to give space to epistemic transformation. As described in cosmopolitan reflections (e.g., Ganter & Ortega, 2019; Waisbord, 2019), this process is complex and requires the additional emotional labor of self-reflection, open listening, and re-positioning—all of which are adverse in the fast-paced, hypercompetitive, and performative, often streamlined knowledge economy (Afonso, 2013). While media and communication governance scholars have started the iteration, it will take some time for the process to come full circle.

Learning from cosmopolitan critique

Points raised through the cosmopolitan critique (Ganter & Badr, 2022) concerning the field have mainly addressed scholarly but also pedagogical and institutional dimensions of academic work. One blind spot that has been highlighted is the lack of attention in media governance studies to transitioning or autocratic systems (Allam, 2022; Sakr, 2022). Scholars have outlined different modes of media governance in the Global South and have emphasized how established ideas, rules, axioms, and postulates of the field are questioned in this context. Specific conceptual points of the critique raised in this context refer, for example, to the need to provide an alternative theorization of AI ethics through emphasizing the "raciality of computing and political economy of technoscience" by providing a "racial and colonial understanding of technoscience" (Hassan, 2023, p. 1430).

Further points of the critique voiced have referred to different interpretations of formality vs. informality and underlined the need to study governance as centered around informalities and oral cultures of governance which shape practices and rules outside of written policies (Raghunath, 2022; Sakr, 2022). While there might be some references to self-governance and co-governance in this approach, the inclusion of locally shaped oral cultures of transmission, shaping, and changing of unwritten rules and practices and their consequences for media and communication, is an aspect worthy of further exploration in the literature. Here, the suggestion is also to consider individuals' negotiation of their freedoms, identities, and opinions in highly restricted public spaces in authoritarian and transformational systems as a governance mode (Matsilele & Mustvairo, 2022; Sakr, 2022) and to consider voice-parity as a value to be upheld in governance processes (Raghunath, 2022). In line with that, the multi-stakeholder approach is frequently critiqued as a contingent concept, a process that is often suggested as an ideal solution to democratic shortcomings, but one that neglects power imbalances and hinders policy innovations. Therefore, scholars who provide perspectives from the Global South offer critical thought and conceptual scrutiny. Shen (2022) points here to the problem of

4 Please see Ganter and Badr (2022) for a detailed explanation of which observations are foundational for cosmopolitan critique and its iteration.

overemphasizing the USA as an influential actor in international multilateral processes and not taking into account other countries such as China and their impact. De Albuquerque and de Matos (2022) criticize how the multi-stakeholder approach has been used to distribute neoliberal ideas and policy programs with the aim of maintaining power relations. Segura and Linares (2022) propose "participatory policy" (p. 215) as an alternative terminology altogether, to provide room for an analysis of multilateral processes which can address and acknowledge different types and stages of participation and how power asymmetries shape access and ability to participate in high-level policy making—as well as to provide encouragement and tools to start reform from below (Segura & Waisbord, 2016). Asthana (2022) similarly highlights the need to enable everyday people's involvement in resistance to established systems through practices of "commoning" (p. 32), thus lowering barriers to access to media infrastructures and goods.

The critique of media governance as a "utopian concept" (Jo & Jin, 2022) is also a powerful reminder that commonplaces from media governance are not widely accepted as evident. This critique points to the increased uneven influence of powerful companies, such as digital platform companies, when it comes to dealing with new practices and rules when citizen involvement in these processes is limited through structural and cultural conditioning. It is interesting to note that the rise of platforms has exposed the impact of similar challenges and their varied consequences for citizens around the world. Different countries negotiate and deal with challenges such as the circulation of content that is potentially harmful to individuals and society at large, economic imbalances and challenges related to the media industry and journalism, or the violation of privacy rights and copyrights in different ways (e.g., Ganter, 2022). Moving forward, the underlying assumption needs to be that all contexts are equally insightful and conceptually and theoretically important and that they all can enrich the development of the field. Overall, cosmopolitan critique brought forward recently have emphasized the need to move beyond conceptualizations of media governance, as based on regulatory or multi-level actions, to consider power relations, particularly those that affect and delimit or enable citizen action, and to reflect more of what is not addressed, studied, or asked about and therefore not conceptually considered.

Challenges to cosmopolitan iteration in media governance studies

The impact that cosmopolitan critique unfolds academically and the results that cosmopolitan iteration will bring to the fore will over time depend on whether new perspectives and voices can obtain good visibility and whether the field engages them. Research suggests that those who manage journals will influence the geographical spread of authors and issues studied (Ganter & Ortega, 2019; Goyanes & Demeter, 2020). Editors in chief of the few current flagship journals of the field are mainly from the US-American or central European context, and editorial board members are similarly distributed (see Table 1). Consequentially, organized special issues and edited volumes that promote perspectives of the Global South or specific regions that are normally underrepresented are still required (e.g., Bizberge et al., 2023; Ganter & Badr, 2022) to promote work on and from within contexts outside the established geographical hubs. However, the risk is that these bundled publications will remain invisible and not engaged with from within hegemonic centers. Comparative perspectives, for example, would offer multiple opportunities for deep engagement and have been popular particularly in the European stream of media governance research (e.g., d'Haenens et al., 2018). However, even studies with large samples are often restricted to more similar cases and are likelier to exclude countries which would contribute experiences different from centered contexts—or adapt outliers to Western understandings (Pies, 2022). The argument that country X is not comparable with country Y is frequently used and often misses the mark when it is about finding new ways to think about and study media governance in the first place (see also Radue et al. on media systems in this book). The question to be asked is not "How comparable is it?" Instead, the important question in cosmopolitan inquiry would be "What are the differences and similarities in the generic thinking and resulting conceptualizations of what constitutes media governance, and why do they exist?" or simply put, "What can we learn from context x or y, and how does it matter?" This, however, requires the courage to "reach beyond the comfort zone" (Ganter, 2020). However, it is not enough to foster cosmopolitan critique and to publish case studies from the Global South or include experiences from the Global South; it will be essential to engage them proactively over time. This engagement will not be successful if constructive critique is mentioned as a counter-hegemonic agenda but then the contributions made are not considered in the emerging conceptualizations of the field (e.g., Puppis et al., 2024)—or if Western-centric conceptualizations, for example, of "algorithmic governance" are being translated and published in contexts of the Global South as if they were also automatically and equally addressing the contexts beyond the West (e.g., see Katzenbach & Ulbricht, 2019).

Pedagogically, embracing epistemic wonderings would also mean training and encouraging students to study scholarly works from a variety of contexts. This particularly includes empowering students from the Global South, regardless of where they chose to study, to engage with concepts and ways of studying from their local contexts and to transport them to their university. However, at times when governmental policies force universities to start reversing from being globalized with a high portion of international students into being more nationalized places, it might be challenging to uphold a cosmopolitan perspective in classrooms—even in universities that currently have a culturally, linguistically, and nationally diverse studentship.[5] Explaining to local students in North America and Europe why findings from studies done in Tanzania, Argentina, China, or Egypt matter and are theoretically, socially, and politically at least as important as studies done in Canada, the USA, or the UK, could require more effort and dialogue in the classroom moving forward. Therefore, to be convincing in this effort, it is even more important to make it a standard that faculty engage with and use scholarly work from outside their own cultural, linguistic, and national contexts. At the same time, it is necessary that efforts for cosmopolitan iteration are institutionally supported and normalized through funding agencies, departmental inclusion of this type of work into evaluations of workload, and its recognition as academic work per se. In particular, funding agencies should engage more open funding schemes that do not preselect a

5 In Canada, for example, the government decided in January 2024 to considerably limit visas for international students for at least two years (https://www.cbc.ca/news/politics/miller-cap-international-students-1.7090779).

limited number of potential cooperation partner countries and enhance and facilitate global partnerships on a level playing field between researchers, as well as support both national and international students with comparable funding schemes. In the field of media governance, this is particularly relevant, as it is a field where policy innovation (in all directions) could emerge from more dialogue between academics from different contexts and with diverse experiences that could be shared widely and benefit not only the academic community but also policymakers and citizens.

Table 1: Journals designated to publishing work in the field of media governance (source: author's compilation)

Journal	Year founded	Editor in Chief/ Managing Board	Affiliation of Editor(s)/ Managing Board Members
Telecommunications Policy	1976	Erik Bohlin	Western University Canada and Chalmers University of Technology, Sweden
International Journal of Cultural Policy	1993	Oliver Bennett,	Centre for Cultural and Media Policy Studies, University of Warwick, Coventry, UK
Communication Law and Policy	1996	Amy Kristin Sanders	University of Texas at Austin, USA
Journal of Digital Media & Policy	2010	Petros Iosifidis	City University, London, UK
Internet Policy Review	2012	Mélanie Dulong de Rosnay Natali Helberger Jeanette Hofmann Martin Kretschmer Vincent Homburg David Megías Jiménez Wolfgang Schulz	CIS-CNRS, Université Paris-Sorbonne IViR, University of Amsterdam HIIG Berlin CREATe, University of Glasgow Johan Skytte Institute, University of Tartu IN3, Universitat Oberta de Catalunya Hans-Bredow Institute, University of Hamburg
Digital Policy, Regulation and Governance*	2017	Anna Visvizi	SGH Warsaw School of Economics – Poland

*Formerly info: The journal of policy, regulation and strategy for telecommunications, information and media

Conclusion

Media governance as a field is built upon three main hegemonic hubs that emerged almost simultaneously and, to some extent, overlapped in thematic areas, conceptual and theoretical considerations, and methodologies but failed to acknowledge the differences that also exist when we think about, approach and study media governance from within different contexts and positionalities. Even though all three hubs have produced self-reflections and calls to reinvigorate or improve the field, little attention has been given to blind spots, which include work done in the field from and on the Global South as well as intersectional perspectives. The temporary renewal of the field through the introduction of new labels, such as social media governance or platform governance, has largely been a testimony of the structural imbalances shaping the field and further cemented the invisibility of de-Westernizing realities and perspectives. This is particularly problematic, as the rise of digital technologies, increased access to the Internet, and the emergence of platform services and AI technologies have also increasingly shown dispersed impact in countries around the world (see also Sarısakaloğlu in this book). There have been varied attempts and opportunities for citizens and the industry to engage in the technological developments and their impact on society. The limited opportunities for agency have led to concrete consequences for access, freedom of expression and privacy rights in countries where powerful actors used new means as a way to control, monitor, and restrict access to unbiased information, their circulation and their production. Regardless, we see emerging forms of resistance and new informal ways to shape media governance, to establish and safeguard rights, and to formulate expectations toward the frameworks and practices that shape media systems. These realities are different from what Western media governance studies have addressed, and they are equally important for a truly broad and open conceptualization and theorization of the field—and publications addressing these realities from a media governance perspective have been slowly increasing since 2014. If the aim is to be inclusive and move the field forward through an epistemic transformation, it is critical to trigger epistemic wonderings in those who have been at the center of the field for many years. These wonderings need to be triggered through critique, the process of cosmopolitan iteration, and the recognition that iteration has no perfect ending, no ideal answers, but that it is ongoing and requires broad engagement and consideration from academics working under all different labels that media governance literature provides us with and from all parts of the world. The creation of spaces where epistemic wonderings are possible enables structural ruptures and changes needed for epistemological transformations. Academic cosmopolitanism provides an encouragement to move out of the established comfort zone and engage in epistemic wonderings. If media governance as a field of study wants to maintain its relevancy and become a more equitable and open area of study, this is where we need to invest the time, labor, and resources to succeed.

References

Afonso, A. (2013, December 11). How academia resembles a drug gang. *LSE Blog*. https://blogs.lse.ac.uk/impactofsocialsciences/2013/12/11/how-academia-resembles-a-drug-gang/

Akpojivi, U. (2022). Cosmopolitan media, contestation, and critique: Assessing international media governance standards from the Nigerian perspective. In S. A. Ganter & H. Badr (Eds.), *Media governance: A cosmopolitan critique* (pp. 103–124). Palgrave Macmillan. https://doi.org/10.1007/978-3-031-05020-6_6

Alhassan, A., & Chakravartty, P. (2011). Postcolonial media policy under the long shadow of empire. In R. Mansell & M. Raboy (Eds.), *The handbook of global media and communication policy* (pp. 366–382). Wiley Blackwell. https://doi.org/10.1002/9781444395433.ch23

Ali, C., & Puppis, M. (2018). When the watchdog neither barks nor bites: Communication as a power resource in media policy and regulation. *Communication Theory*, *28*(3), 270–291. https://doi.org/10.1093/ct/qtx003

Allam, R. (2022). The Egyptian media governance framework: Gains and limitations. In S. A. Ganter & H. Badr (Eds.), *Media governance: A cosmopolitan critique* (pp. 145–164). Palgrave Macmillan. https://doi.org/10.1007/978-3-031-05020-6_8

Asthana, S. (2022). Sovereignty, power, and agency in neoliberal configurations of media and governance in the Global South. In S. A. Ganter & H. Badr (Eds.), *Media governance: A cosmopolitan critique* (pp. 15–38). Palgrave Macmillan. https://doi.org/10.1007/978-3-031-05020-6_2

Bannerman, S. (2020). *Canadian communication policy and law*. Canadian Scholars.

Benhabib, S. (2006). The philosophical foundations of cosmopolitan norms. In S. Benhabib & R. Post (Eds.), *Another cosmopolitanism* (pp. 13–44). Oxford University Press. https://doi.org/10.1093/acprof:oso/9780195183221.003.0002

Bhuiyan, A. (2014). *Internet governance and the Global South: Demand for a new framework*. Palgrave Macmillan. https://doi.org/10.1057/9781137344342

Bizberge, A., Mastrini, G., & Gómez, R. (2023). Discussing Internet platform policy and regulation in Latin America. *Journal of Digital Media & Policy*, *14*(2), 135–148. https://doi.org/10.1386/jdmp_00118_2

Bolaño, C. (2020). Notes on political economy and critical thought in communication studies in Europe and Latin America. In F. O. Paulino, G. Kaplún, M. V. Mariño, & L. Custódio (Eds.), *Research traditions in dialogue: Communication studies in Latin America and Europe* (pp. 107–114). Media XXI.

Braman, S. (2004). Where has media policy gone? Defining the field in the twenty-first century. *Communication Law and Policy*, *9*(2), 153–182. https://doi.org/10.1207/s15326926clp0902_1

Braman, S. (2006). *Change of state: Information, policy and power.* MIT Press.

Braman, S. (2013). Laying the path: Governance in early Internet design. *Journal of Policy, Regulation and Strategy for Telecommunications, Information and Media*, *15*(6), 63–83. https://doi.org/10.1108/info-07-2013-0043

Camou, A. (2020). *Gobernabilidad y Democracia*. Instituto Nacional Electoral. (Original work published 1995)

Coates, K., & Holroyd, C. (2003). *Japan and the Internet revolution*. Palgrave Macmillan. https://doi.org/10.1057/9781403990075

Connell, R. (2007). *Southern theory: The global dynamics of knowledge in social science*. Routledge.

d'Haenens, L., Sousa, H., & Trappel, J. (Eds.). (2018). *Comparative media policy, regulation and governance in Europe: Unpacking the policy cycle*. Intellect.

de Albuquerque, A., & de Matos, L. M. (2022). Media governance and fake news in Brazil. In S. A. Ganter & H. Badr (Eds.), *Media governance: A cosmopolitan critique* (pp. 125–144). Palgrave Macmillan. https://doi.org/10.1007/978-3-031-05020-6_7

de Albuquerque, A., & de Oliveira, T. (2021). Thinking the recolonial in communication studies: Reflections from Latin America. *Comunicação Mídia E Consumo*, *18*(51), 81–100. https://doi.org/10.18568/cmc.v18i51.2521

de Sousa Santos, B. (2005). The future of the World Social Forum: The work of translation. *Development*, *48*(2), 15–22. https://doi.org/10.1057/palgrave.development.1100131

Donders, K., Pauwels, C., & Loisen, J. (Eds.). (2014). *The Palgrave handbook of European media policy*. Palgrave Macmillan. https://doi.org/10.1057/9781137032195

Donges, P. (2007). Medienpolitik und Media Governance. In P. Donges (Ed.), *Von der Medienpolitik zur Media Governance?* (pp. 7–23). Herbert von Harlem.

Flew, T. (2015). Social media governance. *Social Media + Society*, *1*(1), 1–2. https://doi.org/10.1177/2056305115578136

Floridi, L. (Ed.). (2021). *Ethics, governance, and policies in Artificial Intelligence*. Springer. https://doi.org/10.1007/978-3-030-81907-1

Freedman, D. (2008). *The politics of media policy*. Polity Press.

Ganter, S. A. (2016). *Audio-visual media policy and the (re)configuration of audio- visual spaces in the Mercosur: Amplified governance, geographies and sovereignty* [Unpublished doctoral dissertation]. University of Vienna.

Ganter, S. A. (2020, March 9). Academic cosmopolitanism on the conference circuit: Reaching beyond the comfort zone. *Conference Interference: Blogging the World of Conferences*. https://conferenceinference.wordpress.com/2020/03/09/academic-cosmopolitanism-on-the-conference-circuit-reaching-beyond-the-comfort-zone-sarah-anne-ganter/?

Ganter, S. A. (2022). Governance of news aggregators' practices across five emblematic cases: Policy regimes between normative acceptance and resistance. *Information Society*, *38*(4), 290–306. https://doi.org/10.1080/01972243.2022.2076180

Ganter, S. A., & Badr, H. (Eds.). (2022). *Media governance: A cosmopolitan critique*. Palgrave Macmillan. https://doi.org/10.1007/978-3-031-05020-6

Ganter, S. A., & Löblich, M. (2021). Discursive media institutionalism: Assessing Vivien A. Schmidt's framework and its value for media and communication studies. *International Journal of Communication*, *15*, 2281–2300. https://ijoc.org/index.php/ijoc/article/view/16863

Ganter, S. A., & Ortega, F. (2019). The invisibility of Latin American scholarship in European media and communication studies: Challenges and opportunities of de-Westernization and academic cosmopolitanism. *International Journal of Communication*, *13*, 68–91. https://ijoc.org/index.php/ijoc/article/view/8449

Gillespie, T. (2017). Governance of and by platforms. In J. Burgess, A. Marwick, & T. Poell (Eds.), *The Sage handbook of social media* (pp. 254–278). Sage.

Ginosar, A. (2013). Media governance: A conceptual framework or merely a buzz-word? *Communication Theory*, *23*(4), 356–374. https://doi.org/10.1111/comt.12026

Gorwa, R. (2019). What is platform governance? *Information, Communication & Society*, *22*(6), 854–871. https://doi.org/10.1080/1369118X.2019.1573914

Goyanes, M., & Demeter, M. (2020). How the geographic diversity of editorial boards affects what is published in JCR-ranked communication journals. *Journalism & Mass Communication Quarterly*, *97*(4), 1123–1148. https://doi.org/10.1177/1077699020904169

Hamelink, C. J., & Nordenstreng, K. (2007). Towards democratic media governance. In E. de Bens (Ed.), *Media between culture and commerce* (pp. 225–240). Intellect.

Hassan, Y. (2023). Governing algorithms from the South: A case study of AI development in Africa. *AI & Society*, *38*(4), 1429–1442. https://doi.org/10.1007/s00146-022-01527-7

Helberger, N. (2011). Media diversity from the user's perspective: An introduction. *Journal of Information Policy*, *1*, 241–245. https://doi.org/10.5325/jinfopoli.1.2011.0241

Hintz, A. (2016). Policy hacking: Citizen-based policy-making and media reform. In D. Freedman, J. Obar, C. Martens, & R. W. McChesney (Eds.), *Strategies for media reform: International perspectives* (pp. 223–238). Fordham University Press. https://doi.org/10.1515/9780823271672-019

Hong, S. (2022). A new perspective on the importance of the state in global Internet governance: Tracing China's participation. In S. A. Ganter & H. Badr (Eds.), *Media governance: A cosmopolitan critique* (pp. 185–203). Palgrave Macmillan. https://doi.org/10.1007/978-3-031-05020-6_10

Jo, H., & Jin, D. Y. (2022). Media governance as a utopian concept in a local mediascape: Challenges for conceptual development in South Korea. In S. A. Ganter & H. Badr (Eds.), *Media governance: A cosmopolitan critique* (pp. 165–181). Palgrave Macmillan. https://doi.org/10.1007/978-3-031-05020-6_9

Jose, J. (2007). Reframing the 'governance' story. *Australian Journal of Political Science*, *42*(3), 455–470. https://doi.org/10.1080/10361140701513588

Just, N., & Latzer, M. (2017). Governance by algorithms: Reality construction by algorithmic selection on the Internet. *Media, Culture & Society*, *39*(2), 238–258. https://doi.org/10.1177/0163443716643157

Just, N., & Puppis, M. (2018). Moving beyond self-castigation: Let's reinvigorate communication policy research now! *Journal of Communication*, *68*(2), 327–336. https://doi.org/10.1093/joc/jqx012

Just, N., Latzer, M., & Sauerwein, F. (2007). Communications Governance: Entscheidungshilfe für die Wahl des Regulierungsarrangements am Beispiel Spam. In P. Donges (Ed.), *Von der Medienpolitik zur Media Governance?* (pp. 103–126). Herbert von Halem.

Karppinen, K., & Moe, H. (2013). A critique of "media governance". In M. Löblich & S. Pfaff-Rüdiger (Eds.), *Communication and media policy in the era of the Internet: Theories and processes* (pp. 69–80). Nomos. https://doi.org/10.5771/9783845243214-69

Katzenbach, C., & Ulbricht, L. (2019). Algorithmic governance. *Internet Policy Review*, *8*(4). https://doi.org/10.14763/2019.4.1424

Kim, D. (2018). *Media governance in Korea 1980–2017*. Palgrave Macmillan. https://doi.org/10.1007/978-3-319-70302-2

Kleinsteuber, H. J., & Nehls, S. (Eds.). (2011). *Media Governance in Europa: Regulierung – Partizipation – Mitbestimmung*. VS Verlag für Sozialwissenschaften. https://doi.org/10.1007/978-3-531-93265-1

Kleinwächter, W. (2000). ICANN between technical mandate and political challenge. *Telecommunications Policy*, 24(6–7), 553–563. https://doi.org/10.1016/S0308-5961(00)00037-9

Löblich, M (2016). Dissent and political participation: The many faces of communication policy advocacy and activism. *Communication, Culture & Critique*, 9(3), 395–416. https://doi.org/10.1111/cccr.12113

Mansell, R. (2012). *Imagining the Internet: Communication, innovation and governance*. Oxford University Press.

Mansell, R., & Raboy, M. (2011). *The handbook of global media and communication policy*. Wiley Blackwell. https://doi.org/10.1002/9781444395433

Matsilele, T., & Mutsvairo, B. (2022). Understanding the dynamics of social media governance in South Africa. In S. A. Ganter & H. Badr (Eds.), *Media governance: A cosmopolitan critique* (pp. 231–260). Palgrave Macmillan. https://doi.org/10.1007/978-3-031-05020-6_12

Michalis, M. (2010). EU broadcasting governance and PSB: Between a rock and a hard place. In P. Iosifidis (Ed.), *Reinventing public service communication: European broadcasters and beyond* (pp. 36–48). https://doi.org/10.1057/9780230277113_4

Mueller, M. L. (2002). *Ruling the root: Internet governance and the taming of cyberspace*. MIT Press.

Mueller, M. L., Mathiason, J., & Klein, H. (2007). The Internet and global governance: Principles and norms for a new regime. *Global Governance*, 13(2), 237–254. https://www.jstor.org/stable/27800656

Musiani, F. (2013). WSIS+10: The self-praising feast of multi-stakeholderism in Internet governance. *Internet Policy Review*, 2(2). https://doi.org/10.14763/2013.2.121

Opperman, D. (Ed.). (2018). *Internet governance in the Global South: History, theory, and contemporary debates*. University of São Paulo. https://nbn-resolving.org/urn:nbn:de:0168-ssoar-65805-2

Padovani, C. (2018). Gendering media policy research and communication governance. *Javnost – The Public*, 25(1–2), 256–264. https://doi.org/10.1080/13183222.2018.1423941

Padovani, C., Musiani, F., & Pavan, E. (2010). Investigating evolving discourses on human rights in the digital age. *International Communication Gazette*, 72(4–5), 359–378. https://doi.org/10.1177/1748048510362618

Paulino, F. O., & Kaplún, G. (2020). Transatlantic dialogue for future communication. In F. O. Paulino, G. Kaplún, M. V. Mariño, & L. Custódio (Eds.), *Research traditions in dialogue: Communication studies in Latin America and Europe* (pp. 11–14). Media XXI.

Picard, R. G. (2016). Isolated and particularised: The state of contemporary media and communications policy research. *Javnost – The Public*, 23(2), 135–152. https://doi.org/10.1080/13183222.2016.1162991

Pies, J. (2022). Media accountability in a non-democratic context: Conceptual challenges and adaptations. In S. A. Ganter & H. Badr (Eds.), *Media governance: A cosmopolitan critique* (pp. 81–100). Palgrave Macmillan. https://doi.org/10.1007/978-3-031-05020-6_5

Price, M. E. (2002). *Media and sovereignty: The global information revolution and its challenge to state power*. MIT Press. https://doi.org/10.7551/mitpress/4533.001.0001

Puppis, M. (2010). Media governance: A new concept for the analysis of media policy and regulation. *Communication, Culture & Critique, 3*(2), 134–149. https://doi.org/10.1111/j.1753- 9137.2010.01063.x

Puppis, M., Mansell, R., & Van den Bulck, H. (Eds.). (2024). *Handbook of media and communication governance*. Edward Elgar Publishing.

Raboy, M. (Ed.). (2002). *Global media policy in the new millennium*. University of Luton Press.

Raghunath, P. (2022). Deliberative policy ecology approach: Media policy studies from South Asia. In S. A. Ganter & H. Badr (Eds.), *Media governance: A cosmopolitan critique* (pp. 261–281). Palgrave Macmillan. https://doi.org/10.1007/978-3-031-05020-6_13

Reitman, R. (2016). Lessons from the SOPA Fight. In D. Freedman, J. Obar, C. Martens, & R. W. McChesney (Eds.), *Strategies for media reform: International perspectives* (pp. 92–99). Fordham University Press. https://doi.org/10.1515/9780823271672-007

Roberts, H., Cowls, J., Morley, J., Taddeo, M., Wang, V., & Floridi, L. (2021). The Chinese approach to Artificial Intelligence: An analysis of policy, ethics, and regulation. In L. Floridi (Ed.), *Ethics, governance, and policies in Artificial Intelligence* (pp. 47–79). Springer. https://doi.org/10.1007/978-3-030-81907-1_5

Sakr., N. (2022). Media governance as diagnostic lens to probe hidden dimensions of authoritarian decision-making in the Arab Middle East. In S. A. Ganter & H. Badr (Eds.), *Media governance: A cosmopolitan critique* (pp. 39–58). Palgrave Macmillan. https://doi.org/10.1007/978-3-031-05020-6_3

Schulz, W., & Dankert, K. (2016). 'Governance by Things' as a challenge to regulation by law. *Internet Policy Review, 5*(2). https://doi.org/10.14763/2016.2.409

Segura, S. M., & Linares, A. (2022). Democratic governance of media and public communication: Latin American participatory institutions created in the twenty-first century. In S. A. Ganter & H. Badr (Eds.), *Media governance: A cosmopolitan critique* (pp. 205–230). Palgrave Macmillan. https://doi.org/10.1007/978-3-031-05020-6_11

Segura, M. S., & Waisbord, S. (2016). *Media movements: Civil society and media policy reform in Latin America*. Zed Books.

Shah, N., Rajadhyaksha, A., & Hasan, N. A. (2022). *Overload, creep, excess: An Internet from India*. Institute of Network Cultures.

Shen, H. (2022). A new perspective on the importance of the State in global Internet governance: Tracing China's participation. In S. A. Ganter & H. Badr (Eds.), *Media governance: A cosmopolitan critique* (pp. 185–203). Palgrave Macmillan. https://doi.org/10.1007/978-3-031-05020-6_10

Siles, I. (2020). *A transnational history of the Internet in Central America, 1985–2000: Networks, integration, and development*. Palgrave Macmillan. https://doi.org/10.1007/978-3-030-48947-2

Siochrú, S., Girard, B., & Mahan, A. (2002). *Global media governance: A beginner's guide*. Rowman & Littlefield.

Smith, J. A., & Craig, R. T. (Eds.). (2023). *Racializing media policy*. Emerald. https://doi.org/10.1108/9781804557365

Strycharz, J., Ausloos, J., & Helberger, N. (2020). Data protection or data frustration? Individual perceptions and attitudes towards the GDPR. *European Data Protection Law Review, 6*(3), 407–421. https://doi.org/10.21552/edpl/2020/3/10

Tankard, K. (1998). History, the Internet and South Africa. *South African Historical Journal*, *38*(1), 20–33. https://doi.org/10.1080/02582479808671310
van Eeten, M. J. G., & Mueller, M. L. (2013). Where is the governance in Internet governance? *New Media & Society*, *15*(5), 720–736. https://doi.org/10.1177/1461444812462850
Waisbord, S. (2019). *Communication: A post-discipline*. Polity Press.
Waisbord, S., & Mellado, C. (2014). De-Westernizing communication studies: A reassessment. *Communication Theory*, *24*(4), 361–372. https://doi.org/10.1111/comt.12044
Warf, B., & Vincent, P. (2007). Multiple geographies of the Arab Internet. *Area*, *39*(1), 83–96. https://www.jstor.org/stable/20004588
Wasserman, H. (2017). African histories of the Internet. *Internet Histories*, *1*(1–2), 129–137. https://doi.org/10.1080/24701475.2017.1308198
Willems, W. (2023). The reproduction of canonical silences: Re-reading Habermas in the context of slavery and the slave trade. *Communication, Culture & Critique*, *16*(1), 17–24. https://doi.org/10.1093/ccc/tcac047
Woods, N. (1999). Good governance in international organizations. *Global Governance*, *5*(1), 39–61. https://www.jstor.org/stable/27800219
Yang, F., & Mueller, M. L. (2014). Internet governance in China: A content analysis. *Chinese Journal of Communication*, *7*(4), 446–465. https://doi.org/10.1080/17544750.2014.936954

Reimagining risk and crisis communication research through a cosmopolitan lens

Pauline Gidget Estella, Martin Löffelholz & Yi Xu

Introduction

The complexity of modern risks and crises is evident in everyday headlines: the severe impacts of climate emergencies, worsening refugee crises, and multifaceted health crises, such as the COVID-19 pandemic. Such crises transcend borders, mandates, and cultures. Today's transnational risks and protracted crises require global perspectives and solutions, although nationally focused efforts to address these overlapping "systemic events" are no less important (Kunelius, 2020, p. 1; see also Sellnow-Richmond & Lukacovic, 2024).

Against this backdrop, and in an era of increasing globalization and geopolitical shifts, there is a need to discuss what constitutes "cosmopolitan" risk and crisis communication, particularly in terms of research. Risk and crisis communication is highly relevant in light of the current circumstances, as risk communication ideally aims to help people understand potential hazards and manage them through informed decisions, while crisis communication aims to minimize harm and maintain trust during and after critical events. Because communicators are organizations of different types and purposes (e.g., governments and corporations), the field has become broad and heterogeneous, encompassing different disciplines ranging from public health to communication studies to political science.

Risk and crisis communication research has long been concerned with various issues, including identifying effective communication strategies during crises, how these strategies influence audience behavior or crisis perception, and the role of environmental factors (e.g., state–media dynamics) in shaping these outcomes, to name a few. To the extent that risk and crisis perceptions are influenced by an array of contextual factors, transboundary crises can therefore be viewed from different cultural perspectives and studied through multiple theoretical and methodological lenses (e.g., see Schwarz et al., 2016). Such risks and crises characterize what Beck (2011) called a modern "world risk society," where the spatiality and "incalculableness" of risks and crises require perspectives beyond previous international collaborations and epistemological advances in research (p. 665). The COVID-19 pandemic, for instance, hit both developing and industrialized nations (albeit disproportionately), prompting scholars from different settings to examine how governments, organizations, and other entities communicated within and across borders. If anything, the pandemic and comparable crises underscore the need for comparative approaches in research, primarily to understand how differences in macro-level characteristics influence communication, and for studies of environments that are rarely represented in scholarship (see also Esser & Hanitzsch, 2012).

Moreover, these crises also raise the stakes for global or intercultural communication practices, such as cross-border journalism, especially in light of the criticisms of the practice (to be discussed further in the subsequent sections). For instance, research has shown that during the height of the COVID-19 pandemic in 2020, some news media outlets promoted problematic nationalist narratives and even misinformation, which may have contributed to fueling racist sentiments (e.g., Richter et al., 2021; Zhang & Trifiro, 2022).

What has been said so far raises the question: What can be said about risk and crisis communication research and practice concerning phenomena that have both universal and unique repercussions? This chapter tackles such questions, including critical issues related to the state of the art in risk and crisis communication research, particularly the persisting Western, or more precisely, Euro-American ethnocentrism in scholarship. We contend that this narrow focus does not serve a world fraught with protracted, complex crises and increasingly multicultural societies. The first section describes the state of research, followed by a discussion of the nature of complex crises and the corresponding need for effective intercultural crisis communication. The final section unpacks the nature of cosmopolitan risk and crisis communication research as a normative concept and develops suggestions for further conceptual and empirical studies.

Persisting Euro-American ethnocentrism in research

The last few decades have witnessed remarkable growth in risk and crisis communication research. This is evident in the increasing number of studies on (and from) settings beyond Europe and North America (the so-called "mainstream West"), the adoption of international comparative approaches, and the development of cross-country research communities (Löffelholz et al., 2023; see also Diers-Lawson & Meißner, 2021). The COVID-19 pandemic has also led to a surge in the number of studies in the field over the past three to four years; in China and the USA, for instance, the number of publications dealing with risk and crisis communication jumped to about three times the number of studies before the pandemic (Löffelholz et al., 2023, p. 22).

However, much remains to be done as the scholarship on risk and crisis communication is hardly "global" in the true sense of the word. It retains its strong Western bias, with "geographic and perspective blind spots" (Diers-Lawson & Meißner, 2021, p. 165; see also Diers-Lawson, 2017). Much of the knowledge production is still concentrated in institutions of the Western mainstream (or under the contested label of the "Global North"), albeit scholars from developing regions and industrialized Asian countries are publishing more and more (Che et al., 2022; Diers-Lawson, 2017; Diers-Lawson & Meißner, 2021;

Goerlandt et al., 2020; Löffelholz et al., 2023). The research has provided longitudinal evidence illustrating this deeply entrenched imbalance. Goerlandt et al. (2020), for example, conducted a scientometric analysis of the risk communication literature from 1985 to 2019 and found that approximately 75% of all the publications originated from five Western countries (USA, UK, the Netherlands, Germany, and Canada). Their findings are congruent with those of a previous study mapping the crisis communication literature from 1953 to 2015: more than 86% of the publications over the past six decades have been about crisis communication in Europe and North America, especially the USA, which accounts for more than 60% of the literature (Diers-Lawson, 2017). In terms of citation trends in the field, the most cited scholars are based at institutions in the Western mainstream, and most international cooperative research is also concentrated in these regions (Che et al., 2022).

Diers-Lawson (2017) described this state of research as a "state of emergency," given the fact that the most vulnerable regions are also the most underrepresented, with much less research production (p. 1). For instance, African countries have collectively contributed only 1.2% of the crisis communication literature over the last six decades, while South and Southeast Asia account for 3.2% (Diers-Lawson, 2017). It is not surprising, therefore, that researchers have questioned the extent to which practice is informed or enhanced by research in these areas, and whether known frameworks or theoretical assumptions in scholarship are actually applicable across settings. As Diers-Lawson (2017) wrote, "Functionally, crisis communication scholars and practitioners know very little about the topic as it applies to the majority of the world's population" (p. 11).

In recent decades, numerous studies have discussed the potential consequences of Western centrism in knowledge production, with repeated calls for the "de-Westernization" or "decolonization" of research (e.g., Ganter & Ortega, 2019; Glück, 2018; Waisbord & Mellado, 2014; Willems, 2014). The significant disparity in research development between the Western mainstream and many other parts of the world can lead to what Waisbord and Mellado (2014) called "universalistic pretensions based on a narrow slice of context-specific cases" (p. 365). Such "pretensions" can take the form of popular theoretical assumptions or established frameworks in the field, or even research epistemologies in general, that cannot fully account for circumstances in areas beyond the Western mainstream. For example, frameworks that are often used to underpin empirical research, particularly the Situational Crisis Communication Theory (SCCT; Coombs, 2007) and the Image Repair Theory (Benoit, 1995), have been criticized for their "lack of non-Western perspectives," leading to the exclusion of some strategies commonly used in other settings (Hu & Pang, 2018, p. 108). In the case of mainland China, for example, some of the commonly used corporate crisis response strategies were either inconsistent with the prescriptions of the SCCT or were not mentioned at all in the framework (Hu & Pang, 2018). For example, public apologies or defensive response strategies, which are prescribed in SCCT when appropriate, are rarely used in China, which can be explained by the prevalence of the "face-saving" mindset as a cultural value (Wang & Lauder, 2020, p. 7). In fact, risk and crisis communication is context sensitive, and any inquiry should therefore consider macro-level contextual factors, such as cultural influences and the role of the government (Löffelholz et al., 2023; Schwarz, 2016; Wu et al., 2016).

Furthermore, despite the existence of transboundary crises, crises as objects of analysis still vary across countries, as they experience unique crisis events, and crisis events capture media and public attention (or the public imagination) in different ways. This is evident in the historical development of the field and the recurring research themes that correspond to crises impacting specific regions: the scholarly attention to natural disasters and large-scale crises in many parts of Asia, the Chernobyl disaster and its impact in directly affected European countries, and the "weakness of civil society" and political opposition in the Middle East and North Africa (Löffelholz et al., 2023; see also Frandsen & Johansen, 2016; Huang et al., 2016; Saleh, 2016, p. 316). Even the conceptualization of risk and crisis can be influenced by political and cultural contexts. In the Soviet Union, for example, "risk" and "crisis" were generally viewed as "foreign notions" that were inevitable consequences of the Western capitalist order, in line with the propagandistic rhetoric of the time (Samoilenko, 2016, p. 399).

However, it would not be fair to say that the research landscape in regions beyond the Western mainstream is entirely barren. After all, as Diers-Lawson (2017) noted, there is a growing trend toward international comparative studies and a significant increase in the number of publications originating from and about Chinese societies (see also Che et al., 2022), with some scholars even proposing a Chinese model of crisis communication (Wu et al., 2016) and a framework for Chinese crisis communication research (Huang, 2010). Despite the dearth of research in low- and middle-income countries (LMICs), empirical studies on risk and crisis communication still exist, many of which are qualitative or descriptive case studies (e.g., Adamu et al., 2018; Arora, 2022). The COVID-19 pandemic also triggered increased knowledge production in these countries (e.g., George et al., 2021; Sakhiyya et al., 2024; Tam et al., 2021).

What might explain the disparity in knowledge production that threatens the plurality of academic discourse, not just in risk and crisis communication but in the field of communication studies in general? We argue that it can be attributed to five factors: 1) a research and development infrastructure that is insufficiently developed, which is tied to broader chronic societal dilemmas such as poor governance and systemic poverty, 2) the historical development of the field, where institutions in the USA and Europe gained footing much earlier than those in the rest of the world, 3) constraints or external influences on research production, such as pressures in tightly controlled political systems, 4) the English language as the "academic lingua franca," which contributes to the invisibility of work in other languages, regardless of the intellectual rigor involved (Lillis et al., 2010, p. 111), and 5) deeply embedded academic structures that favor scholarship and research traditions originating from and focused on high-income, predominantly Western countries, which in turn marginalize scholars from underrepresented regions and communities (Barreto de Souza Martins et al., 2023; Chakravartty et al., 2018; Fox Tree & Vaid, 2022; Glück, 2018).

The first four factors mentioned above can be briefly explained, but the last one requires further elaboration. The first relates to the lack of adequate support mechanisms for research, such as underinvestment and poor working conditions in academia in many LMICs (e.g., Heng et al., 2023; Quiroga-Garza et al., 2022). The limited investment in research can also hinder scholars from pursuing studies that require substantial financial resources, such as large-sample surveys. The second point concerns the relative advantage of US-American and European institutions in research production, partly due to the earlier institutionalization of risk and crisis communication research, or even communication studies in general, in these regions (Löffelholz et al., 2023; see also Barreto de Souza Martins et al., 2023; Waisbord & Mellado, 2014). The third factor can be seen in how political influence and tight control over public information can limit research or academic freedom in certain settings, such as Russia, where the academic community has limited access to primary data on crises (Samoilenko, 2016). The fourth concerns the dominance of English as the medium in knowledge production, as seen, for instance, in citation practices in international journals, as was found by Lillis et al. (2010). By analyzing citations, reviewer comments, and ethnographic data, they found that there is a "pressure" to cite English-medium works and to publish in English, and that the journal "gatekeepers" are even "challenging citations in other languages" (Lillis et al., 2010, p. 131). In an academic landscape marked by these "Anglophone gatekeeping practices," there is a significant risk that valuable knowledge published in other languages may be overlooked by the broader research community (Lillis et al., 2010, p. 111).

The last factor is largely about the uneven accumulation of "academic capital" that is supposed to be gained from one's publication record, number of citations, networks, and the like (Barreto de Souza Martins et al., 2023, p. 481). Again, empirical research has provided evidence of this uneven playing field. For example, there is a severe lack of representation from the Global South in the editorial boards of top communication journals, with more than 79% of board members coming from Australia, Canada, the USA, the UK, and Germany (Goyanes, 2020; see also Chakravartty et al., 2018). Demeter (2019), who studied the "career paths" of 426 scholars from the Global South, found that it was "almost impossible" for Global South researchers to become internationally recognized scholars in the field of communication without some capital from the Global North (e.g., education or affiliation) (p. 578). Having capital from the Global North is also associated with increased chances of publishing in top-tier communication journals, as 85% of the Global South authors who have published in these journals have some form of Global North capital. In fact, there are even some top-ranked journals that have almost never published work from Global South authors (Demeter, 2019, p. 592).

There are also other research traditions that differ from those of the Western mainstream and are more often sidelined in academic discourse. For example, in the case of scholarship on pandemic crisis communication, researchers from LMICs tend to focus on "social epidemiology and collective health," as opposed to mainstream approaches focusing on social psychology, media effects, and individual behavior and risk factors (Barreto de Souza Martins et al., 2023, p. 488). Many researchers from LMICs also seem to prefer qualitative methods, as seen in Brazil during the pandemic, despite the international focus on quantitative approaches (Barreto de Souza Martins et al., 2023).

From the discussion, it is evident that the persisting Euro-American ethnocentrism in research is inextricably linked with systemic conditions that shape environments and fields of knowledge, such as underinvestment in research resulting from problematic governance. These conditions are extremely difficult to change, but in the face of prolonged crises and the increasing need for effective transnational and transcultural crisis communication, steps toward decentering knowledge production are necessary.

Prolonged crises and the challenges of transcultural crisis communication

For several scholars, the COVID-19 pandemic exemplified the new, more complex type of crisis that characterizes today's global risk society (Auld et al., 2021; Jin et al., 2021; Löffelholz et al., 2023; Sellnow-Richmond & Lukacovic, 2024). For what seems like countless times, the pandemic crisis has been described as unprecedented, or unlike any other crisis in recent history, because it is a health crisis that has generated political, economic, and other types of social crises of immense magnitude. It has been referred to by various names in the academic literature: "mega crisis" (Sellnow-Richmond & Lukacovic, 2024), "sticky crisis" (Coombs et al., 2021), "prolonged crisis" (Diers-Lawson & Omondi, 2024), and "super wicked problem" (Auld et al., 2021), all of which share several conceptual overlaps. As both a prolonged crisis and a sticky crisis, the pandemic has been characterized by 1) multiple sub-crises affecting entire societies, 2) different forms of uncertainties, 3) the need for complex solutions, 4) transcending cultures and borders, and 5) prolonged duration (see Coombs et al., 2021; Diers-Lawson & Omondi, 2024; Jin et al., 2021).

As a super wicked problem, the pandemic crisis can be viewed as a problem that can exacerbated by the very actors attempting to solve it (Auld et al., 2021). For example, the disease spreads rapidly due to international flows of knowledge, labor, and capital, from which people also benefit. Furthermore, measures to mitigate the effects are often shortsighted or address only part of the broader consequences (Auld et al., 2021, p. 711). For the purposes of the discussion in this chapter, we have chosen to refer to these crises as "prolonged crises," but it should be noted that these crises also exhibit the characteristics of super wicked problems and sticky crises.

However, the pandemic is neither the first nor the last of its kind. The climate emergency, for example, was already described as a super wicked problem more than a decade before the pandemic (e.g., Lazarus, 2009; Levin et al., 2012). The world has experienced crises that had transnational consequences and were widely debated in the global public sphere, such as the Russian invasion of Ukraine, the global financial crisis, and others. The modern "world risk society" faces such transnational and complex risks and crises, and modern societies are "shaped by new kinds of risks, that their foundations are shaken by the global anticipation of global catastrophe" (Beck, 2011, p. 665). For Yuan (2021), the COVID-19 pandemic exemplified the "perfect storm of modern risk society," stemming from "man-made ecological disturbance" and spreading through the infrastructure of intensified globalization, such as international migration flows and greater economic and political interdependence among countries (p. 323).

In the context of a global risk society, Beck and Levy (2013) argued that "cosmopolitanized risk collectivities" arise largely through the communication of risk in the global media landscape and are therefore an important "facet" of the global risk society (p. 3). This serves as a theoretical starting point for the discussion of cosmopolitan risk and crisis communication. These cosmopolitanized risk collectives, such as transnational advocacy publics, emerge through the international mediatization of risk and constant exposure to risk-related information, even those coming from the other side of the world. An important example of this is the global civil society movement centered on climate change advocacy. The collectivities are partly anchored in the presence of a cosmopolitan mindset, which, in a normative sense, is characterized by a commitment to the virtues of global justice and welfare, an awareness of the interconnectedness of the world, and a dialogue with different cultures, especially the marginalized identities (Cheah, 2006; Delanty, 2012; Waisbord, 2015; Wardle, 2015). The term "partly" is essential here, as the collectivities may or may not be informed by the ideal form of a cosmopolitan mindset. This normative and critical cosmopolitanism also informs research on cosmopolitan risk and crisis communication, which we will discuss in more detail in the next section.

While the concepts of nationhood, national identities, and even parochial attitudes remain intact, the communication of complex risks and prolonged crises cannot be viewed solely through nation-centric lenses, and even the concept of nationhood is being "reimagined" through cosmopolitanization (Beck & Levy, 2013, p. 6). As Beck and Levy (2013) noted, "Globalization provides a new context for the transformation of national identifications" (p. 6). In this regard, they asserted that contrary to popular assumptions in literature, cosmopolitanism is not the "antithesis to the essentialized version of the national" (Beck & Levy, 2013, p. 5). Instead, these two seemingly polar opposites (global and local or universal and particular) are "interconnected and reciprocally interpenetrating" (Beck, 2006, p. 72, as cited in Beck & Levy, 2013, p. 6). For example, exposure to other settings with different political systems can cause individuals to reconsider their views about national government.

The rise of cosmopolitan risk collectives and prolonged crises underscores the increasing importance of intercultural risk and crisis communication, the main forms of which include 1) organizations or public institutions communicating about transboundary crises and to transnational audiences, 2) organizations or public institutions communicating to a multicultural society, such as state government communication in large cities, and 3) media coverage of transboundary crises or news reporting for international audiences. However, many practitioners feel that they are not equipped to deal with multicultural publics, nor do they adapt their strategies according to the needs of different cultures (de Fatima Oliveira, 2013). In fact, there have been many cases where multinational organizations failed to implement an appropriate crisis response strategy in an international context, primarily due to a lack of understanding of cultural characteristics (An et al., 2010).

A number of studies have described the challenges and complexities of intercultural communication during crises (e.g., Kharbat et al., 2024; Lehmberg & Hicks, 2018; de Fatima Oliveira, 2013). For instance, research has shown how crisis perceptions and the effectiveness of communication strategies can vary widely across countries and cultures, such that collectivist cultures may perceive messages quite differently from individualist cultures (Claeys & Schwarz, 2016; see also An et al., 2010). At the time of the COVID-19 pandemic, government communicators in Europe and the USA, particularly in culturally diverse cities such as New York or Amsterdam, reported that tailoring communications to specific audience segments was one of the major challenges in their work, including translating messages into multiple languages and identifying opinion leaders (Löffelholz et al., 2023).

When it comes to news coverage, two major ethical challenges include representing cultural identities without peddling ethnocentric narratives and resisting instrumentalization for propaganda, both of which become more of a normative ideal when one takes into account the various influences on the journalist. Several authors have criticized the lapses in reporting on international crises, international coverage on crises, or reporting on crises related to different cultures, as seen, for instance, in journalistic framing during conflicts as crises (e.g., Baden & Meyer, 2018; Nohrstedt, 2016). As Baden and Meyer (2018) noted, there are cases of "biased, ethnocentric interpretations" against an out-group, sometimes even justifying a warmongering agenda (p. 33; see also Baden & Tenenboim-Weinblatt, 2018; Löffelholz, 2004). There are also documented cases of sensationalized reporting on violent conflicts, often at the expense of necessary contextualization and accuracy, presumably to capture audience attention (e.g., Ishaku, 2021; see also Schleicher & Sarısakaloğlu in this book). The media, at times, facilitated the "othering" of vulnerable groups, such as Muslims and immigrants, creating an "us vs. them" dichotomy through the "discourse of fear" (Nohrstedt, 2010, p. 38). In the case of the COVID-19 pandemic, some conservative media outlets in the USA spread misinformation and conspiracy theories, blaming China for the spread of the disease, and potentially contributing to "racial tensions" in the country in 2021 (Zhang & Trifiro, 2022, p. 1044). Vilifying the "Other" or an outgroup is also a strong theme in COVID-19-related news coverage across several countries in the Middle East and North Africa, all while reinforcing nationalist identities (Richter et al., 2021).

Given the rise of complex risks and prolonged crises, the challenges (and lapses) in intercultural risk and crisis communication, and the ongoing Western-centrism in knowledge production, it is crucial to discuss the elements of what can be called cosmopolitan risk and crisis communication research.

Principles of cosmopolitan risk and crisis communication research

In this section, we will propose some guiding principles for cosmopolitan risk and crisis communication research, inspired by the elements of normative cosmopolitanism mentioned in the previous section (e.g., see Cheah, 2006; Waisbord, 2015), Badr and Ganter's (2021) work on "academic cosmopolitanism" (see also Alves & Medeiros, 2021), Waisbord and Mellado's (2014) dimensions in de-Westernizing communication studies, and Diers-Lawson's (2017) intercultural communication research agenda. For some of the principles, we will provide examples from our experience with the DECIPHER project, an international research consortium funded by the German Research Foundation that examined COVID-19-related risk and crisis communication in the USA and six European countries.

As a normative framework, cosmopolitan risk and crisis communication research is guided by the following principles: 1) it recognizes the nature of complex risks and prolonged crises, especially their transnational nature; 2) it strives for a truly cosmopolitan perspective, one that is informed by critical de-Westernization, thereby including perspectives and realities from underrepresented environments; 3) it involves attempts to address long-standing inequities in academic spaces; 4) it accommodates the particularities of cultures and environments yet seeks to apply an inclusive yet unified framework for mature comparative studies; and 5) it is devoid of "epistemic hierarchization" and involves academic dialogue on equal footing (Alves & Medeiros, 2021, p. 12). Furthermore, following the United Nations Sendai Framework for Disaster Risk Reduction 2015–2030

(United Nations, 2015) and other agreements at the United Nations level, cosmopolitan risk and crisis studies should be people-centered. While many existing theories focus on image enhancement and reputation management, a cosmopolitan approach incorporates a humanitarian perspective that focuses more on stakeholders and the diversity of people affected by complex risks and protracted crises.

The first principle takes into account the characteristics of prolonged crises, as explained in the first section. Because the effects and causes of such crises can no longer be located within national borders, studies with an international or comparative perspective offer insights into how risk and crisis communication is conducted or perceived in different settings. As Esser and Hanitzsch (2012) noted, international comparative research allows us to see our own communication patterns in a different light by "[contrasting] them critically with those prevalent in other societies" (p. 4), thus answering the question as to how particular political, cultural, or historical contexts explain the variance in message reception or choice of communication strategies. Moreover, such studies allow scholars to question theoretical frameworks or ways of thinking that are considered commonsensical or taken for granted as "natural." In terms of practical implications, comparative studies show which solutions work in which contexts and why and which best practices from other environments can be adapted to the current setting (Esser & Hanitzsch, 2012).

The abovementioned reasons were also the reasons for adopting a comparative approach in the DECIPHER project. In this project, we compared pandemic-related government communication, news coverage, citizen perceptions, and online discourse in Italy, Germany, the Netherlands, Spain, Sweden, the UK, and the USA. The countries were selected according to differences in system-level characteristics; for example, some countries represented individualistic risk cultures while others represented state-oriented and fatalistic risk cultures (Cornia et al., 2016). The rationale was to identify similarities or differences in communication and message reception patterns and how these related to system-level characteristics.

The nature of prolonged crises as multiplex crises also necessitates careful consideration of their cross-sectoral impact on society and the different considerations or "trade-offs" that influence crisis communication. It can be argued, for instance, that COVID-19-related risk and crisis communication is not only health communication but also a form of strategic government communication and is therefore also driven by political considerations. This raises the questions: How do communicators in different countries "juggle" these sometimes competing considerations (e.g., communicating safety vs. communicating uncertainty), or which considerations take precedence over others? And what particular features of the environment might explain this?

The second principle, meanwhile, is predicated on the elements of a cosmopolitan mindset, as mentioned in the previous section: commitment to virtues of equality and justice that transcend borders, strong knowledge foundations on global interconnectedness, and a critical understanding of the historical and contemporary dynamics of the Global North and the Global South (e.g., Cheah, 2006; Delanty, 2012; Waisbord, 2015). This means encouraging the inclusion of underrepresented environments (LMICs and industrialized countries beyond the Western mainstream) in comparative studies or encouraging research from and about these settings. This principle is crucial in addressing what Diers-Lawson (2017) described as the US-centric state of research discussed in the first section of this chapter. This effort is best guided by the critical de-Westernization paradigm, which requires an awareness of the status quo in knowledge production and the factors that have led to persisting Euro-American ethnocentrism in research, and, more importantly, a willingness to help correct such disparities.

But first, we must ask what exactly de-Westernization means. The term has several meanings. For many scholars, especially in the so-called West, de-Westernization is more about inclusiveness or "considering experiences, research findings, and theoretical frameworks developed in the rest of the world," but for scholars beyond the mainstream West, it is more about "a necessary shift to reorient intellectual work against academic Eurocentrism" (Waisbord & Mellado, 2014, p. 362). The more radical tone, so to speak, of the second conceptualization can be understood in the context of the deeply embedded structures of Euro-American ethnocentrism, which was discussed in depth in the first section. These structures are linked to systemic conditions in the LMICs or in academia in general, and while it is extremely difficult to change these, research communities can start by de-Westernizing important dimensions of their research. They can de-Westernize the "subject of study" (conducting studies on the circumstances of underrepresented settings), the "body of evidence" (e.g., including LMICs or the Global South countries in comparative studies to arrive at truly global conclusions), the "analytical frameworks" (e.g., using theoretical frameworks that accommodate realities of the Global South, or modifying them for this purpose, or using indigenous frameworks from these countries), and "academic cultures" (understanding the differences in the practices, norms, and beliefs of scholars around the world and why certain academic cultures are more "privileged" than others) (Waisbord & Mellado, 2014, pp. 363–370).

De-Westernizing academic cultures involves a defining characteristic of academic cosmopolitanism: "intellectual and structural critique towards academia" (Badr & Ganter, 2021, p. 2). This structural critique underpins the third principle: helping correct long-standing inequities in academic spaces, as seen in the experiences and numbers of women and people of color in academia, especially in tenured positions and bodies that have a role in the accumulation of academic capital (see Fox Tree & Vaid, 2022). This can take the form of fostering collaborations with scholars from the Global South as principal investigators in comparative projects, including more scholars from the Global South on the editorial boards of risk and crisis communication journals, supporting scholars from the Global South in conducting research in their own settings, and promoting more collaborations among countries in the Global South.

In practice, however, such collaborations are more complex. In fact, collaborations in comparative studies in general are quite complex. Part of what makes it so is related to the fourth principle: capturing the particularities of the environments while still using an inclusive yet unified framework for comparative studies, which entails functional equivalence or the "equivalence of concepts" (Hanitzsch & Esser, 2012, p. 504). This principle is also in line with another aspect of academic cosmopolitanism: "room for differentiation" in the process of creating "common spaces" (Badr & Ganter, 2021, p. 2). Seeking functional equivalence, meanwhile, means ensuring an "identical, or at least similar, definition of core concepts across all investigated cultures in a comparative study" (Hanitzsch & Esser, 2012, p. 504). This also requires the construction of a theoretical framework that can be applied across environments without ignoring important environmental particularities.

The fourth principle can be illustrated by describing our sampling strategy in the DECIPHER subproject on government communication. In selecting our respondents, we needed to identify representatives who could speak on behalf of executive, public health, and communication functions in the national and subnational governments. However, we encountered some complications due to system-level differences. For example, in some countries, those who perform key functions do not always meet the sampling criteria or work in units with mandates different from their supposed counterparts in other countries. In one case, the equivalent subnational government did not have the crisis management competencies that its counterparts in other countries had. Addressing these issues was, of course, a complicated task and required a literature-based justification, but this was largely resolved through regular discussions with our international partners, who were professors from the countries in our sample and had a long history of research on these environments. This is the "assembling strategy" in establishing equivalence: experts from different cultural origins take the time to discuss concepts and approaches and arrive at "universally applicable" but still context sensitive concepts (Hanitzsch & Esser, 2012, p. 504).

In the workshops and regular conversations with international partners, we were also able to extensively discuss issues of research epistemologies, all the while trying to maintain an open-minded attitude in a dialogue of equals, which illustrates the fifth guiding principle. The fifth principle is based on another facet of academic cosmopolitanism: an "open-minded and impartial attitude toward scholars and their work and seeks dialogue on even grounds" (Badr & Ganter, 2021, p. 2). As the consortium was composed of scholars from different disciplines, we were able to gain insights into the norms of knowledge production in other related disciplines as well as reflect on our own. This was also a step toward avoiding "epistemic hierarchization," borrowing Alves and Medeiros's (2021, p. 12) term. Epistemic hierarchization refers to the dominance of some research traditions that remain unchallenged by virtue of their position in the structures that provide academic capital, as discussed in a previous section.

The discussion in this section suggests that cosmopolitan risk and crisis communication research goes beyond the mere internationalization of scholarship. It addresses not only the deficiencies in research epistemologies but also the structural dilemmas of academia when it comes to the representation of Global South scholars or people of color in research communities. Addressing the latter is crucial because the themes, approaches, and overall research traditions in the field are linked to such issues of representation.

Conclusion

This chapter examined the persisting Euro-American ethnocentrism in risk and crisis communication research, exploring its causes and implications. We also discussed the nature of prolonged crises and cosmopolitan risk collectivities that have emerged due to the increasing interconnectedness of the world, as well as the challenges and critical issues surrounding intercultural risk and crisis communication. Finally, we proposed a set of guiding principles for research on cosmopolitan risk and crisis communication. Our aim was not only to improve the study of global crises as the world risk society faces one protracted crisis after another, but also to address long-standing disparities in knowledge production that prevent the field from moving forward.

While the principles are grounded on normative cosmopolitanism, we provided illustrative examples from research experience. Nevertheless, it remains to be seen if these principles can actually be applied in cosmopolitanizing research (which one can test through future empirical studies). There is hope, as there are notable ongoing efforts to address the disparities in knowledge production. Compared to a couple of decades ago, the field has had substantial gains in terms of internationalization and institutionalization (Löffelholz et al., 2023). As mentioned in an earlier section, the academic community is incorporating more and more perspectives from beyond the mainstream West, as seen in the increasing number of comparative projects involving Global South environments and scholars publishing on realities in these areas.

To truly advance the field, it is essential to commit to a cosmopolitan approach in risk and crisis communication research, which involves not only seeking diverse perspectives and methodologies but also challenging Euro-American ethnocentrism in knowledge production and structures of academia. Only then can a genuine cosmopolitan turn take place in risk and crisis communication research. This cosmopolitan turn can lead to a better understanding of risks and crises, most which impact developing and underrepresented regions much more than industrialized ones, ultimately translating to more effective practice.

References

Adamu, A. A., Mohamad, B. B., & Abdul Rahman, N. A. B. (2018). Towards measuring internal crisis communication: A qualitative study. *Journal of Asian Pacific Communication*, 28(1), 107–128. https://doi.org/10.1075/japc.00006.ada

Alves, C. N. R., & Medeiros, D. (2021). Towards cosmopolitanism in German academia? Shedding light on colonial underpinnings of communication research in a globalized world. *Global Media Journal – German Edition*, 11(1). https://doi.org/10.22032/dbt.49167

An, S.-K., Park, D.-J., Cho, S., & Berger, B. (2010). A cross-cultural study of effective organizational crisis response strategy in the United States and South Korea. *International Journal of Strategic Communication*, 4(4), 225–243. https://doi.org/10.1080/1553118X.2010.515543

Arora, S. (2022). Post-disaster communities on social media: Citizen participation in crisis communication after the Nepal earthquake, 2015. *Journal of Applied Communication Research*, 50(1), 1–18. https://doi.org/10.1080/00909882.2021.1964572

Auld, G., Bernstein, S., Cashore, B., & Levin, K. (2021). Managing pandemics as super wicked problems: Lessons from, and for, COVID-19 and the climate crisis. *Policy Sciences*, 54(4), 707–728. https://doi.org/10.1007/s11077-021-09442-2

Baden, C., & Meyer, C. O. (2018). Dissecting media roles in conflict: A transactionist process model of conflict news production, dissemination and influence. In R. Fröhlich

(Ed.), *Media in war and armed conflict: The dynamics of conflict news production and dissemination* (pp. 23–48). Routledge.

Baden, C., & Tenenboim-Weinblatt, K. (2018). The search for common ground in conflict news research: Comparing the coverage of six current conflicts in domestic and international media over time. *Media, War & Conflict, 11*(1), 22–45. https://doi.org/10.1177/1750635217702071

Badr, H., & Ganter, S. A. (2021). Towards cosmopolitan media and communication studies: Bringing diverse epistemic perspectives into the field. *Global Media Journal – German Edition, 11*(1). https://doi.org/10.22032/dbt.49164

Barreto de Souza Martins, F., Yu, J., & Domahidi, E. (2023). A global health crisis with divided research traditions? A comparative review of Brazilian and international research in communication on the COVID-19 pandemic. *Annals of the International Communication Association, 47*(4), 479–496. https://doi.org/10.1080/23808985.2023.2242817

Beck, U. (2011). Clash of risk cultures or critique of American universalism. *Contemporary Sociology, 40*(6), 662–667. https://doi.org/10.1177/0094306111425017

Beck, U., & Levy, D. (2013). Cosmopolitanized nations: Re-imagining collectivity in world risk society. *Theory, Culture & Society, 30*(2), 3–31. https://doi.org/10.1177/0263276412457223

Benoit, W. L. (1995). *Accounts, excuses, apologies: A theory of image restoration strategies*. State University of New York Press.

Chakravartty, P., Kuo, R., Grubbs, V., & McIlwain, C. (2018). #CommunicationSoWhite. *Journal of Communication, 68*(2), 254–266. https://doi.org/10.1093/joc/jqy003

Che, S., Kamphuis, P., Zhang, S., Zhao, X., & Kim, J. H. (2022). A visualization analysis of crisis and risk communication research using CiteSpace. *International Journal of Environmental Research and Public Health, 19*(5), Article 2923. https://doi.org/10.3390/ijerph19052923

Cheah, P. (2006). Cosmopolitanism. *Theory, Culture & Society, 23*(2–3), 486–496. https://doi.org/10.1177/026327640602300290

Claeys, A.-S., & Schwarz, A. (2016). Domestic and international audiences of organizational crisis communication: State of the art and implications for cross-cultural crisis communication. In A. Schwarz, M. W. Seeger, & C. Auer (Eds.), *The handbook of international crisis communication research* (pp. 224–235). Wiley Blackwell. https://doi.org/10.1002/9781118516812.ch21

Coombs, W. T. (2007). Attribution Theory as a guide for post-crisis communication research. *Public Relations Review, 33*(2), 135–139. https://doi.org/10.1016/j.pubrev.2006.11.016

Coombs, W. T., Holladay, S. J., & White, R. (2021). Corporate crises: Sticky crises and corporations. In Y. Jin, B. H. Reber, & G. J. Nowak (Eds.), *Advancing crisis communication effectiveness: Integrating public relations scholarship with practice* (pp. 35–51). Routledge.

Cornia, A., Dressel, K., & Pfeil, P. (2016). Risk cultures and dominant approaches towards disasters in seven European countries. *Journal of Risk Research, 19*(3), 288–304. https://doi.org/10.1080/13669877.2014.961520

de Fatima Oliveira, M. (2013). Multicultural environments and their challenges to crisis communication. *Journal of Business Communication, 50*(3), 253–277. https://doi.org/10.1177/0021943613487070

Delanty, G. (Ed.). (2012). *Routledge handbook of cosmopolitanism studies*. Routledge. https://doi.org/10.4324/9780203837139

Demeter, M. (2019). So far, yet so close: International career paths of communication scholars from the Global South. *International Journal of Communication, 13*, 578–602. https://ijoc.org/index.php/ijoc/article/view/10181

Diers-Lawson, A. (2017). A state of emergency in crisis communication an intercultural crisis communication research agenda. *Journal of Intercultural Communication Research, 46*(1), 1–54. https://doi.org/10.1080/17475759.2016.1262891

Diers-Lawson, A., & Meißner, F. (2021). Editor's essay: Moving beyond Western Corporate Perspectives: On the need to increase the diversity of risk and crisis communication research. *Journal of International Crisis and Risk Communication Research, 4*(1), 165–176. https://doi.org/ https://doi.org/10.30658/jicrcr.4.1.6

Diers-Lawson, A., & Omondi, G. (2024). Ripples, waves, and riptides: Reconceptualizing wicked, novel, and ongoing crises as prolonged crises. In B. F. Liu & A. M. Mehta (Eds.), *Routledge handbook of risk, crisis, and disaster communication* (pp. 179–196). Routledge.

Esser, F., & Hanitzsch, T. (2012). On the why and how of comparative inquiry in communication studies. In F. Esser & T. Hanitzsch (Eds.), *The handbook of comparative communication research* (pp. 3–22). Routledge.

Fox Tree, J. E., & Vaid, J. (2022). Why so few, still? Challenges to attracting, advancing, and keeping women faculty of color in Academia. *Frontiers in Sociology, 6*, Article 792198. https://doi.org/10.3389/fsoc.2021.792198

Frandsen, F., & Johansen, W. (2016). Crisis communication research in Northern Europe. In A. Schwarz, M. W. Seeger, & C. Auer (Eds.), *The handbook of international crisis communication research* (pp. 373–383). Wiley Blackwell. https://doi.org/10.1002/9781118516812.ch34

Ganter, S. A., & Ortega, F. (2019). The invisibility of Latin American scholarship in European media and communication studies: Challenges and opportunities of de-Westernization and academic cosmopolitanism. *International Journal of Communication, 13*, 68–91. https://ijoc.org/index.php/ijoc/article/view/8449

George, T. J., Shafqat, N., Verma, R., & Bali, S. (2021). Handling the pandemic our way: A qualitative content analysis of the guidelines issued by Apex Institutes of National Importance (INIs) of India to combat COVID-19 crisis. *Journal of Family Medicine and Primary Care, 10*(11), 4054–4065. https://doi.org/10.4103/jfmpc.jfmpc_263_21

Glück, A. (2018). De-Westernization and decolonization in media studies. In M. Powers (Ed.), *Oxford encyclopedia of communication*. Oxford University Press. https://doi.org/10.1093/acrefore/9780190228613.013.898

Goerlandt, F., Li, J., & Reniers, G. (2020). The landscape of risk communication research: A scientometric analysis. *International Journal of Environmental Research and Public Health, 17*(9), Article 3255. https://doi.org/10.3390/ijerph17093255

Goyanes, M. (2020). Editorial boards in communication sciences journals: Plurality or standardization? *International Communication Gazette*, *82*(4), 342–364. https://doi.org/10.1177/1748048518825322

Hanitzsch, T., & Esser, F. (2012). Challenges and perspectives of comparative communication inquiry. In F. Esser & T. Hanitzsch (Eds.), *The handbook of comparative communication research* (pp. 501–516). Routledge.

Heng, K., Hamid, M. O., & Khan, A. (2023). Research engagement of academics in the Global South: The case of Cambodian academics. *Globalisation, Societies and Education*, *21*(3), 322–337. https://doi.org/10.1080/14767724.2022.2040355

Hu, Y., & Pang, A. (2018). The indigenization of crisis response strategies in the context of China. *Chinese Journal of Communication*, *11*(1), 105–128. https://doi.org/10.1080/17544750.2017.1305978

Huang, Y.-H. (2010). Theorizing Chinese communication research: A holistic framework for comparative studies. *Chinese Journal of Communication*, *3*(1), 95–113. https://doi.org/10.1080/17544750903528880

Huang, Y.-H. C., Wu, F., Cheng, Y., & Lyu, J. C. (2016). Crisis communication research in the Chinese Mainland. In A. Schwarz, M. W. Seeger, & C. Auer (Eds.), *The handbook of international crisis communication research* (pp. 269–282). Wiley Blackwell. https://doi.org/10.1002/9781118516812.ch25

Ishaku, J. (2021). Peace journalism or war journalism? An analysis of newspaper coverage of ethno-religious conflicts in Southern Kaduna, Nigeria (2020–2021). *International Journal of Communication Research*, *11*(4), 295–313.

Jin, Y., Choi, S. I., & Diers-Lawson, A. (2021). Special issue editor's essay: Advancing public health crisis and risk theory and practice via innovative and inclusive research on COVID-19 communication. *Journal of International Crisis and Risk Communication Research*, 4(2), 177–192. https://jicrcr.com/index.php/jicrcr/article/view/44

Kharbat, F. F., Kannan, Y., Gleason, K., & Qasim, A. (2024). Corporate communication during the COVID-19 crisis in a multicultural environment: Culture and tweet impact. *Electronic Commerce Research*, 24(1), 675–709. https://doi.org/10.1007/s10660-023-09777-3

Kunelius, R. (2020). On the overlap of systemic events: Covid-19, climate, and journalism. *Social Media + Society*, 6(3). https://doi.org/10.1177/2056305120948197

Lazarus, R. J. (2009). Super wicked problems and climate change: Restraining the present to liberate the future. *Cornell Law Review*, 94(5), 1153–1234. https://scholarship.law.cornell.edu/clr/vol94/iss5/8/

Lehmberg, D., & Hicks, J. (2018). A 'glocalization' approach to the internationalizing of crisis communication. *Business Horizons*, *61*(3), 357–366. https://doi.org/10.1016/j.bushor.2018.01.002

Levin, K., Cashore, B., Bernstein, S., & Auld, G. (2012). Overcoming the tragedy of super wicked problems: Constraining our future selves to ameliorate global climate change. *Policy Sciencies*, 45(2), 123–152. https://doi.org/10.1007/s11077-012-9151-0

Lillis, T., Hewings, A., Vladimirou, D., & Curry, M. J. (2010). The geolinguistics of English as an academic lingua franca: Citation practices across English-medium national and English-medium international journals. *International Journal of Applied Linguistics*, 20(1), 111–135. https://doi.org/10.1111/j.1473-4192.2009.00233.x

Löffelholz, M. (2004). Krisen- und Kriegskommunikation als Forschungsfeld: Trends, Themen und Theorien eines hoch relevanten, aber gering systematisierten Teilgebiets der Kommunikationswissenschaft. In M. Löffelholz (Ed.), *Krieg als Medienereignis II: Krisenkommunikation im 21. Jahrhundert* (pp. 13–55). VS Verlag für Sozialwissenschaften. https://doi.org/10.1007/978-3-322-90833-9_1

Löffelholz, M., Estella, P. G., & Xu, Y. (2023). COVID-19 as a catalyst of global risk society: Institutionalization, de-Westernization, and datafication of crisis communication research. In N. Yamamoto (Ed.), *The COVID-19 pandemic and risks in East Asia* (pp. 14–40). Routledge.

Löffelholz, M., Estella, P., Radechovsky, J., Schleicher, K., & Xu, Y. (2023, October 5–7). *How governments in Europe and the United States understand and deal with citizens' perceptions of government crisis communication: Lessons from a seven-country study* [Paper presentation]. 7th International Crisis Communication Conference, Gothenburg, Sweden.

Nohrstedt, S. A. (2010). Threat society and the media. In S. A. Nohrstedt (Ed.), *Communicating risks: Towards the threat society* (pp. 17–52). Nordicom.

Nohrstedt, S. A. (2016). The role of the media in the discursive construction of wars. In A. Schwarz, M. W. Seeger, & C. Auer (Eds.), *The handbook of international crisis communication research* (pp. 133–144). Wiley Blackwell. https://doi.org/10.1002/9781118516812.ch13

Quiroga-Garza, A., Garza-Cisneros, A. N., Elizondo-Omaña, R. E., Vilchez-Cavazos, J. F., Montes-de-Oca-Luna, R., Villarreal-Silva, E., Guzman-Lopez, S., & Gonzalez-Gonzalez, J. G. (2022). Research barriers in the Global South: Mexico. *Journal of Global Health*, *12*, Article 03032. https://doi.org/10.7189/jogh.12.03032

Richter, C., Al-Shami, A., Khalifa, S., Osman, S., & Mundua, S. (2021). The virus of the 'others'? Corona and discursive othering in Arab media. *Journal of Arab & Muslim Media Research*, *14*(1), 3–24. https://doi.org/10.1386/jammr_00022_1

Sakhiyya, Z., Dewi Saraswati, G. P., Anam, Z., & Azis, A. (2024). What's in a name? Crisis communication during the COVID-19 pandemic in multilingual Indonesia. *International Journal of Multilingualism*, *21*(2), 1169–1182. https://doi.org/10.1080/14790718.2022.2127732

Saleh, I. (2016). Crisis communication research in the Middle East & North Africa (MENA): Echoes of normalizing historical crisis. In A. Schwarz, M. W. Seeger, & C. Auer (Eds.), *The handbook of international crisis communication research* (pp. 313–326). https://doi.org/10.1002/9781118516812.ch29

Samoilenko, S. A. (2016). Crisis management and communication research in Russia. In A. Schwarz, M. W. Seeger, & C. Auer (Eds.), *The handbook of international crisis communication research* (pp. 397–410). Wiley Blackwell. https://doi.org/10.1002/9781118516812.ch36

Schwarz, A. (2016). Crisis communication research in Germany. In A. Schwarz, M. W. Seeger, & C. Auer (Eds.), *The handbook of international crisis communication research* (pp. 357–372). Wiley Blackwell. https://doi.org/10.1002/9781118516812.ch33

Schwarz, A., Seeger, M. W., & Auer, C. (2016). Significance and structure of international risk and crisis communication research: Toward an integrative approach. In A. Schwarz, M. W. Seeger, & C. Auer (Eds.), *The handbook of international crisis commu-*

nication research (pp. 1–10). Wiley Blackwell. https://doi.org/10.1002/9781118516812.ch1

Sellnow-Richmond, D. D., & Lukacovic, M. N. (2024). The COVID-19 pandemic as exemplar of the chaos of mega-crises. In T. L. Sellnow & D. D. Sellnow (Eds.), *Communicating risk and safety* (pp. 97–108). De Gruyter Mouton. https://doi.org/10.1515/9783110752427-006

Tam, L. T., Ho, H. X., Nguyen, D. P., Elias, A., & Le, A. N. H. (2021). Receptivity of governmental communication and its effectiveness during COVID-19 pandemic emergency in Vietnam: A qualitative study. *Global Journal of Flexible Systems Management*, *22*(1 Supplement), 45–64. https://doi.org/10.1007/s40171-021-00269-7

United Nations. (2015). *Sendai framework for disaster risk reduction 2015–2030*. https://www.undrr.org/publication/sendai-framework-disaster-risk-reduction-2015-2030

Waisbord, S. (2015). De-Westernization and cosmopolitan media studies. In C.-C. Lee (Ed.), *Internationalizing "International Communication"* (pp. 178–200). University of Michigan Press. https://doi.org/10.2307/j.ctv65sxh2.11

Waisbord, S., & Mellado, C. (2014). De-Westernizing communication studies: A reassessment. *Communication Theory*, *24*(4), 361–372. https://doi.org/10.1111/comt.12044

Wang, Y., & Laufer, D. (2020). How does crisis management in China differ from the West?: A review of the literature and directions for future research. *Journal of International Management*, *26*(1), Article 100708. https://doi.org/10.1016/j.intman.2019.100708

Wardle, H. (2015). Cosmopolitanism. In J. D. Wright (Ed.), *International encyclopedia of the social & behavioral sciences* (2nd ed., pp. 41–46). Elsevier. https://doi.org/10.1016/B978-0-08-097086-8.12043-4

Willems, W. (2014). Beyond normative dewesternization: Examining media culture from the vantage point of the Global South. *Global South*, *8*(1), 7–23. http://eprints.lse.ac.uk/id/eprint/61882

Wu, F., Huang, Y.-H. C., & Kao, L. (2016). East meets West: A new contextual perspective for crisis communication theory. *Asian Journal of Communication*, *26*(4), 350–370. https://doi.org/10.1080/01292986.2016.1148186

Yuan, E. J. (2021). Governing risk society: The socio-technological experiences of China and South Korea in the COVID-19 pandemic. *Asian Journal of Communication*, *31*(5), 322–336. https://doi.org/10.1080/01292986.2021.1913620

Zhang, Y., & Trifiro, B. (2022). Who portrayed it as "the Chinese virus"? An analysis of the multiplatform partisan framing in U.S. news coverage about China in the COVID-19 pandemic. *International Journal of Communication*, *16*, 1024–1050. https://ijoc.org/index.php/ijoc/article/view/17916

Prefix journalisms

Selected concepts and their cosmopolitan potentials and pitfalls

Melanie Radue, Thomas Eckerl, Oliver Hahn & Beate Illg[1]

Introduction

Concepts of journalism(s) have always been subject to change worldwide. Although global moments of journalism occur more and more due to technological change, digitalization, and globalization, journalism in different countries and world regions still remains very sensitive to cultures and languages (Hahn, 2008). In international comparative and collaborative research on journalistic cultures (Hanitzsch et al., 2019), different contextual understandings of concepts of journalisms must be taken into serious and permanent consideration. This theoretical and methodological awareness is the *conditio sine qua non* for the endeavor of a cosmopolitan approach to journalism studies that might help to avoid inter- and transcultural pitfalls and to overcome othering through constructed boundaries.

Journalism studies tend to think in typologies, where one can easily define what, for example, peace journalism, tabloid journalism, Ubuntu journalism, advocacy journalism, or independent journalism characterizes and which one is the most desirable of those journalisms. This packing into boxes and their assessment is mostly done from a Western perspective (Hanitzsch & Vos, 2018), where distinguished experience in diverse journalistic contexts from around the world does not necessarily exist. "In so doing, journalism scholarship privileges a vision of journalism that is narrower than reality, and it fails to account for distinctive approaches in non-democratic and non-Western contexts, as well as for forms of journalism beyond political news," according to Hanitzsch and Vos (2018, p. 159). Here, we want to critically raise the question concerning who defines which journalisms for whom and who is judging which journalism is more valuable than the other?

We summarize different concepts of journalisms as "prefix journalisms". In the literature, terms like "pioneer journalism" or "X journalism" can also be found to describe

1 The authors would like to thank Marlon Hahn (University of Passau) for his support in the preparation of this chapter.

various new and old reporting practices with certain specificities and novelties (Hepp & Loosen, 2021; Loosen et al., 2022). Prefix journalisms do not necessarily refer to the topics reported on, such as news, political, economic, financial, cultural, sports, or science journalism, etc. Neither are they necessarily inscribed into the debate about quality (aka "expensive") journalism vs. tabloid journalism, nor do they speculate about the future of traditional journalism. Rather, we suggest investigating prefix journalisms according to whether their concepts are normative, method-driven, technology-driven, outlet-driven or art-driven, as we grouped them in Table 1.

Table 1: Typology of prefix journalisms (source: authors' compilation)

Normative	**Method-driven**	**Technology-driven**	**Outlet-driven**	**Arts-driven**
- Development journalism - Peace journalism - Conflict-sensitive journalism - Constructive journalism - Advocacy journalism - Exile journalism - Ubuntu journalism - Civic/public journalism - Slow journalism - Patriotic journalism	- Investigative journalism - (Investigative) (collaborative) cross-border journalism - Data(-driven) journalism - Precision journalism	- Robotic journalism - Computational journalism - Algorithmic (automated) journalism - Immersive journalism - AI journalism - Drone journalism	- Online journalism - Mobile journalism - Video journalism - Citizen journalism - Corporate journalism	- New journalism/ literature journalism - Gonzo journalism

This collection of prefix journalisms, of course, does not exclude the overlapping or combination of concepts; some concepts maybe even be seen as general concepts including subordinated concepts. Some of these concepts are fluid.

Many normative concepts are discussed controversially due to their permeability for political instrumentalization. In contrast, there is more of a global consensus on many method-driven concepts. Some of the technology-driven concepts are even questioned whether or not they are still journalism or something else completely. Whereas digital media outlet-driven concepts seem to be widely accepted as journalistic concepts today, arts-driven concepts are oftentimes criticized as not being journalism at all. While research in this important area of prefix journalisms is still lacking, some relevant research in the area of "pioneer journalism as those forms of journalism that aim to redefine the field through experimental practices and imaginaries of journalism's possible future" (Hepp et al., 2021, p. 551) has been already undertaken.

In our review of three prefix journalisms, we shed light on their cultural bonds, normative entanglements, and transnational practices. There exist many roles and practices of journalism around the world, each having their contextual justification. Hence, we show the cosmopolitan potential arising from a review of different journalisms and a critical analysis of academic typologies. The concepts we are looking at are particularly interesting because they not only characterize a certain method or relate to certain outlets but also are particularly important for a cosmopolitan understanding of journalisms (research).

In this chapter, we take a closer look at three distinctive prefix journalisms that are relevant and timely for cosmopolitan journalism research. Therefore, we discuss the functions and normative and structural foundations of advocacy journalism, Ubuntu journalism, and cross-border journalism as three different prefix journalisms. The review of the three prefix journalisms highlights the varying functions of different journalisms within changing settings and the takeaways for a more cosmopolitan journalism research from those examples. We take different view angles to examine the normative assumptions of such prefix journalisms, their cultural bonds, situational and structural practices, and upcoming ambiguities. Since both advocacy journalism and Ubuntu journalism base their practices on a normative foundation, we take a more conceptual and normative perspective in our analysis in order to show whether and how such journalisms reveal their social functions and journalistic roles. We investigate these two prefix journalisms from two distinct viewpoints: advocacy journalism coming from a Western-centric perspective and Ubuntu journalism being based on the African Ubuntu philosophy. Specifically, this means that we first contrast advocacy journalism with the normativity of the Western influenced paradigms of objectivity and neutrality and explore how different perspectives on these paradigms may impact how we evaluate their functions in societies. For example, our perception of advocacy changes when it comes to exile journalism. Second, we investigate the concept of Ubuntu journalism. Here we discuss the community-centered perspective of Ubuntu, which forms an African normative approach to reporting. Third, our analysis of cross-border journalism takes a slightly different perspective on this specific form of investigative journalism. As we have categorized transnational cross-border journalism as a method-driven prefix journalism (see Table 1), the investigation takes its practices and methods as a starting point to underpin cosmopolitan aspects in research and reporting.

Overall, our aim is to zoom into these prefix journalisms to highlight cosmopolitan perspectives and potentials, rather than to paint a complete picture of these journalistic practices and their academic rootedness. Another aim in this chapter is to show how the perceptions of functions deviate when journalism is viewed from different viewpoints. Therefore, we adopt a normative conceptual perspective. The aim is to show how different normative assumptions of such prefix journalisms, their cultural bonds, and situational and structural practices lead to ambiguities in the evaluation of functions. Those ambiguous spaces hold the potential to shift our understanding of prefix journalisms considerably within a cosmopolitan perspective. Furthermore, a practice-orientated perspective also enables us to show ways in which journalisms can be better studied in order to recognize potentials in different situational and contextual settings.

This review of prefix journalisms must be seen against the background that all authors involved in this chapter are trained and do research at European academic institutions. This means our access to non-Western perspectives is admittedly limited. For us, research traditions outside this perspective are only accessible through articles published in English that take critical perspectives toward those concepts. This is one reason why the sources upon which this chapter is based are not as diverse and global as we would wish. Another reason is that most of the accessible knowledge production in journalism research is circulated within Western academic publications and citations.

This chapter opens perspectives to the ambiguity of (Western) theories and concepts that were mostly developed in the West or with reference to Western contexts. All the authors have acquired knowledge in research and practice that is not centered on the Global North, for example, the Middle East, Southeast Asia, South Asia, and Eastern Europe. Such research and practical experiences are often in contradiction to the modalities of Western-orientated academic knowledge production.

Advocacy journalism: The crux of objectivity and neutral reporting

The valuation of advocacy journalism seems to be ambiguous due to its perspectives and functions. In the following, we will mainly shed light on the Western perception of advocacy journalism and how it might deviate in its positionality. In media contexts in which a so-called free and independent reporting is the norm (e.g., in Germany, Europe, and the USA), advocacy journalism is considered problematic due to its non-objective and non-neutral reporting. It is therefore seen as an irregularity in terms of the concept of independent journalism, which is often seen as universal ideal. In contrast, non-objective advocacy journalism is supported and funded by governments from such contexts (e.g., Germany and the USA) through international media assistance programs designed to support, for example, exile journalists from countries such as Myanmar, Belarus, Russia, Afghanistan, or Ukraine, as non-objective advocacy journalism pursues foreign policy interests. "Journalists forced into exile have become primary targets for support by press freedom groups because they are the personification of a threatened virtue," according to Skjerdal (2010, p. 50). As a result, prefix journalism, such as advocacy journalism, may vary with its perspective of political ends and media contexts.

Therefore, we will briefly discuss if these two styles of journalism have a different base for assessment and how research can tackle such ambiguities. Although no consensus about the functions of journalism exists, neither in Western academia nor in journalism itself, most agree that the media should provide access to information for its society. In which style this information is disseminated is intensely discussed and it is without question context dependent.

When we talk about the ideal of neutral, objective, and so-called independent news reporting, most think of Western media reporting. This is based on "a normalization of Western ideals and practices of journalism as the 'professional' standard against which journalism in the non-Western world was gauged" (Hanitzsch & Vos, 2018, p. 150), although the reality in newsrooms in many "Western" media systems differs essentially. Still, in most journalism textbooks and often cited academic theories, this news style is presented as a universal ideal without questioning the Eurocentric contextual bond. This leads to the convention to define and assess all other journalistic styles in relation to a so-called independent/information journalism marked by objectivity and neutrality and leads to the unreflective assessment of other journalism styles, such as advocacy. "Western journalism scholarship has reproduced this hierarchy, privileging a journalistic world that is narrower than that which resides in practice," noted Hanitzsch and Vos (2018, p. 150) Thus, the buzzwords "journalistic objectivity" and "neutrality" are the crux of the matter. Generally, a Eurocentric understanding of independent/information journalism is formed by the idea that neutral and objective information is essential to form a pluralistic democratic society in which the journalist is an "independent public-spirited verifier of factual information as the superego of the news industry" (Hanitzsch & Vos, 2018, p. 150). As liberal societies all over the world become more and more under pressure from undemocratic forces, the perception of advocacy styles of journalism starts to change slightly. For example, journalism for the common good gets more attention as a support for democratic structures (Kramp & Weichert, 2023).

There might exist different understandings not only of journalistic styles around the world but also of how information serves different forms of societies and political practices. Examples from Nepal and Bhutan support the idea that journalists in different world regions follow the "Western" standard of objectivity and neutrality (this could be because of Eurocentric journalism textbooks or Western financed journalism training). At the same time, journalists in those contexts explain that they have to find "their own way" (see Illg, 2018, 2019; see also Keßler et al. in this book). This refers to specific social, cultural, and political circumstances within those countries and reflects different experiences with forms of democratic practices as the studies of Illg (2018, 2019) have shown. Both states—Nepal as well as Bhutan—have only short experiences with democracy and follow quite different pathways. They try to find their own way because neighboring democratic countries are not seen as a positive example to follow.

Even though different political frameworks exist around the world, no modern society is free of power imbalances and oppressions; hence, media has to find a way how to handle such factors. The credo of the independent journalist would be to report these in a neutral and objective manner. However, as journalists always have to focus on some specific aspects of a story, framing is unavoidable. With a constructivist perspective on media and journalism, there is no doubt about the impossibility of neutrality and objectivity. Plurality of realities is what radical constructivists state is the case. Neutrality and objectivity are not their relevant criteria, but functionality is the primary criterion. Even if one does not refer to a constructivist perspective, it is obvious and unavoidable that journalists always have to shorten and compress and establish a focus (Merten et al., 1994).

Weischenberg (2004, p. 80) took a critical stance toward a neutral and objective presentation of news that showcases injustices. Presenting injustices in a neutral way may legitimize or virtually absolve/excuse oppression, Weischenberg (2004) argued. He concluded that "such objectivity is inauthentic and violates humanity and truth, because it is quiet where one ought to be angry and abstains from accusation where it is contained in the facts themselves. The tolerance expressed in such impartiality serves to minimize or even absolve the prevailing intolerance and oppression. But if objectivity has anything to do with truth, and if truth is more than a matter of logic and science, then this kind of objectivity is false and this kind of tolerance is inhuman" (Weischenberg, 2004, p. 80, original quote in German). His plea basically screams for an advocacy style of journalism to fulfill the function of the media to detect grievances in society. This function of the media to critique most researchers and practitioners would be considered important for pluralistic, liberal democratic societies. One could—or perhaps should—ask whether objective and neutral reporting is always the best way to fulfill such social functions.

In Nepal and Bhutan, nearly all journalists who have been asked in qualitative interviews about their societal role, see their main function as "giving a voice to the voiceless" and helping to improve people's lives (Illg, 2019). This includes the function of a watchdog as media are considered to be the fourth estate in a democratic country. So far, it fits to an Anglo-Saxon understanding of journalism and is probably influenced by it. But "giving a voice to the voiceless" seems to have its limits when journalists avoid reporting social grievances like child abuse in order not to harm the Gross National Happiness (GNH) ranking of Bhutan. Here, it seems to deviate from a Western norm. The GNH is an important issue in Bhutan, and it also influences journalism. From a Bhutanian perspective, this could indeed be considered advocacy journalism, while from a "Western" perspective, one could call it censorship.

But is there something that the "Western" perspective and research could learn from this? The negativity of news factors is seen as problem in "Western" societies as well. Guided by the assumptions of Herman and Chomsky's Propaganda Model (2002) and Bennett's Indexing Theory (2016), we could argue that a so-called independent journalism in democratic societies exists to support the powers that be. McNair (2009) stated the following within a Marxian critique that:

> ... "freedom of the press," and the "bourgeois" notion of freedom in general, is essentially an ideological hoax, a form of false consciousness which merely legitimizes the status quo and distracts the masses from serious scrutiny of a system which exploits and oppresses them. The media are structurally locked into pro-systemic bias, and will rarely give "objective" coverage to anything which seriously threatens the social order of capitalism. The aspirations of objectivity, and of independence from the state, are masks for the production by the media of dominant ideology, or bourgeois hegemony, in the sphere of political coverage as elsewhere. (p. 240)

Prinzing (2018) emphasized how journalistic styles, such as advocacy journalism, public journalism, and constructive journalism realize integration functions for societies. She argued that those styles of journalism can help to integrate society in times when globalization and digitization create problems which lead to the disintegration and polarization of societies.

From a Western, idealistic, and normative point of view, journalists are no activists and cannot fulfill both roles at the same time. Most so-called independent journalists do not want to be put in the same box with advocacy journalists or do not want to be considered biased. As a privileged journalist working in a stable and secure environment, it might be easy to take the liberty of being "impartial." But what is the role of journalists in crisis situations and wars where partisanship and civic engagement is required? Journalists usually become active participants in the conflict they are reporting on (Ruigrok, 2010). Impartiality becomes a luxury good. But, even in times of crisis, organizations with a Western, idealistic understanding of journalistic norms, like Free Press Unlimited (FPU), claim: "While most journalists in exile will have suffered from the actions of the government of the home country from which they are in exile, it is important to maintain a clear line between journalism and activism" (Uiterkamp, 2022, p. 13).

During times of crisis, we observe exoduses of people, including journalists, fleeing their home countries. Often, exile journalism is the only way to keep up with journalistic practices which would be punished under authoritarian rules. Those journalists in exile are often seen as activists, and therefore, they often do not find jobs as journalists in their new home countries. Then, they either leave their profession or work for exiled media houses (Mugabo, 2023).

An example of an enormous exile news media industry is the exiled media houses of Myanmar. The Democratic Voice of Burma is one example how exile media works in a way that is both professional and partial at the same time. Although the Western norm of independent journalism would like to distinguish itself from such forms of news reporting, Western governments put a lot of effort and money into the support of such media houses. We have learned from examples like Afghanistan that most media houses only existed because of foreign money (Page & Siddiqi, 2012). Recently, many journalists have been supported after fleeing from Ukraine, Belarus, and Russia, not to mention other exiled journalists fleeing injustice around the world. Their transnational efforts in conflict mediation involve documenting arrests and human rights violations as well as sharing evidence of oppression and election irregularities with international media (Arafat, 2021). Exile media, such as the exile media industry of Myanmar, promote "this inside-out and outside-in channeling of voices for democracy" (Oo, 2006, p. 238). In her study of Syrian exile journalism, Arafat (2021) defined "diaspora advocacy journalism" as follows:

> The purposive involvement of exiled subjects in transnational news gathering and production where new forms of collaboration/networking with people, civil society actors, and human rights defenders from the homeland and host country are introduced to promote underrepresented voices and mobilize a democratic political reformation and or social change. (p. 2191)

Within this definition and her analysis that "Syrian opposition diaspora journalists do not see a contradiction between being advocates and independent professional journalists at the same time and do not believe they risk the quality of their media work. They rather believe the two roles go hand in hand to promote democracy" (Arafat, 2021, p. 2191). Thus, exile media do often report on the injustices of the oppressors in their home countries (Balasundaram, 2019). They mostly do that in an advocacy style of news presentation. A Nepali journalist emphasized this, based on his manifold international experience in general as well as war journalist in Asia, saying that you cannot report in a neutral and objective way about your home country (Illg, 2018, 2019).

As people who have been socialized in a certain, perhaps conflict ridden, context, journalists can become activists. This is driven by the will to go beyond writing news and the experience that there is often more needed to point out grievances: namely action. We notice that journalism research separates journalistic writing very strongly from the journalistic physicality—the body and its actions itself. We see that journalistic work can often be impacted by physical events and actions, such as being part of social movements or being detained for reporting. Journalists and journalism (research) could therefore, for a more cosmopolitan perspective, also investigate the actions of journalists.

We can see that there are good reasons to not follow the ideal of an objective journalism, and because journalism has to perform criticism, it should not always be fair in the sense of being equal. As journalism is seen as the "fourth estate," partial (sometimes biased) reporting is just and reasonable in some cases. Journalists who take their social responsibility seriously, but in their work, criticize the powerful and expose scandals, abuse of power, and corruption will find it difficult to provide balanced reporting that treats all parties fairly. So, the assessment of whether advocacy journalism is good or bad depends on whether or not it supports one's own interest. The key takeaway from our discussion of the normative foundations and evaluation of advocacy journalism is that we should examine and treat this prefix journalism as a normative concept in its own right. Advocacy journalism should not be treated as an abnormal practice of information journalism and should not be evaluated only against the background of objectivity and neutrality. In a similar fashion, Ubuntu journalism represents an independent normative journalistic concept which facilitates active interpretation and participation and is geared toward society or rather understood as part of the community.

Ubuntu journalism: A community focused approach to journalism

Unlike the Western normative entanglements of advocacy journalism, Ubuntu journalism is rooted in the ancient (South) African philosophy of Ubuntu. Here, we first have to clarify what Ubuntu means. Essentially it means that "I am because we are" (Mutwarasibo & Iken, 2019) and builds the foundation of a community focused approach "to rediscover and re-establish idealised values of traditional African culture(s) and traditional African communities" (Fourie, 2008, p. 53). The collectivistic nature of African culture is reflected in the value system of Ubuntu, which emphasizes humanity. Ubuntu means that an individual's existence impacts those around them, focusing on the collective rather than the individual (Poovan et al., 2006, p. 23). Therefore, the Ubuntu philosophy represents a contrast to the concept of individualism as it emphasizes "togetherness, where an individual does not survive alone but with others" (Khan & Ntakana, 2023, p. 218).

The Ubuntu normative framework may prioritize journalistic practices that focus on voicing the community's concerns, ideas, and opinions. The media's main role would be to stimulate citizen and community participation, seeking consensus through broad consultation. The media would emphasize collective well-being over individual rights with the aim of social transformations, acting as a catalyst for moral agency (Fourie, 2008, p. 64). Metz (2011) noted that Ubuntu offers an orientation on how to resolve disputes about justice and morality. This is based on the assumption that people derive their dignity from their ability to form a community by identifying and showing solidarity with others. At the media level, this means that news outlets must be empathetic and that they have to minimize the potential level of suffering of the people being reported on (Rodny-Gumede, 2015). In comparison to other journalistic media concepts, which are generally shaped by Western traditions, the concept of Ubuntu journalism emphasizes that journalists should take on a decidedly participatory role. In particular, this means that they see themselves first and foremost as part of the community and less as neutral observers.

Furthermore, the political-democratic dimension of the implications of this must be taken into account. In the course of newly emerging democracies, which are based on the legacy of long-lasting slavery, colonialism, and apartheid, this is to be aligned according to a new principle that differs from Western normative understandings of journalism. The media should help democracies stand on a stable foundation. The idea is that the media should offer people a platform to express different opinions; they should convey knowledge about political participation processes and simply inform citizens about their rights and obligations. In this light, there are authors who advocate for the "Africanization" of the news media. This refers to a type of public service ethos that is based on the ideas and values of the Ubuntu philosophy (Rodny-Gumede, 2015).

Rediscovering the value system based on Ubuntu is a fundamental part of the renegotiation of cultural identities in post-apartheid South Africa (Rao & Wasserman, 2007). While Ubuntu journalism "may encourage action towards civic transformation and community renewal," it is not without challenges (Fourie, 2008, p. 64). Economic pressures, political biases, and societal prejudices can hinder the practice of compassionate journalism. Yet, countless journalists and media organizations around the world are embracing the Ubuntu philosophy, highlighting the triumph of empathy over apathy and unity over division. In times when media houses are dying and cost-intensive investigative journalism is under economic and social pressure, journalism for the common good is on everyone's lips as the third pillar for ensuring social and political stability. Commercial media reporting and public service media no longer seem to be able to fulfill this task sufficiently (Kramp & Weichert, 2023). As we move forward in this ever-changing world, Ubuntu journalism stands as a testament to the enduring power of compassion. It underscores the notion of collective interdependence, even within the complexities of news cycles. By incorporating the principles of Ubuntu, journalism has the potential to contribute to a future where understanding prevails over ignorance, empathy mitigates prejudice, and the narratives produced reflect a shared human experience. Understood as journalism for the common good, it might also ensure democracy on a local level.

In the end, the question remains as to whether Ubuntu journalism is an originally new concept or whether there are strong similarities and overlaps with existing ideas, such as public journalism or civic journalism, from which Ubuntu journalism does not appear to be essentially different. Nevertheless, it should be noted that many authors who subscribe to the normative concept refer to historical and social framework conditions. These are to be understood as elementary.

Proponents of the Ubuntu approach have argued that reporters should break away from the Western paradigm of "truth-seeking" associated with the principles of independence and objectivity. In contrast, their duties should be oriented toward the needs of communities and societies (Christians, 2015). When Ubuntu journalism is explained in practical detail, it seems that Ubuntu and civic journalism are very similar. This can be seen, among other things, in the assumption that journalists should focus less on statements by prominent people or other high-profile events. Rather, the aim should be an integrative actor in a community, communicating the reality of life and the concerns of the community to the outside world (Fourie, 2007). It seems, and here there are also similarities to the concept of civic journalism, that an Ubuntu-inspired approach to journalism shows weaknesses when it comes to more than idealized scenarios. Beyond the idealized scenarios mentioned above, the approach offers no indications as to how journalism can be changed or improved in concrete terms (Benequista, 2016). The values anchored in the philosophy of Ubuntu, such as homogeneity and communitarianism, tend to paint a strongly romanticized picture of Africa. It also seems unclear whether and when the individual journalist should break away from the majority opinion and truly act independently. On an abstract level, this means that values such as freedom of opinion or journalistic autonomy would be jeopardized (Fourie, 2008). Nevertheless, an Ubuntu style of news reporting that is geared toward community can fulfill media functions for democracies that we often think of when we are talking about the norms and functions of so-called independent information journalism.

Overall, the concept of Ubuntu offers an alternative perspective for assessing the competing values of community orientation and individualism in news reporting and the impact these may have on societies in a more cosmopolitan manner.

(Investigative) (collaborative) cross-border journalism: A transnational way of doing journalism

Along with our zoom into the method-driven concept of cross-border journalism, we would like to highlight a more practice-oriented perspective on cosmopolitan potentials in journalistic practice and research. For journalism per se, the past few years have been characterized above all by numerous transformations. Against the ever-present backdrop of digitization, these have been primarily phenomena such as declining revenue models and diminishing returns in the sales business of publishing houses as well as changing media usage behavior (Lobigs, 2016, pp. 70–71). One particular aspect of journalism has become increasingly important in light of this development: (investigative) collaborative cross-border journalism. This development seemed to be taking shape even before the publicity milestone of the Panama Papers (Konow-Lund et al., 2019). This transnational and cross-cultural approach make investigative cross-border journalism particularly interesting for journalism research with a cosmopolitan perspective. The Migrant Files, Paradise Papers, FINCEN Files, and Football Leaks—all these journalistic titles, as well as the Panama Papers—were the result of international research cooperation. Journalists from all over the world have contributed their extensive research to realize a joint and enormously extensive journalistic output. The image of the journalist in intensive investigations, often portrayed as a "lone wolf," has become a thing of the past, at least since the Panama Papers were published. They were published simultaneously around the world on April 3, 2016. This journalistic achievement was aggregated from an incredibly large amount of raw data, which would have far exceeded the output of a single investigative newsroom. In the case of the Panama Papers, 11.5 million individual documents were analyzed and appropriately processed together with around 400 journalists and around 100 media partners across 80 countries over a period of one year (Obermayer et al., 2016).

Journalistic collaborations have become extremely popular, especially in investigative journalism. Carson and Farhall (2018) described this as a "shift in investigative reporting practice from the 'old model' . . . to a collaborative model of multiple newsrooms (and countries) sharing information to expose wrongdoing" (p. 1901). It is not surprising that in a global networked world, data volumes are growing, and corruption and crime do not stop at national borders, making it necessary for investigative journalism to form teams and work together to hold the powerful accountable.

Accordingly, it is obvious that the "big" problems are no longer unique topics of a community or a nation. For example, climate change and digitalization with all its implementation on information flow, working processes, etc. are problems that have effects all over the world and can only be handled in cooperation. As a logical consequence, it makes sense that journalists also cooperate across national boundaries (Berglez, 2013).

Investigative journalism has always had deep sociopolitical significance and underscores its importance as the fourth estate in a democracy. Walton (2010) even credits full-time investigative journalists with the title of "elite special forces" (para. 19) of the fourth estate. Such metaphors for the traditional investigative journalist sometimes better outline their work and approach. It is precisely investigative journalism that is currently experiencing a fundamental structural transition. The portrayal of investigative journalists as "muckrakers" or "junkyard dogs" underlines this structural transition (Donsbach, 1995).

Seymour Hersh, one of the most important investigative journalists in the USA, is considered a classic "lone wolf" in his profession. Revelations such as the torture scandal in the Iraqi prison Abu Ghraib and war crimes in the Vietnam War can be traced back to him (Hersh, 2018). If we look at other and older investigative publications such those related to the Watergate affair (Seibert, 2017) or the *Boston Globe's* Spotlight revelations (Fürst, 2016), they represent the achievements of individual investigative journalists or small investigative and local teams. A publication like the Panama Papers would not have been feasible in this setting. The high complexity of the topics and the increasing quantity of data can hardly be evaluated or summarized by individual journalists or editorial teams. Another characteristic of earlier individual investigative efforts was the highly competitive way of working among journalists (Alfter, 2019; Graves & Konieczna, 2015).

But what makes investigative journalism collaborative and transnational now? Strictly speaking, journalism can always be described as collaborative insofar as more than one person is involved. The literature likes to point out that journalistic collaborations between reporters and newsrooms have always existed. An example of this is the founding of the Associated Press (AP), an alliance of New York newspapers, to get news about the Mexican War to the North faster and to minimize costs by sharing labor (AP, n.d.; Lewis, 2017; Stonbely, 2017). In terms of collaborative journalism, "it is now being practiced on a scale that constitutes a revolution in journalism," according to Stonbely (2017, p. 59). For Stonbely (2017), collaborative journalism is defined as "a cooperative arrangement (formal or informal) between two or more news and information organizations, which aims to supplement each organization's resources and maximize the impact of the content produced" (p. 14), and it can be stated that investigative journalism benefits from collaborations (p. 17).

According to Alfter (2019), the murder of journalist Don Bolles in 1976 in Phoenix, Arizona, described by her as an "iconic event" (p. 5), is considered a major milestone in the history of cross-border journalism. At the time, Don Bolles was a founding member of the young association Investigative Reporters and Editors (IRE) (Monge Duarte, 2018). The murder of their colleague Bolles thus became the impetus for the Arizona Project (IRE, 2006). Forty journalists from 28 different media outlets across the USA traveled to Phoenix to look into the murder of Don Bolles and complete his research on corruption in Arizona (Duarte, 2018). The further development of the collaborative approach was significantly driven by digitalization. However, not only has computing power increased, but the amount of data journalists have to deal with has also continuously increased. While the WikiLeaks/Cablegate scandal in 2010 involved only 1.7 gigabytes of data, this figure increased 1,500-fold six years later with the Panama Papers, with more than 2.6 terabytes (or 11.5 million individual documents) of data (Obermayer et al., 2016). The Panama Papers have shown what is possible. In addition to establishing and further developing data journalism techniques, the Panama Papers have established a culture of collaboration and sharing (Baack, 2016).

Today, we already know that journalistic collaborations can exist in many different forms and vary in the duration, complexity, and diversity of teams (Alfter, 2017; Heft et al., 2019; Stonbely, 2017). Basically, the phenomenon of collaborative cross-border journalism best describes the changed understanding of roles in investigative journalism. In her standard work, Alfter (2019) described collaborative cross-border journalism via the presence of the following characteristics: "[1] journalists from different countries, [2] decide on an idea of mutual interest, [3] gather and share material," and "[4] publish to their own audiences" (p. 18).

This fundamental shift in the way journalists work simultaneously provides the basis for professional networks or organizations that have made it their mission to provide resources and know-how to connect journalists worldwide and enable them to work collaboratively at a professional level. Examples on a global level are the International Consortium of Investigative Journalists (ICIJ) and the Organized Crime and Corruption Reporting Project (OCCRP). However, such networks are also forming in smaller contexts as it is becoming increasingly worthwhile to collaborate, especially against the backdrop of the worsening economic situation in journalism (Stonbely, 2017). Since the 2016 Panama Papers in 2016 at the latest, collaborative and cross-border journalism has become an integral part of investigative journalism (Leihs, 2019). It can be noted that "[t]he era of the lone wolf is over. . . . A borderless world needs watchdogs who can transcend borders" (Coronel, 2016, paras. 20–23).

However, despite this cosmopolitan potential of a global phenomenon, the literature shows a strong Western focus in comparative journalism research (Hanusch & Vos, 2020). Clear evidence regarding cross-border practices in countries outside North America and Europe is significantly underrepresented (Estella, 2023, p. 68). This leads to the question of the extent to which large and financially strong established networks are dominant, vis-à-vis organizations in non-Western regions, and possibly impose a Western-style reporting model. Alfter (2019) spoke less about the transfer of a specific journalistic practice but rather raised the question of a homogenization of journalistic practices.

We argue that the application of collaborative practices is more about the creation of a uniform framework than the transfer of a Western practice. The uniform framework conditions are the lowest common denominator, which builds the basis for a set of shared rules and norms that should not curtail cultural and social practices in different contexts (Lück & Schultz, 2019, p. 111). The people involved in such projects write for their own national audience, which may counteract the assumption of a generally imposed Western journalistic practice. However, the fact that working conditions can deteriorate must be viewed critically. This was shown in a study by Heft and Baack (2022), which looked at cross-border projects in Europe. Therefore, it is crucial that journalism research deals specifically with networks outside the Europe and the North American contexts and asks those involved in other world regions about their perceptions and to see whether (working) conditions are impacted negatively or even improve. In addition, structural differences in relation to large and small networks must be considered in the further course of events, which also means analyzing the content of the journalistic output of a particular project abroad (see also Alfter, 2019; Alfter & Cândea, 2019; Heft & Baack, 2022).

From a practical perspective, while the ICIJ's pioneering role in coordinating the Panama Papers may give the impression of Western dominance, the quality and impact of such efforts are shaped by the diverse partners and journalistic influences involved.

Some reflections and concluding remarks

With our zoom into the prefix journalisms advocacy journalism, Ubuntu journalism, and cross-border journalism, we have looked into norms and practices and their contextual and situational foundations. On the one hand, we see that different normative understandings of journalism exist and that manifold practices can have different impacts on societies and understandings of roles and functions. On the other hand, we have shown within a more practical/method-driven perspective how digitalization, transnationalization, and collaborative work bring more cosmopolitan practices of investigative journalism in the form of cross-border practice to the front.

From our investigation into the normative concepts of advocacy journalism and Ubuntu journalism, we learned that studies of journalism usually emphasize the Eurocentric paradigms of neutrality and objectivism as the ideal for which every media should strive. Other journalistic practices are often assessed as less desirable and seen as an abnormity of the Western role model.

These two different forms of journalism, which we discussed here as normative prefix journalisms, follow autonomous normative perspectives with their own advantages and disadvantages; they are not just modifications of one universal normative Western understanding and its paradigms (see also Thomass in this book).

In contrast to the praise of paradigms of objectivity and neutrality, we have argued that neutral objective reporting can promote inequality and injustice in societies instead of denouncing them, as it does not question prevailing power structures but instead equalizes them. How we evaluate the practices, functions, and roles of different prefix journalisms is impacted by contextual norms and the definition of media functions in the respective societies. Additionally, the conclusion about the paradigms of "journalism's independence from the state and the political elite, and its capacity to be objective, ... tend to be premised on one's views about the nature of capitalism itself, its viability as a system, and the scope for serious alternatives" (McNair, 2009, p. 242). This emphasizes the concept that cosmopolitan journalism research is looking not only into media functions in relation to politics but also into media-society and media-economy relations. Altogether, one could get the impression that classifying different styles of journalism as advocacy, exile, Ubuntu, or civic journalism is more important for those outside than those inside the respective society. Actually, we should question the promotion of neutral reporting and its legitimacy in times when a neo-colonialist perspective asserts that such leading values must be found in the various contexts of operations rather than in universal norms. Overall, we question whether something like objective and neutral can exist in news content created by journalists, in other words subjects. We argue that everything is information journalism, but nothing is neutral or objective about what is created by a subjective mind. This does not mean that distinctive prefix journalisms, such as investigative or information journalism, do not follow specific rules and norms in their daily practice.

Within a practice-oriented analysis of cross-border journalism, we have shown that another paradigm seems to vanish: the paradigm of the competition-loving "lone wolf" has been partly overcome and has given way to a culture of information sharing and collaboration across borders. Certainly, external factors, such as the worsening economic situation for the media, have accelerated this transformation. However, it seems undisputed that the multinational influence on journalistic investigations can specifically increase journalistic impact on societies around the globe, for example, the resignation of ministers in Iceland, Pakistan, and Spain after the publication of the Panama Papers. (Investigative) (collaborative) cross-border journalism thrives on the diversity and different journalistic approaches of its participants and therefore exemplifies a cosmopolitan understanding of transnational and cross-cultural journalistic collaboration, holding the powerful accountable even across borders. In a globalized world, crime, social grievances, and (political) misuse of power do not stop at national borders; hence, networks such as ICIJ are a valuable response to such a development (Berglez, 2013).

Still, a cosmopolitan perspective should critically assess how such collaborations may perpetuate global power imbalances. We see that research on this issue does not adequately cover all world regions. Therefore, cosmopolitan journalism research should look more closely into interactions within cross-border journalism networks and their impacts on journalism cultures, functions, and practices. Network analyses would be a fitting inductive approach to tackling such questions in all world regions. Overall, we need to establish a research culture in journalism studies that promotes local or Indigenous perspectives from different world regions to form a non-Western critique of journalism and journalism research.

References

Alfter, B. (2017). New method, new skill, new position? Editorial coordinators in cross-border collaborative teams. In R. Sambrook (Ed.), *Global teamwork: The rise of collaboration in investigative journalism* (pp. 41–58). Reuters Institute for the Study of Journalism.

Alfter, B. (2019). *Cross-border collaborative journalism: A step-by-step guide*. Routledge.

Alfter, B., & Cândea, S. (2019). Cross-border collaborative journalism: New practice, new questions. *Journal of Applied Journalism & Media Studies*, *8*(2), 141–149. https://doi.org/10.1386/ajms.8.2.141_1

AP. (n.d.). *Our story: The Associated Press*. https://www.ap.org/about/our-story/

Arafat, R. (2021). Examining diaspora journalists' digital networks and role perceptions: A case study of Syrian post-conflict advocacy journalism. *Journalism Studies*, *22*(16), 2174–2196. https://doi.org/10.1080/1461670X.2021.1990110

Baack, S. (2016). What big data leaks tell us about the future of journalism – and its past. *Internet Policy Review*. https://policyreview.info/articles/news/what-big-data-leaks-tell-us-about-future-journalism-and-its-past/413

Balasundaram, N. (2019). Exiled journalists as active agents of change: Understanding their journalistic practices. In I. S. Shaw & S. Selvarajah (Eds.), *Reporting human rights, conflicts, and peacebuilding: Critical and global perspectives* (pp. 265–280). Palgrave Macmillan. https://doi.org/10.1007/978-3-030-10719-2_16

Benequista, N. (2016). *The moral dilemmas of journalism in Kenya's politics of belonging* [Doctoral dissertation, London School of Economics and Political Science]. LSE Theses Online. https://etheses.lse.ac.uk/3458/

Bennett, W. L. (2016). Indexing theory. In *International encyclopedia of political communication*. https://doi.org/10.1002/9781118541555.wbiepc180

Berglez, P. (2013). *Global journalism: Theory and practice*. Peter Lang. https://doi.org/10.3726/978-1-4539-1088-7

Carson, A., & Farhall, K. (2018). Understanding collaborative investigative journalism in a "post-truth" age. *Journalism Studies*, *19*(13), 1899–1911. https://www.doi.org/10.1080/1461670X.2018.1494515

Christians, C. G. (2015). Introduction: Ubuntu for journalism theory and practice. *Journal of Media Ethics*, *30*(2), 61–73. https://doi.org/10.1080/23736992.2015.1020158

Coronel, S. (2016, June 15–19). *A golden age of global muckraking at hand* [Keynote]. 2016 Conference of Investigative Reporters and Editors, New Orleans, LA, United States. https://gijn.org/stories/a-golden-age-of-global-muckraking/

Donsbach, W. (1995). Lapdogs, watchdogs and junkyard dogs. *Media Studies Journal*, *9*(4), 17–30.

Estella, P. G. (2023). Cross-border journalism and de-Westernization. In L. Rothenberger, M. Löffelholz, & D. H. Weaver (Eds.), *The Palgrave handbook of cross-border journalism* (pp. 67–81). Palgrave Macmillan. https://doi.org/10.1007/978-3-031-23023-3_5

Fourie, P. J. (2007). Moral philosophy as the foundation of normative media theory: The case of African Ubuntuism. *Communications*, *32*(1), 1–29. https://doi.org/10.1515/COMMUN.2007.001

Fourie, P. J. (2008). Ubuntuism as a framework for South African media practice and performance: Can it work? *Communicatio*, *34*(1), 53–79. https://doi.org/10.1080/02500160802144520

Fürst, D. (2016, February 29). Der Skandal, um den es in "Spotlight" geht. *Süddeutsche Zeitung*. https://www.sueddeutsche.de/panorama/boston-der-skandal-um-den-es-in-spotlight-geht-1.2885087

Graves, L., & Konieczna, M. (2015). Sharing the news: Journalistic collaboration as field repair. *International Journal of Communication*, *9*, 1966–1984. https://ijoc.org/index.php/ijoc/article/view/3381

Hahn, O. (2008). Journalismus an der Kreuzung der Kulturen und Sprachen: Grundlagen der anthropologischen interkulturellen Kommunikationsforschung und Ethnolinguistik. In O. Hahn & R. Schröder (Eds.), *Journalistische Kulturen: Internationale und interdisziplinäre Theoriebausteine* (pp. 31–55). Herbert von Halem.

Hanitzsch, T., & Vos, T. P. (2018). Journalism beyond democracy: A new look into journalistic roles in political and everyday life. *Journalism*, *19*(2), 146–164. https://doi.org/10.1177/1464884916673386

Hanitzsch, T., Hanusch, F., Ramaprasad, J., & de Beer, A. S. (Eds.). (2019). *Worlds of journalism: Journalistic cultures around the globe*. Columbia University Press. https://doi.org/10.7312/hani18642

Hanusch, F., & Vos, T. P. (2020). Charting the development of a field: A systematic review of comparative studies of journalism. *International Communication Gazette*, *82*(4), 319–341. https://doi.org/10.1177/1748048518822606

Heft, A., & Baack, S. (2022). Cross-bordering journalism: How intermediaries of change drive the adoption of new practices. *Journalism*, *23*(11), 2328–2346. https://doi.org/10.1177/1464884921999540

Heft, A., Alfter, B., & Pfetsch, B. (2019). Transnational journalism networks as drivers of Europenanisation. *Journalism*, *20*(9), 1183–1202. https://doi.org/10.1177/1464884917707675

Hepp, A., & Loosen, W. (2021). Pioneer journalism: Conceptualizing the role of pioneer journalists and pioneer communities in the organizational re-figuration of journalism. *Journalism*, *22*(3), 577–595. https://doi.org/10.1177/1464884919829277

Hepp, A., Loosen, W., Kühn, H., Solbach, P., & Kramp, L. (2021). Die Figuration des Pionierjournalismus in Deutschland: Akteure und Experimentierbereiche. *Medien und Kommunikationswissenschaft (M&K)*, *69*(4), 551–577. https://doi.org/10.5771/1615-634X-2021-4-551

Herman, E. S., & Chomsky, N. (2002). *Manufacturing consent: The political economy of the mass media*. Pantheon.

Hersh, S. M. (2018). *Reporter: A memoir*. Penguin Books.

Illg, B. (2018). Freiheit durch Demokratie – Freiheit in der Demokratie? Chancen und Grenzen des Journalismus in Neapel: Eine empirische Analyse zum Selbstverständnis nepalesischer Journalist_innen. In A. Czepek, M. Hellwig, B. Illg, & E. Nowak (Eds.), *Freiheit und Journalismus* (pp. 173–188). Nomos. https://doi.org/10.5771/9783845280851-173

Illg, B. (2019). Self-perception of Nepalese journalists within the democratization process of the country. In B. Dernbach & B. Illg (Eds.), *Journalism and journalism education in developing countries* (pp. 116–129). Manipal University Press.

Investigative Reporters and Editors. (2006, March 5). *The Arizona project*. https://web.archive.org/web/20060305044342/http://www.ire.org:80/history/arizona.html

Khan, S., & Ntakana, U. M. (2023). Theorising a theory of Ubuntu – the divide between individualism compared to a socialistic understanding of African society. *African Journal of Social Work*, *13*(4), 217–223. https://dx.doi.org/10.4314/ajsw.v13i4.5

Konow-Lund, M., Gearing, A., & Berglez, P. (2019). Transnational cooperation in journalism. In M. Powers (Ed.), *Oxford research encyclopedia of communication*. https://doi.org/10.1093/acrefore/9780190228613.013.881

Kramp, L., & Weichert, S. (2023). *Whitepaper Non-Profit-Journalismus: Handreichungen für Medien, Politik und Stiftungswesen* (OBS-Arbeitsheft 112). Otto Brenner Stiftung. https://www.otto-brenner-stiftung.de/fileadmin/user_data/stiftung/02_Wissenschaftsportal/03_Publikationen/AH112_Nonprofit_Journalismus.pdf

Leihs, N. (2019, September 4). *Partnerschaft statt Wettbewerb*. European Journalism Observatory. https://de.ejo-online.eu/qualitaet-ethik/partnerschaft-statt-wettbewerb

Lewis, C. (2017). Tear down these walls: Innovations in collaborative accountability research and reporting. In R. Sambrook (Ed.), *Global teamwork: The rise of collaboration in investigative journalism* (pp. 5–25). Reuters Institute for the Study of Journalism.

Lobigs, F. (2016). Finanzierung des Journalismus – von langsamen und schnellen Disruptionen. In K. Meier & C. Neuberger (Eds.), *Journalismusfoschung* (2nd ed., pp. 69–138). Nomos.

Loosen, W., Ahva, L., Reimer, J., Solbach, P., Deuze, M., & Matzat, L. (2022). 'X Journalism.' Exploring journalism's diverse meanings through the names we give it. *Journalism*, *23*(1), 39–58. https://doi.org/10.1177/1464884920950090

Lück, J., & Schultz, T. (2019). Investigative data journalism in a globalized world. *Journalism Research*, 2(2), 93–114. https://doi.org/10.1453/2569-152X-22019-9858-en

McNair, B. (2009). Journalism and democracy. In K. Wahl-Jorgensen & T. Hanitzsch (Eds.), *The handbook of journalism studies* (pp. 237–249). Routledge. https://doi.org/10.4324/9780203877685

Merten, K., Schmidt, S. J., & Weischenberg, S. (Eds.). (1994). *Die Wirklichkeit der Medien: Eine Einführung in die Kommunikationswissenschaft*. VS Verlag für Sozialwissenschaften Wiesbaden. https://doi.org/10.1007/978-3-663-09784-6

Metz, T. (2011). Ubuntu as a moral theory and human rights in South Africa. *African Human Rights Law Journal*, *11*(2), 532–559. https://www.ahrlj.up.ac.za/metz-t

Monge Duarte, J. (2018, September 19). Kollaboratives Arbeiten im Investigativjournalismus: Chancen und Herausforderungen. *Fachjournalist*. https://www.fachjournalist.de/kollaboratives-arbeiten-im-investigativjournalismus/

Mugabo, L. E. (2023, June 14). Journalism in exile: Lessons from Latin America and East Africa [Video]. *Reuters Institute for the Study of Journalism*. https://reutersinstitute.politics.ox.ac.uk/calendar/journalism-exile-lessons-latin-america-and-east-africa

Mutwarasibo, F., & Iken, A. (2019). I am because we are – the contribution of the Ubuntu philosophy to intercultural management thinking. *Interculture Journal*, *18*(32), 15–32. https://doi.org/10.24403/jp.1267962

Obermayer, B., Obermaier, F., Wormer, V., & Jaschensky, W. (2016). Das sind die Panama Papers. *Süddeutsche Zeitung*. https://panamapapers.sueddeutsche.de/articles/56ff9a28a1bb8d3c3495ae13/

Oo, Z. (2006). Exit, voice and loyalty in Burma: The role of overseas Burmese in democratizing their homeland. In T. Wilson (Ed.), *Myanmar's long road to national reconciliation* (pp. 231–259). Institute of Southeast Asian Studies, Asia Pacific Press.

Page, D., & Siddiqi, S. (2012). *The media of Afghanistan: The challenges of transition* [Policy briefing #5]. BBC Media Action. https://downloads.bbc.co.uk/mediaaction/policybriefing/bbc_media_action_afghanistan_is_in_transition.pdf

Poovan, N., Du Toit, M. K., & Engelbrecht, A. S. (2006). The effect of the social values of Ubuntu on team effectiveness. *South African Journal of Business Management*, *37*(3), 17–27, https://doi.org/10.4102/sajbm.v37i3.604

Prinzing, M. (2018). Bewusst alte Muster durchbrechen? Anwaltschaftlicher und konstruktiver Journalismus etc. aus ethischer Perspektive. In M. Prinzing, N. Köberer, & M. Schröder (Eds.), *Migration, Integration, Inklusion: Medienethische Herausforderungen und Potenziale für die digitale Mediengesellschaft* (pp. 105–120). Nomos. https://doi.org/10.5771/9783845285085-105

Rao, S., & Wasserman, H. (2007). Global media ethics revisited: A postcolonial critique. *Global Media and Communication*, *3*(1), 29–50. https://doi.org/10.1177/1742766507074358

Rodny-Gumede, Y. (2015). An assessment of the public interest and ideas of the public in South Africa and the adoption of Ubuntu journalism. *Journal of Media Ethics*, *30*(2), 109–124. https://doi.org/10.1080/23736992.2015.1020379

Ruigrok, N. (2010). From journalism of activism towards journalism of accountability. *International Communication Gazette*, *72*(1), 85–90. https://doi.org/10.1177/1748048509350340

Seibert, T. (2017, June 16). 45 Jahre Watergate: Die schmutzigen Tricks des Richard Nixon. *Tagesspiegel*. https://www.tagesspiegel.de/politik/die-schmutzigen-tricks-des-richard-nixon-6615751.html

Skjerdal, T. S. (2010). How reliable are journalists in exile? *British Journalism Review*, *21*(3), 46–51. https://doi.org/10.1177/0956474810383773

Stonbely, S. (2017). *Comparing models of collaborative journalism*. Center for Cooperative Media | Montclair State University. https://collaborativejournalism.org/wp-content/uploads/2017/09/Models-for-Collaborative-Journalism-research-paper.pdf

Uiterkamp, T. S. (Ed.). (2022). *Exile Media: Mapping the challenges faced by independent media in exile*. Free Press Unlimited. https://kq.freepressunlimited.org/wp-content/uploads/2022/05/FPU_Exile-Media.pdf

Walton, M. (2010, September). Investigative shortfall. *American Journalism Review*, *32*(3), 18. https://ajrarchive.org/article.asp?id=4904

Weischenberg, S. (2004). *Journalistik. Medienkommunikation: Theorie und Praxis. Vol. 1. Mediensysteme – Medienethik – Medieninstitutionen* (3rd ed.). VS Verlag für https://doi.org/10.1007/978-3-322-80843-1

Cosmopolitanization of war coverage research

Empowering non-Western narratives

Kathrin Schleicher & Aynur Sarısakaloğlu

Beyond borders: War as events of cosmopolitan relevance

From the Russian–Ukrainian War to tensions between Israel and Iran, from the Gaza War to the Houthi attacks on international shipping, to the friction between China and Taiwan—these wars and conflicts are currently dominating the global media headlines.

Throughout history, the media have been drawn to the captivating and influential nature of wars and armed conflicts, which have always characterized news cycles and public attention. Their attraction lies in their inherent news value, which includes several key factors, such as harm, negativity, and personalization, that strongly influence journalistic choices in topic selection (Eilders & Hagen, 2005). In addition, wars attract public interest because, from a Western European perspective, they mostly take place in distant regions beyond our immediate experience. We therefore rely on media coverage as a source of information (Wolff, 2018)—it shapes our understanding of wars (Jungblut, 2023) and often contributes to the escalation of conflicts (Hoxha & Hanitzsch, 2018).

However, empirical studies reveal a prevailing tendency within the media to focus on a select few high-profile conflicts, leaving many others unnoticed (e.g., Zerback & Holzleitner, 2018). This results in selective and disproportionate media war coverage, which either leads to significant attention or minimal coverage of a given conflict (Maier et al., 2011; Zerback & Holzleitner, 2018). In particular, Western media tend to focus their attention on conflicts in countries often situated at the periphery of mainstream news (Löffelholz, 2004), and they are criticized for "excessively focusing on "bad news" about developing countries while neglecting stories about positive events and progress" (Sobel et al., 2020, p. 172). This phenomenon is notably widespread in media war coverage and foreign news reporting, where the media selectively and crisis-drivingly cover non-Western regions, such as Africa, the Arab world, or Southeast Asia (Hafez, 2002). This presumably explains why "Western media repeatedly fail to report on the world's most significant (in terms of casualties) conflicts and crises" (Robinson, 2017, p. 189).

A parallel trend can be observed in communication research. The field of war communication has received considerable scholarly attention (Wolff, 2018), particularly between 1990 and 2005, resulting in an extensive corpus of case studies that continues to dominate the existing literature (Löffelholz, 2004). However, academic studies focus less frequently on wars and armed conflicts that have gained limited visibility in Western media (Löffelholz, 2004). Instead, existing research predominantly centers on a few very salient and high-profile conflicts, notably the Gulf War (1990), the Kosovo conflict (1999), the Afghanistan War (2001), the Iraq War (2003), the ongoing Israeli–Palestinian conflict, and the war in Ukraine (Jungblut, 2023; Löffelholz, 2004). In addition, research focuses on the analysis of media war coverage—the main research attention is on the topics, patterns, and quality of war reporting with most studies' attention on current conflicts (Eilders & Hagen, 2005; Löffelholz, 2004). Despite some revealing individual findings, systematic research into the forms and content of war reporting in the media remains largely incomplete (Jungblut, 2022). This concentration on media war coverage leads to a large number of content analyses, while studies in the field of communicator research, such as the strategic communication of political elites and the role of war correspondents or journalists, remain comparatively scarce. Similarly, research on media platforms, audience reception, and media effects, including the impact of such reporting on war-related policy, is less prevalent. Due to the dominance of media content research, this article examines the state of cosmopolitanization in research on media war and conflict coverage. In doing so, we consider conflicts to be the basis for wars and crises. Conflicts that break out between states or parties within a state can be described as wars or armed conflicts if they escalate, as the latter do not require a formal declaration (Zerback & Holzleitner, 2018).

Research about war is a deeply cosmopolitan field of study because when presented in the news worldwide, wars and global conflicts can also influence processes of globalization, deepen our sense of globality (Zhang & Luther, 2020), and potentially contribute to what Beck (2009) called a cosmopolitan perspective: in a connected world, globally mediated crises like wars and armed conflicts transcend the traditional boundaries that define the local and the global by fostering a cosmopolitan perspective (Zhang & Luther, 2020, p. 405; see also Estella et al. in this book for a similar discussion).

Zhang and Luther (2020) argued that apart from a few cases, there is a notable absence in the research regarding the interplay and diffusion of global conflicts and their reliance on the global news media and new forms of global journalism. This points to a gap in the field of communication studies' understanding of wars and conflicts as cosmopolitan events. The question therefore arises as to whether the observations that apply to communication studies as a whole also apply to academic research on the media representation of wars and conflicts. Data from Demeter (2018) showed that around 90% of the publication output of the *Journal of Communication* comes from the West, without a significant increase in output from the Global South. Within the Global South, inequalities persist, with the developing countries in Asia (especially China) and South America making some contributions, while other regions, such as the Middle East, Africa, and Eastern Europe, remain largely ignored (Demeter, 2018). In addition, Hanusch and Vos (2020) presented the results of a systematic review of comparative journalism studies from 2000 to 2015. Analyzing 441 articles from 22 major journals, they found consistent patterns, including a focus on Western authorship, Western countries, and elite media (Hanusch & Vos, 2020).

The need to rethink cultural differences in academic discourse is obvious (Demeter, 2020). Proponents of de-Westernization criticize the dominance of Western perspectives and call for revising power relations in global knowledge production and dissemination (Glück, 2018): it aims to expand academic knowledge by incorporating diverse global perspectives and theoretical frameworks (Waisbord & Mellado, 2014). Consequently, de-Westernization aims to overcome Anglo-American and Eurocentric perspectives, while deep internationalization (Badr et al., 2020) advocates open approaches to representative academic knowledge production from around the world (Badr & Ganter, 2021). This is not a simple internationalization, that is, a whole range of case studies worldwide in English that do not sufficiently address common issues and questions or do not build on theoretical concepts, arguments, or conclusions of Western sciences or can be linked to them (Waisbord, 2015).

Therefore, de-Westernization should promote the development of a cosmopolitan perspective that raises awareness of global differences and similarities, develops a broader understanding of global issues, and fosters a scholarly environment that is sensitive to comparative and global studies (Waisbord, 2015). In essence, cosmopolitanization requires an epistemic paradigm shift toward theoretical and methodological inclusivity in order to overcome the marginalization of certain voices (Ganter & Badr, 2022). To advance the de-Westernization of communication studies in the sense of cosmopolitanism, Waisbord (2015) proposed three research strategies: (1) ask questions that are absent in Western research and search for "empirical and theoretical blindspots that reflect the powerful influence of Western concerns and categories," (2) conduct comparative studies with non-Western cases and have an "interest in exploring whether and why 'context matters' to explain similarities and differences," and (3) analyze global, transnational phenomena and "study 'global media events'" (pp. 187–188).

To explore the de-Westernization and associated cosmopolitanization of research on war coverage, we examine three key areas of global media coverage of war in the following sections. First, we delve into studies analyzing the visibility of different conflicts. Second, we investigate studies examining how conflicts are portrayed in the news, including the framing process and the factors influencing media war coverage. Third, we review studies that assess the overall quality of reporting. Within each section, we provide a research overview and identify shortcomings in the cosmopolitanization of the research field. We discuss approaches to adopting a cosmopolitan perspective and explain the prerequisites for cultivating such a viewpoint in the research field. In conclusion, we address the initial question regarding the extent of cosmopolitanism in war reporting within the research media and propose recommendations for advancing the cosmopolitanization of the research field on war reporting.

Selective choice: Research on the visibility of conflicts in media coverage

One strategy for de-Westernization proposed by Waisbord (2015) is to conduct research that transcends national and regional boundaries and focuses on global issues. This approach emphasizes both common and diverse characteristics and examines causes and effects. A genuinely global research topic is the international flow of news, which concerns investigating global patterns of war and conflict reporting and which wars play a role in the media and why. According to Hafez (2018), this research field not only attempts to capture news geographies worldwide but also to explore the causes of similarities and differences in a global comparison.

Wars and armed conflicts are intensively researched, but little is known about the factors that influence their visibility in the news media. Empirical studies on news values shedding light on the factors that influence the visibility of conflicts in the media have shown that contextual factors such as a country's population, economy, and political influence and event-related factors, in particular high casualty figures, influence media attention (Jungblut, 2022; Zerback & Holzleitner, 2018). Event-related factors are directly linked to the conflict or war itself, whereby it is often assumed that conflicts with a high number of victims receive more media attention, although empirical evidence questions this assertion (Hawkins, 2011; Zerback & Holzleitner, 2018). On the other hand, contextual factors are intertwined with the characteristics of the nations involved in the conflict. While the importance of these factors in the context of war is still not fully understood (Zerback & Holzleitner, 2018), observations so far suggest that conflicts involving nations with larger populations, greater economic power, or greater political influence most often receive greater media attention (Jones et al., 2011). This trend unveils the reasoning behind a country's attractiveness and how this impacts the amount of foreign media coverage it receives, arising from factors such as its global power, cultural affinity, economic ties, or conversely, its marginalization and geographical distance (Hafez, 2018). Additionally, the relationship between the countries facing conflict and those reporting on it is crucial. Media outlets exhibit a higher tendency to cover conflicts occurring in countries that share geographical, cultural, political, or economic proximity to their own (Heimprecht, 2017).

However, it is problematic that these empirical studies largely look at Western media and their selection mechanisms. For example, Zerback and Holzleitner (2018) limited themselves to analyzing German news coverage, while Jones et al. (2011), Hawkins (2011), and Sobel et al. (2020) mainly examined US-American media and interpreted them from a US-centric perspective. Specifically, Sobel et al. (2020) analyzed the relationship between visibility in reporting and US-American interests, thus creating a very one-sided perspective on news values from a global perspective.

In order to understand the dynamics of international news flows, comprehensive international comparative studies on news selection in conflict and war reporting are essential—another strategy to de-Westernize media studies in the sense of cosmopolitan scholarship, according to Waisbord (2015). The author also explained that comparative research is motivated by the need to overcome the prevailing "US-centrism" in the literature and develop evidence-based, theoretically informed arguments based on diverse cases (Waisbord, 2015). International comparative studies offer added value by providing theoretical insights into the drivers of global media discourses that go beyond the assumption of universal news factors (Hafez, 2018).

However, comparative studies on international media war coverage are rare, with only a few exceptions. For example, Fengler et al. (2020) analyzed media coverage of the Ukraine conflict in 13 countries, including Western, Central, and Eastern European countries as well as Russia. They noted that while the structures of foreign reporting in Western Europe have been extensively researched, Central and Eastern European countries have been comparatively little studied. Their study found that the representation of conflicts varies from country to country, possibly influenced by factors such as geographical and cultural proximity to the conflict area or economic relevance (Fengler et al., 2020). Jungblut (2020) also revealed that media coverage of international conflicts often depends on ethnocentrism and elite status.

Heimprecht (2017) selected an even more global sample in her study on foreign news coverage in television news from 16 countries, including Chile, China, Egypt, and Singapore. The aim was to identify overarching patterns by including countries with significant differences in state structures and media systems. The main findings confirmed earlier observations that country characteristics influence foreign reporting and that conflicts are not automatically newsworthy, but depend on the relevance of a country to the reporting media system and are influenced by country-specific interests (Heimprecht, 2017). Furthermore, the study showed that journalists consider the interests of their own countries when reporting on political conflicts, which is reflected in the relevance of the crisis status and proximity to the conflict.

Overall, further cross-cultural studies are essential to fully understand how news factors, together with event- and context-related factors, influence news selection processes. However, Hafez (2018, p. 142) emphasized that effective news value analysis is primarily applicable only to "free" media systems, which poses a challenge when transferring this concept to non-Western contexts, especially authoritarian media systems. Therefore, the feasibility of applying the news value concept in these contexts needs to be assessed. It is important not to transfer all theoretical concepts and approaches to non-Western contexts uncritically. Nevertheless, there remains a need to extend the study of media coverage of war beyond the currently dominant case study designs. There is an increasing need to systematically compare "how different types of conflicts from different regions of the world are covered in different types of media outlets in the context of different journalistic working environments" (Jungblut, 2023, p. 131): "In doing so, studies will be able to identify generalizable patterns of news coverage and determinants that shape how conflict is covered, for instance by influencing the power relations between the press and politics or by posing a challenge for journalistic research" (Jungblut, 2023, p. 131).

Research should also focus more on conflicts without major Western involvement, as these conflicts have so far been largely ignored by researchers (Jungblut, 2023). The selective choice of salient and high-profile conflicts as research objects described above favors the thematic interests of academics based in the USA and Europe (Waisbord, 2016). Scholarship has also largely focused only on those wars that play a role in the Western media and on issues that are of interest to Western academia, and it has marginalized wars and related issues that are relevant in other regions. These priorities may have been influenced by third-party funding and publication attention, leading to a distorted perspective on conflict issues and "the danger is that Western thematic preoccupations overshadow non-Western questions in the globalized academia" (Waisbord, 2016, p. 879).

However, this bias could be balanced by deliberately including other previously overlooked conflicts in the research agenda. Advancing the cosmopolitanization of research requires the development of novel research agendas that encompass non-Western subjects (Waisbord, 2015). Studies outside the mainstream of communication studies focusing on media from non-Western nations, such as Frère's (2007) work on media and conflicts in Central Africa, could leverage these insights, potentially by linking up with Western studies that deal with media selection mechanisms. Similar considerations extend to the research agendas of other regions, as highlighted by Kozman (2021), in the context of Arab media research. She noted that "not surprisingly, the studies dealt with issues pertinent to the region and reflected topical areas that are most relevant to Arab audiences and journalists" (Kozman, 2021, p. 244). These areas include topics such as the attribution of blame and mistrust of journalistic reporting on conflict negotiations as well as the politicization of photos. However, in the perspective of cosmopolitan research, case studies on individual wars or countries neglected by communication studies scholarship should not be limited to issues and arguments that are relevant only to specific regions. Instead, it is essential that the research questions and findings are linked to larger theoretical and empirical debates, contributing to the overall body of knowledge within the discipline (Waisbord, 2015).

Context matters: The role of the global and the local

Empirical studies scrutinizing war communication primarily focus on analyzing the characteristics of media war coverage—mainly on topics, patterns, and the quality of war coverage (for an overview, see Jungblut, 2023).

Empirical studies show that news stories, framed by details, myths, and shared values, influence audience understanding and evaluative attitudes toward wars and conflicts (Baden & Meyer, 2018). This framing process emphasizes specific aspects of reality: "a particular problem definition, causal interpretation, moral evaluation, and/or treatment recommendation for the item described" (Entman, 1993, p. 52). Framing studies, for example, examine which actors have been given a voice in conflict reporting (e.g., Golan, 2013; Jungblut, 2020). Here, the focus remains on Western media coverage, with studies analyzing the framing in Western European media (e.g., Ojala et al., 2017).

"Given the West-centric nature of research agendas, important questions in the Global South have not been at the forefront or even discussed in the West" (Waisbord, 2015, p. 187). However, it is important to learn more about authoritarian regimes because "in countries dominated by authoritarian regimes or where governments exert powerful influence on media economics and access to information, journalists' strategies to navigate government restrictions are different from the West" (Waisbord, 2015, pp. 187–188). Studies by Liu (2022) or Chernov (2023), which dealt with the Russian–Ukrainian conflict, or Baden and Stalpouskaya (2020), which examined the Syrian Chemical Weapons crisis, explored ways to de-Westernize research on war media coverage. Going beyond the commonly studied Western democracies, they analyzed the media framing of Russian media in comparison to liberal democracies, such as the US-American and British news media.

However, overall, there is a lack of existing comparative studies that include the countries of the Global South. One example that includes the perspective of the South is the analysis of the Ukraine conflict by Kuisz and Wigura (2023), who identified four competing narratives on the conflict: First, the Russian narrative portrays Russia's actions as a defensive war against the West, referred to as a "Russian fairy tale." Second, the "Western European fox" narrative aims to avoid a military confrontation between NATO and Russia. Third, the "Ukrainian hedgehog" emphasizes the struggle for national existence. Finally, another view from the Global South sees the conflict as a "white man's war" in which people would instead not get involved (Kuisz & Wigura, 2023, pp. 67–70).

In addition to framing patterns in media coverage, studies have also dealt with the influence of journalism cultures (e.g., Baden & Stalpouskaya, 2020) and different political and media systems (e.g., Chernov, 2023) on the journalistic framing of wars. The study by Baden and Stalpouskaya (2020) revealed that established news frames, rooted in cultural perceptions and elite groups' strategic framing, continue to influence journalists' framing practices. These findings imply the substantial impact of journalists' preexisting attitudes on news selection and framing (Baden & Stalpouskaya, 2020). This highlights the significance of cultural influences in news production, where journalists' selection of specific frames often carries cultural nuances (von Sikorski & Merz, 2022). Also, different media cultures are expected to be framed differently (e.g., Peng, 2008). The research emphasizes the role of normative pressures stemming from political and social systems, as well as factors like political climate, public opinion, and journalistic practices (Nygren et al., 2018; Peng, 2008).

To sum up, media war coverage cannot be separated from the influences of their production conditions as well as the political and cultural environment (Wolff, 2018). A contextual approach is crucial when analyzing media coverage of war and conflict; for example, political and media systems play an essential role in understanding the research subject, as each war and conflict has its own unique dynamics (Fröhlich, 2018). Such comparative research designs should also include countries and interpretations from the Global South. Waisbord (2015) emphasized the importance of cross-national analyses to understand how media and political systems interact and influence war reporting. This raises the question of the factors that influence media war coverage. Such comparative studies are intended to explore whether and why context matters to explain similarities and differences in media war coverage.

A comprehensive understanding of global journalism requires a systematic comparison of coverage of different types of conflict in different regions of the world that take the diverse media landscape and journalistic environment into consideration (Jungblut, 2023). As Waisbord (2015, p. 184) argued, "global" thinking requires broadening analytical perspectives in media studies. Given the international nature of war reporting and the influence of globalized news, especially through social media platforms, the role of journalists and new media in conflicts cannot be adequately captured from a national perspective alone (Fröhlich, 2018; Zahoor & Sadiq, 2021). Therefore, de-Westernization also involves the analysis of communication phenomena that transcend traditional geopolitical boundaries and encompass events, processes, and trends that go beyond specific countries and regions (Waisbord & Mellado, 2014).

To better understand the complex relationships between media and conflicts, Gilboa et al. (2016) proposed six levels for media analysis, drawing from Gilboa (2009): local (e.g., urban television and radio stations, local newspapers), national (e.g., The New York Times, CNN-US), regional (e.g., Al Jazeera Arabic), international (e.g., Al Jazeera English, RT), global (e.g., BBC World, CNN International), and glocal media (e.g., Facebook, Twitter)—"glocal" is a portmanteau coined from combining global and local. According to Gilboa et al. (2016), glocal media allow for the analysis of interactions between media, conflict parties, and other stakeholders, such as non-governmental organizations (NGOs) and the United Nations. They noted that the key distinguishing features among media are geography, content, audience type, and ownership structures. Using this approach, the authors explained, future research can answer these critical questions: How do global and local media dimensions empower actors inside and outside conflict zones, such as the diaspora, to advance their agendas, and how does this influence global and international responses?

Even if war and conflicts are translocal, transnational, and globally embedded, studies should not ignore the cultural specificity of conflict dynamics and the existence of local means of conflict resolution. For example, the recourse to anthropological approaches that add local perspectives to national and international interventionist approaches to conflict resolution could be useful (Bräuchler & Budka, 2020). At this point, anthropologically informed communication research could make an important contribution to the further development of conflict research due to its cross-cultural and context sensitive approach with its ethnographic methods, such as participant observation and qualitative interviews. However, studies that analyze conflicts and the media tend to rely solely on media content and quantitative datasets to determine the role of the media in conflicts (Bräuchler & Budka, 2020).

To understand the role of the media in conflicts and to ensure an inclusive view of it, studies need to combine insights into the case-specific characteristics of each individual conflict and the contextual factors that shape media roles. These factors can be summarized as follows: "a) culturally embedded conflict perceptions and belief systems; (b) audiences' media reliance and trust; (c) the configuration, diversity and freedom of relevant media; (d) the constellation of conflict actors, including regional and international powers; and (e) the course and intensity of conflict" (Baden & Meyer, 2018, p. 45).

Moving away from negativity: Studies on assessing the quality of media war coverage

Wars often elude journalistic quality standards, such as objectivity (Imhof, 1995). This raises the question of the quality of reporting on wars. Empirical studies focus, in particular, on the partiality or impartiality of the media. Springer et al. (2022) examined patriotic journalism in reporting on the Ukraine conflict in Sweden and Ukraine. They analyzed how journalists in both countries adhered to the standard of objectivity. The results indicated that Swedish journalists showed value-based loyalty to allies, while in Ukraine, official views were disseminated. However, the researchers found that the latter were not generally framed in a positive light, as one might expect from patriotic journalists (Springer et al., 2022).

The universal application of such quality criteria to all research topics is a challenge, especially in the context of wars and conflicts, where the media function according to their own internal dynamics, according to Stremlau et al. (2016). The authors illustrated this with the example of Somalia, where the media landscape, characterized by owners, business interests, and journalists seeking political and financial gain, operates according to its own incentives and rules. Efforts to promote typical journalistic standards, such as neutrality and objectivity, face significant hurdles due to the deeply entrenched political economy of Somali media. While many owners and journalists recognize ethical standards, ongoing conflict hinders their implementation. Therefore, the de-Westernization of media studies should also include a critique of research conventions, such as those of news value research, which often measure results from other environments against Western contexts.

In addition to patriotic journalism, other forms of journalistic war reporting that are frequently examined and criticized by empirical studies are those that involve othering and stereotyping. Research shows that war and conflict reporting often adopts a one-sided ethnocentric perspective (Baden & Tenenboim-Weinblatt, 2018), which is influenced by deeply rooted socialization and enculturation mechanisms (Wolff, 2018). Opinions of ingroups are presented uncritically, while outgroups, such as opposing parties, are downplayed and sometimes dehumanized (Baden & Tenenboim-Weinblatt, 2018). State sources dominate reporting, suggesting limited press independence (Bjørge & Kalnes, 2021; Robinson et al., 2009). In conflict situations, journalists may orient themselves toward the government's narrative and report in a one-sided manner (Robinson et al., 2009). In the ongoing Global War on Terror, for example, the media often perpetuate an Orientalist narrative in which Muslims are portrayed as the "other," and this conflation of xenophobia and war propaganda reflects a worrying trend (Nohrstedt, 2016). Stereotypical friend–enemy images are constructed with content analyses focusing on language to uncover prejudices and opposing representations (Löffelholz, 2004).

Study findings also indicate a prevalent tendency in war coverage to prioritize narratives steeped in violence and negativity (e.g., Baden & Tenenboim-Weinblatt, 2018). This emphasis persists even when the reported events do not necessarily involve violent conflict. Simultaneously, reports on peace negotiations are at the periphery of media attention (Baden & Tenenboim-Weinblatt, 2018). An automated content analysis of coverage of four conflicts in 17 different media outlets revealed a dominant focus on violent actions and their repercussions within conflict reporting (Jungblut, 2020), which may lead to a diminished public perception of efforts aimed at conflict resolution (Wolfsfeld, 2004). The same seems to apply to research on media war coverage. Despite clear indications of the power of the media for peacebuilding (Bräuchler, 2020), media research has so far focused more on conflict (Bräuchler & Budka, 2020).

A pioneering way to promote cosmopolitanization in the research field of war reporting could, therefore, be to direct a positive focus on media war reporting. In this context, communication studies already offer established concepts, such as constructive journalism or peace journalism, which help not only to assess the quality of reporting appropriately but also to measure it. Closely linked to cosmopolitanization is the research question of the extent to which journalism can invite cosmopolitan solidarity, which constructive journalism can generate. Specifically, Morse (2018) analyzed how Al Jazeera English reported on the 2008–2009 Gaza War and how media representation can promote global solidarity by making distant suffering morally significant. Using the concept of "mediatized grievability," the study showed how Al Jazeera constructed narratives of Palestinian suffering, bridging emotional and geographical divides to create a sense of shared humanity and solidarity (Morse, 2018, p. 385).

The normative concept of peace journalism (Galtung, 2002; Kempf, 2003) emphasizes a non-violent approach to conflict reporting, which proves to be more supportive. This approach promotes proactive media coverage that focuses on humanitarian issues, balanced reporting, conflict resolution, and peace (Gouse et al., 2019; Zahoor & Sadiq, 2021). War journalism and peace journalism are two competing frameworks for conflict reporting, with war journalism propagating violence, victory, and an elitist orientation, while peace journalism highlights cause and solutions, giving all parties a voice (Gouse et al., 2019). However, this concept has faced criticism in communication research (e.g., Hanitzsch, 2007), as it attributes political and military responsibilities to the media and neglects the common structural constraints and traditional processes of journalism (Zillich et al., 2011). Balabanova (2017) outlined how increased human rights reporting in peace journalism can cultivate a cosmopolitan global consciousness. Her study thus examined a global research subject that serves to promote a cosmopolitan research agenda (Waisbord, 2015). While some see the global media as a tool for building a cosmopolitan culture (Szerszynski & Urry, 2006), others warn of its potential exploitation by authoritarian forces and news oligopolies (Chouliaraki, 2008). However, there is a risk of graphic media content becoming "disaster pornography," and constant exposure to human rights violations can lead to "compassion fatigue" (Balabanova, 2017, p. 236).

Peace journalism research primarily relies on content analyses of conflict reporting using war and peace frames (e.g., Gouse et al., 2019; Ishaku, 2021). A study by Neumann and Fahmy (2016) proposed an index for conflict reporting that combined 18 practices of war and peace journalism, including aspects such as emphasizing elites linked with war journalism or covering psychological damage associated with peace journalism. So, despite the effects of globalization, research has revealed significant disparities in normative values among journalists worldwide, particularly concerning the principle of professional detachment (Waisbord, 2012).

Conclusion: Implications and future research directions

To answer the initial question of whether the genuinely cosmopolitan research subject of media war coverage is appropriately studied by communication and media research, it can be concluded that much of the literature on the relationship between media and conflict focuses on state-centric Western political communication and the performance of journalism and how Western media coverage of conflict has influenced Western governments (Gilboa et al., 2016). The academic US- and Eurocentrism discussed in previous sections are deemed insufficient to explain the realities in many environments outside the West. As de-Westernization is a multi-layered process, it requires changes in four dimensions (Waisbord & Mellado, 2014). First, de-Westernization involves examining research subjects that are either understudied or absent in Western contexts or extend beyond traditional geographical boundaries (Waisbord & Mellado, 2014). To de-Westernize research is not only to broaden the analytical scope but also to address topics that might be overlooked by Western scholars but are significant in the non-Western world. For instance, examining how non-Western media cover proxy wars waged by superpowers adds layers of analysis to understanding how global events are represented in different contexts. De-Westernization aims for a more inclusive and balanced portrayal of conflicts, challenging dominant Western-centric perspectives in reporting conflicts (Wasserman & de Beer, 2009).

As a second point, de-Westernization refers to expanding the body of evidence in communication research: it highlights the importance of considering non-Western countries to arrive at more nuanced conclusions (Waisbord & Mellado, 2014, p. 364; see also Cazzamatta in this book). Transitioning from single-case studies to emphasizing comparative research is crucial. To propel future theoretical advancements in the realm of media war coverage and communication during conflicts, it is imperative to validate concepts through comparative approaches across diverse national and international conflict cases, media productions, practitioners, and political contexts (Fröhlich, 2018). Recognizing the inadequacy of understanding the role of journalists and (new) media in conflicts solely from a national perspective, there is a crucial need to shift toward a more global viewpoint (Fröhlich, 2018). In addressing this, there is an urgent call to de-Westernize the field by expanding both theoretical and empirical engagement with research beyond the Euro-Atlantic sphere (Kaneva, 2023). However, many comparative studies have disproportionately focused on the West compared with other parts of the world (Hanusch & Vos, 2020). Waisbord (2022) further underscored the significance of expanding cross-national comparative research—by broadening the scope of comparative projects to encompass countries from the Global South, researchers can gain deeper insights into how local knowledge shapes conventional forms of communication. Kotišová (2024) examined the collaboration between "parachute journalists"[1] and local journalists, known as "fixers," to explore how the personal proximity between these two groups influences the quality of reporting. Interviews with journalists, including those from Ukraine, revealed that local journalists are sometimes trusted less due to their proximity to events (Kotišová, 2024). Nevertheless, the proximity of local journalists is viewed in practice as valuable knowledge for fact-based and ethical reporting. Examining the relationship between parachute journalists and fixers in war zones provides valuable insights into these dynamics.

Third, reconsidering Western theories and concepts is a crucial step for the discipline, necessitating a critical examination of normative assumptions, especially regarding the universality of liberal-democratic values (Kaneva, 2023). Robinson et al. (2017) emphasized "the need to expand our theorising beyond the traditional focus on elite Western media, publics and their associated foreign policies so that we recognise more fully the truly global and multi-level nature of the contemporary media environment" (p. 337). A cross-cultural perspective on the media's role in conflicts, as proposed by Gilboa et al. (2016), can be utilized as a theoretical starting point to operationalize this expanded

1 Financial constraints often compel journalists, especially those in countries not directly involved in the conflict, to travel to conflict zones without possessing sufficient knowledge (Jungblut, 2022; Sobel et al., 2020).

focus in empirical studies and contribute to the de-Westernization of research by involving various actors.

As a fourth point, a primary measure for advancing research in the field of media war coverage and communication is the adoption of a cross-cultural research perspective. However, comparative studies in war reporting predominantly focus on the Western context, often comparing arguments from the USA in a European setting. The widespread knowledge of English among European academics and the incentives to publish in English favor comparative research (Waisbord, 2015). Müller and Knieper (2019) underscored the importance of international exchange among research groups, advocating for the establishment of collaborative research consortia that leverage diverse language and cultural competencies (see also Kozman, 2021). Nygren et al. (2018) emphasized ongoing cooperation between researchers from conflicting sides, fostering a common understanding and providing insights beyond nationalistic discourse.

The research consortium should also incorporate researchers from the countries under study. Based on a review of the most reputable European journals and academic conferences on communication studies from 2010 to 2016, Ganter and Ortega (2019) demonstrated that discussions of Latin American contexts in European media and communication studies often fail to incorporate perspectives from within the continent. They advocated for a critical implementation of de-Westernization, which necessitates "more geographically diverse editorial boards, greater international cooperation, and comparative accounts to capture diversity in regional contexts" (Ganter & Ortega, 2019, p. 68).

Integral to the analysis of war coverage is the incorporation of historical, cultural, and sociopolitical contexts, as this enhances contextual awareness and understanding (Kaneva, 2023). International collaboration and a cooperative approach to research play vital roles in including diverse perspectives, combining experiences from Western and non-Western countries and ensuring more representative and inclusive narratives. This collaboration facilitates resource sharing, including data, access to local sources, and journalistic networks, enabling a comprehensive analysis of conflicts from multiple perspectives (Fröhlich, 2018). Of course, this claim is often countered by the argument that this collaboration would not be feasible in current international comparative research practice. The main reasons for this are the lack of or unequal access to research funding worldwide and differing levels of methodological expertise (Heimprecht, 2017).

Furthermore, existing conflict research should extend beyond moments of escalation, encompassing other conflict phases, such as peace processes and fragile peace situations (Jungblut, 2023). Badr et al. (2020) emphasized the importance of overcoming the one-way knowledge transfer from democracies to non-democratic states, advocating for learning from non-Western regions and embracing innovative communication models. For instance, Frère's work exemplifies the expansion of knowledge beyond traditional perspectives. In particular, her contributions have improved the understanding of Francophone Africa in journalism studies, as Mutsvairo and de Bruijn (2021) wrote in their obituary for Frère. She promoted research from and about Francophone Africa within English-speaking academia through publications and editorial roles, underscoring the importance of cross-cultural comparisons to dismantle barriers in African journalism studies (Mutsvairo & de Bruijn, 2021).

References

Baden, C., & Meyer, C. O. (2018). Dissecting media roles in conflict: A transactionist process model of conflict news production, dissemination and influence. In R. Fröhlich (Ed.), *Media in war and armed conflict: The dynamics of conflict news production and dissemination* (pp. 23–48). Routledge.

Baden, C., & Stalpouskaya, K. (2020). Maintenance of news frames: How US, British and Russian news made sense of unfolding events in the Syrian chemical weapons crisis. *Journalism Studies*, *21*(16), 2305–2325. https://doi.org/10.1080/1461670X.2020.1843066

Baden, C., & Tenenboim-Weinblatt, K. (2018). The search for common ground in conflict news research: Comparing the coverage of six current conflicts in domestic and international media over time. *Media, War & Conflict*, *11*(1), 22–45. https://doi.org/10.1177/1750635217702071

Badr, H., & Ganter, S. A. (2021). Towards cosmopolitan media and communication studies: Bringing diverse epistemic perspectives into the field. *Global Media Journal – German Edition*, *11*(1). https://doi.org/10.22032/dbt.49164

Badr, H., Behmer, M., Fengler, S., Fiedler, A., Grüne, A., Hafez, K., Hahn, O., Hamidi, K., Hanitzsch, T., Horz, C., Illg, B., Litvinenko, A., Löffelholz, M., Radue, M., Richter, C., Thomaß, B., & Töpfl, F. (2020). Kosmopolitische Kommunikationswissenschaft: Plädoyer für eine "tiefe Internationalisierung" des Fachs in Deutschland: Ein wissenschaftliches Positionspapier. *Publizistik*, *65*(3), 295–303. https://doi.org/10.1007/s11616-020-00576-6

Balabanova, E. (2017). Media and human rights. In P. Robinson, P. Seib, & R. Fröhlich (Eds.), *Routledge handbook of media, conflict and security* (pp. 230–240). Routledge.

Beck, U. (2009). *World at risk*. Polity Press.

Bjørge, N. M., & Kalnes, Ø. (2021). Cultures of anarchy: Images of Russia in the narrative of Norwegian mainstream news media during the Ukraine crisis 2014. *Media, War & Conflict*, *14*(2), 150–173. https://doi.org/10.1177/1750635219864024

Bräuchler, B. (2020). From war to peace in Indonesia: Transforming media and society. In P. Budka & B. Bräuchler (Eds.), *Theorising media and conflict* (pp. 295–318). Berghahn Books.

Bräuchler, B., & Budka, P. (2020). Introduction: Anthropological perspectives on theorising media and conflict. In P. Budka & B. Bräuchler (Eds.), *Theorising media and conflict* (pp. 3–31). Berghahn Books.

Chernov, G. (2023). The Russian – Ukrainian war: Persistence of frames and the media issue-cycles. *Athens Journal of Mass Media and Communications*, *9*, 1–13. https://doi.org/10.30958/ajmmc.X-Y-Z

Chouliaraki, L. (2008). The symbolic power of transnational media: Managing the visibility of suffering. *Global Media and Communication*, *4*(3), 329–351. https://doi.org/10.1177/1742766508096084

Demeter, M. (2018). Theorizing international inequalities in communication and media studies: A field theory approach. *KOME*, *6*(2), 92–110. https://doi.org/10.17646/KOME.75692.94

Demeter, M. (2020). Power relations in global knowledge production: A cultural/critical approach. *Journal of Multicultural Discourses*, *15*(1), 1–17. https://doi.org/10.1080/17447143.2019.1657124

Eilders, C., & Hagen, L. M. (2005). Kriegsberichterstattung als Thema kommunikationswissenschaftlicher Forschung: Ein Überblick zum Forschungsstand und den Beiträgen in diesem Themenheft. *Medien und Kommunikationswissenschaft (M&K)*, *53*(2–3), 205–221. https://doi.org/10.5771/1615-634x-2005-2-3-205

Entman, R. M. (1993). Framing: Toward clarification of a fractured paradigm. *Journal of Communication*, *43*(4), 51–58. https://doi.org/10.1111/j.1460-2466.1993.tb01304.x

Fengler, S., Kreutler, M., Alku, M., Barlovac, B., Bastian, M., Bodrunova, S. S., Brinkmann, J., Dingerkus, F., Hájek, R., Knopper, S., Kus, M., Láb, F., Lees, C., Litvinenko, A., Medeiros, D., Orlova, D., Ozolina, L., Paluch, A., Nicoleta Radu, R., Stefanikova, S., Veldhoen, H., & Zguri, R. (2020). The Ukraine conflict and the European media: A comparative study of newspapers in 13 European countries. *Journalism*, *21*(3), 399–422. https://doi.org/10.1177/1464884918774311

Frère, M.-S. (2007). *The media and conflicts in Central Africa*. Lynne Rienner Publishers. https://doi.org/10.1515/9781685853662

Fröhlich, R. (2018). The integration of findings: Consequences of empirical results for the advancement of theory building. In R. Fröhlich (Ed.), *Media in war and armed conflict: The dynamics of conflict news production and dissemination* (pp. 287–309). Routledge.

Galtung, J. (2002). Peace journalism – A challenge. In W. Kempf & H. Luostarinen (Eds.), *Journalism and the new world order: Studying the war and the media* (Vol. 2, pp. 260–280). Nordicom.

Ganter, S. A., & Badr, H. (2022). Introduction: Re-examining media governance through cosmopolitan critique. In S. A. Ganter & H. Badr (Eds.), *Media governance: A cosmopolitan critique* (pp. 1–12). Palgrave Macmillan. https://doi.org/10.1007/978-3-031-05020-6_1

Ganter, S. A., & Ortega, F. (2019). The invisibility of Latin American scholarship in European media and communication studies: Challenges and opportunities of de-Westernization and academic cosmopolitanism. *International Journal of Communication*, *13*, 68–91. https://ijoc.org/index.php/ijoc/article/view/8449

Gilboa, E. (2009). Media and conflict resolution: A framework for analysis. *Marquette Law Review*, *93*(1), 87–110. https://scholarship.law.marquette.edu/mulr/vol93/iss1/9/

Gilboa, E., Jumbert, M. G., Miklian, J., & Robinson, P. (2016). Moving media and conflict studies beyond the CNN effect. *Review of International Studies*, *42*(4), 654–672. https://doi.org/10.1017/S026021051600005X

Glück, A. (2018). De-Westernization and decolonization in media studies. In M. Powers (Ed.), *Oxford encyclopedia of communication*. Oxford University Press. https://doi.org/10.1093/acrefore/9780190228613.013.898

Golan, G. J. (2013). The gates of op-ed diplomacy: Newspaper framing the 2011 Egyptian revolution. *International Communication Gazette*, *75*(4), 359–373. https://doi.org/10.1177/1748048513482264

Gouse, V., Valentin-Llopis, M., Perry, S., & Nyamwange, B. (2019). An investigation of the conceptualization of peace and war in peace journalism studies of media coverage of national and international conflicts. *Media, War & Conflict*, *12*(4), 435–449. https://doi.org/10.1177/1750635218810917

Hafez, K. (2002). *Die politische Dimension der Auslandsberichterstattung*. Nomos.

Hafez, K. (2018). Heimprecht, Christine: Determinanten der Auslandsberichterstattung. Eine Mehrebenenanalyse des internationalen Nachrichtenflusses. *Publizistik*, *63*(1), 141–143. https://doi.org/10.1007/s11616-017-0388-7

Hanitzsch, T. (2007). Situating peace journalism in journalism studies: A critical appraisal. *Conflict & Communication Online*, *6*(2), 1–9.

Hanusch, F., & Vos, T. P. (2020). Charting the development of a field: A systematic review of comparative studies of journalism. *International Communication Gazette*, *82*(4), 319–341. https://doi.org/10.1177/1748048518822606

Hawkins, V. (2011). Media selectivity and the other side of the CNN Effect: The consequences of not paying attention to conflict. *Media, War & Conflict*, *4*(1), 55–68. https://doi.org/10.1177/1750635210396126

Heimprecht, C. (2017). *Determinanten der Auslandsberichterstattung: Eine Mehrebenenanalyse des internationalen Nachrichtenflusses*. Springer VS. https://doi.org/10.1007/978-3-658-14820-1

Hoxha, A., & Hanitzsch, T. (2018). How conflict news comes into being: Reconstructing 'reality' through telling stories. *Media, War & Conflict*, *11*(1), 46–64. https://doi.org/10.1177/1750635217727313

Imhof, K. (1995). Kriegskommunikation im sozialen Wandel. In K. Imhof & P. Schulz (Eds.), *Medien und Krieg — Krieg in den Medien* (pp. 123–135). Seismo.

Ishaku, J. (2021). Peace journalism or war journalism? An analysis of newspaper coverage of ethno-religious conflicts in Southern Kaduna, Nigeria (2020–2021). *International Journal of Communication Research*, *11*(4), 295–313.

Jones, T. M., van Aelst, P., & Vliegenthart, R. (2011). Foreign nation visibility in U.S. news coverage: A longitudinal analysis (1950–2006). *Communication Research*, *40*(3), 417–436. https://doi.org/10.1177/0093650211415845

Jungblut, M. (2020). *Strategic communication and its role in conflict news: A computational analysis of the international news coverage on four conflicts*. Springer VS. https://doi.org/10.1007/978-3-658-29122-8

Jungblut, M. (2022). Kriegs- und Konfliktberichterstattung im digitalen Zeitalter. In I. Borucki, K. Kleinen-von-Königslow, S. Marschall & T. Zerback (Eds.), *Handbuch Politische Kommunikation* (pp. 313–326). Springer VS. https://doi.org/10.1007/978-3-658-26233-4_23

Jungblut, M. (2023). Content analysis in the research field of war coverage. In F. Oehmer-Pedrazzi, S. H. Kessler, E. Humprecht, K. Sommer & L. Castro (Eds.), *Standardisierte Inhaltsanalyse in der Kommunikationswissenschaft – Standardized content analysis in communication research* (pp. 125–136). Springer VS. https://doi.org/10.1007/978-3-658-36179-2_11

Kaneva, N. (2023). "Brave Like Ukraine": A critical discourse perspective on Ukraine's wartime brand. *Place Branding and Public Diplomacy*, *19*(2), 232–236. https://doi.org/10.1057/s41254-022-00273-3

Kempf, W. (2003). Konstruktive Konfliktberichterstattung – ein sozialpsychologisches Forschungs- und Entwicklungsprogramm. *Conflict & Communication Online*, *2*(2), 1–15.

Kotišová, J. (2024). The epistemic injustice in conflict reporting: Reporters and 'fixers' covering Ukraine, Israel, and Palestine. *Journalism*, *25*(6), 1290–1309. https://doi.org/10.1177/14648849231171019

Kozman, C. (2021). Reconceptualizing Arab media research: Moving from centrism toward inclusiveness and balance. *Journalism & Mass Communication Quarterly*, *98*(1), 241–262. https://doi.org/10.1177/1077699020942924

Kuisz, J., & Wigura, K. (2023, April 24). Kampf der Narrative. *Internationale Politik*, *3*, 66–70. https://internationalepolitik.de/de/kampf-der-narrative

Liu, Z. (2022). News framing of the Euromaidan protests in the hybrid regime and the liberal democracy: Comparison of Russian and UK news media. *Media, War & Conflict*, *15*(4), 407–426. https://doi.org/10.1177/1750635220953445

Löffelholz, M. (2004). Krisen- und Kriegskommunikation als Forschungsfeld: Trends, Themen und Theorien eines hoch relevanten, aber gering systematisierten Teilgebietes der Kommunikationswissenschaft. In M. Löffelholz (Ed.), *Krieg als Medienereignis II: Krisenkommunikation im 21. Jahrhundert* (pp. 13–55). VS Verlag für Sozialwissenschaften. https://doi.org/10.1007/978-3-322-90833-9_1

Maier, M., Stengel, K., Ruhrmann, G., Marschall, J., Zillich, A. F., & Göbbel, R. (2011). Synchronisierung von Erhebungsinstrumenten zur Erfassung des dynamischen Nachrichtenprozesses am Beispiel der Krisenkommunikation. In O. Jandura, T. Quandt & J. Vogelgesang (Eds.), *Methoden der Journalismusforschung* (pp. 237–255). VS Verlag für Sozialwissenschaften. https://doi.org/10.1007/978-3-531-93131-9_14

Morse, T. (2018). Mediatized war and the moralizing function of news about disruptive events. *Journalism*, *19*(3), 384–401. https://doi.org/10.1177/1464884917693861

Müller, M. G., & Knieper, T. (2019). Terror der Bilder. In K. Lobinger (Ed.), *Handbuch Visuelle Kommunikationsforschung* (pp. 145–179). Springer VS. https://doi.org/10.1007/978-3-658-06508-9_9

Mutsvairo, B., & de Bruijn, M. (2021). Marie-Soleil Frère. *African Journalism Studies*, *42*(1), 130–131. https://doi.org/10.1080/23743670.2021.1911123

Neumann, R., & Fahmy, S. (2016). Measuring journalistic peace/war performance: An exploratory study of crisis reporters' attitudes and perceptions. *International Communication Gazette*, *78*(3), 223–246. https://doi.org/10.1177/1748048516630715

Nohrstedt, S. A. (2016). The role of the media in the discursive construction of wars. In A. Schwarz, M. W. Seeger & C. Auer (Eds.), *The handbook of international crisis communication research* (pp. 133–144). Wiley Blackwell. https://doi.org/10.1002/9781118516812.ch13

Nygren, G., Głowacki, M., Hök, J., Kiria, I., Orlova, D., & Taradai, D. (2018). Journalism in the crossfire: Media coverage of the war in Ukraine in 2014. *Journalism Studies*, *19*(7), 1059–1078. https://doi.org/10.1080/1461670X.2016.1251332

Ojala, M., Pantti, M., & Kangas, J. (2017). Whose war, whose fault? Visual framing of the Ukraine conflict in Western European newspapers. *International Journal of Communication*, *11*, 474–498. https://ijoc.org/index.php/ijoc/article/view/5917

Peng, Z. (2008). Framing the anti-war protests in the global village: A comparative study of newspaper coverage in three countries. *International Communication Gazette, 70*(5), 361–377. https://doi.org/10.1177/1748048508094293

Robinson, P. (2017). The CNN effect and humanitarian action. In P. Robinson, P. Seib & R. Fröhlich (Eds.), *Routledge handbook of media, conflict and security* (pp. 186–196). Routledge.

Robinson, P., Goddard, P., Parry, K., & Murray, C. (2009). Testing models of media performance in wartime: U.K. TV News and the 2003 invasion of Iraq. *Journal of Communication, 59*(3), 534–563. https://doi.org/10.1111/j.1460-2466.2009.01435.x

Robinson, P., Seib, P., & Fröhlich, R. (2017). Conclusion: Looking ahead. In P. Robinson, P. Seib & R. Fröhlich (Eds.), *Routledge handbook of media, conflict and security* (pp. 334–340). Routledge.

Sobel, M. R., Kim, S., & Riffe, D. (2020). The world at war: Three and a half decades of New York Times conflict coverage. *Media, War & Conflict, 13*(2), 170–187. https://doi.org/10.1177/1750635219828763

Springer, N., Nygren, G., Widholm, A., Orlova, D., & Taradai, D. (2022). Narrating "their war" and "our war": The patriotic journalism paradigm in the context of Swedish and Ukrainian conflict coverage. *Central European Journal of Communication, 15*(2), 178–201. https://doi.org/10.51480/1899-5101.15.2(31).1

Stremlau, N., Fantini, E., & Osman, R. M. (2016). The political economy of the media in the Somali conflict. *Review of African Political Economy, 43*(147), 43–57. https://doi.org/10.1080/03056244.2015.1048795

Szerszynski, B., & Urry, J. (2006). Visuality, mobility and the cosmopolitan: Inhabiting the world from afar. *British Journal of Sociology, 57*(1), 113–131. https://doi.org/10.1111/j.1468-4446.2006.00096.x

von Sikorski, C., & Merz, P. (2022). Die Bedeutung der medialen Kommunikation in Kriegen und Konflikten. In C. Cohrs, N. Knab & G. Sommer (Eds.), *Handbuch Friedenspsychologie* (pp. 3–25). Philipps-Universität Marburg. https://doi.org/10.17192/es2022.0061

Waisbord, S. (2012). Democracy, journalism, and Latin American populism. *Journalism, 14*(4), 504–521. https://doi.org/10.1177/1464884912464178

Waisbord, S. (2015). De-Westernization and cosmopolitan media studies. In C.-C. Lee (Ed.), *Internationalizing "International Communication"* (pp. 178–200). University of Michigan Press. https://doi.org/10.2307/j.ctv65sxh2.11

Waisbord, S. (2016). Communication studies without frontiers? Translation and cosmopolitanism across academic cultures. *International Journal of Communication, 10*, 868–886. https://ijoc.org/index.php/ijoc/article/view/3483

Waisbord, S. (2022). What is next for de-Westernizing communication studies? *Journal of Multicultural Discourses, 17*(1), 26–33. https://doi.org/10.1080/17447143.2022.2041645

Waisbord, S., & Mellado, C. (2014). De-Westernizing communication studies: A reassessment. *Communication Theory, 24*(4), 361–372. https://doi.org/10.1111/comt.12044

Wasserman, H., & de Beer, A. S. (2009). Towards de-Westernizing journalism studies. In K. Wahl-Jorgensen & T. Hanitzsch (Eds.), *The handbook of journalism studies* (pp. 428–438). Routledge.

Wolff, M. A. (2018). *Kriegsberichterstattung und Konfliktsensitivität: Qualitätsjournalismus zwischen Anspruch und Wirklichkeit*. Springer VS. https://doi.org/10.1007/978-3-658-22089-1

Wolfsfeld, G. (2004). *Media and the path to peace*. Cambridge University Press. https://doi.org/10.1017/CBO9780511489105

Zahoor, M., & Sadiq, N. (2021). Media and armed conflicts: An overview. *NUST Journal of International Peace & Stability*, 4(1), 70–80. https://doi.org/10.37540/njips.v4i1.80

Zerback, T., & Holzleitner, J. (2018). Under-cover: The influence of event- and context-traits on the visibility of armed conflicts in German newspaper coverage (1992–2013). *Journalism*, *19*(3), 366–383. https://doi.org/10.1177/1464884916683552

Zhang, X., & Luther, C. A. (2020). Transnational news media coverage of distant suffering in the Syrian civil war: An analysis of CNN, Al-Jazeera English and Sputnik online news. *Media, War & Conflict*, *13*(4), 399–424. https://doi.org/10.1177/1750635219846029

Zillich, A. F., Göbbel, R., Stengel, K., Maier, M., & Ruhrmann, G. (2011). Proactive crisis communication? News coverage of international conflicts in German print and broadcasting media. *Media, War & Conflict*, 4(3), 251–267. https://doi.org/10.1177/1750635211420629

Cosmopolitanization of journalism research in the era of artificial intelligence

Bridging the global divide in algorithm-driven journalism

Aynur Sarısakaloğlu

Introduction

As the communication and media research landscape evolves, a pivotal epistemic shift has emerged, highlighting the need to adopt a cosmopolitan approach in scholarly works, including journalism research. The cosmopolitanization of journalism research can be encapsulated in a series of crucial concepts underscoring the global evolution of scholarly inquiry. At its core, this paradigm seeks to transcend the Western-centric influences prevalent in the journalism research canon (Katz et al., 2003), thus nurturing a more inclusive and diverse perspective in the production of academic knowledge to enhance the relevance of research outcomes on a truly international scale (Badr & Ganter, 2021; Carpentier et al., 2020; Waisbord, 2015). Embracing a cosmopolitan mindset implies an acknowledgment of the increasing globalization and interconnectedness of the academic world, advocating for the inclusion of scholarly voices irrespective of their geographic or linguistic origins, fostering theoretical openness and methodological sensibility, and unearthing the structural reasons leading to underrepresented perspectives (Badr & Ganter, 2021; Beck, 2006; Ganter & Ortega, 2019; Hanitzsch, 2019; Waisbord, 2015).

Therefore, discussions surrounding the cosmopolitanization of journalism research have often been intertwined with calls to de-Westernize, decolonize, and decenter the field to address the imbalance of knowledge production across the globe. More than two decades ago, Curran and Park (2000) encouraged Western scholars to reconsider established academic knowledge to ensure its pertinence and applicability to contexts beyond the Global North.[1] The discourse on the de-Westernization of journalism research is closely linked to the following four dimensions of analysis proposed by Waisbord and

1 In this chapter, countries are classified as being in the Global North or Global South based on the categorization provided at https://worldpopulationreview.com/country-rankings/global-north-countries.

Mellado (2014), which can prove instrumental in guiding the cosmopolitanization of academic research: "the subject of study, the body of evidence, analytical frameworks, and academic cultures" (p. 363). The dedication to a cosmopolitan vision beyond North America and Western Europe is geared toward deepening our understanding of media dynamics and providing universal applicability for theoretical frameworks and research evidence across diverse global contexts while recognizing the particularities of individual regions to avoid possible blind spots (Gunaratne, 2010; Kozman, 2021; Waisbord & Mellado, 2014; Wang, 2011).

In this vein, journalism research strives for a cosmopolitan perspective to challenge the concepts and evidence rooted solely within Western hegemony, aiming to bridge gaps in knowledge and strengthen its emphasis on previously marginalized countries (Hanitzsch, 2019; Wahl-Jorgensen & Hanitzsch, 2009; Wasserman & de Beer, 2009). The importance of cosmopolitan journalism research becomes evident, for instance, when considering the diversity of national journalistic cultures, including differences in the professional views of journalists, journalistic practices and ethics across different countries, and the challenges or limitations they encounter in newsrooms worldwide (Hanitzsch et al., 2019). The impetus behind the cosmopolitanization of journalism research is triggered not only by globalization processes but also by the ongoing digitalization trends and the resulting algorithm-driven transformation of journalism (Löffelholz & Sarısakaloğlu, 2024), reflecting the constant evolution of journalism through the integration of new media and communication technologies (Pavlik, 2000; Zelizer, 2019), such as the use of automation and artificial intelligence (AI) technologies in newsrooms (see also Ganter in this book). In this chapter, the term AI refers to "a collection of ideas, technologies, and techniques that relate to a computer system's capacity" (Brennen et al., 2018, pp. 1–2) to display intelligent behavior by "computationally simulating human activities and skills in narrowly defined domains, typically [through] the application of machine learning approaches" (Simon, 2024, p. 11) that allow machines to learn and enhance their performance from data. This article adopts the term algorithm-driven journalism as an umbrella concept to comprehensively address the integration of automation and AI-driven systems into journalistic practice (Sarısakaloğlu, 2025).

As algorithmic systems are implemented in journalism with greater frequency, a broad array of new opportunities for information and communication processes become possible. However, this also results in a global AI-driven divide, not merely due to the lack of or inadequate physical access to new technologies (i.e., first-level AI-driven divide) between countries in the Global North and the Global South but also due to emerging inequalities in the use of algorithm-driven applications (i.e., second-level AI-driven divide). These disparities are mainly due to the insufficient and imbalanced availability of journalistic competencies (Jamil, 2023). Furthermore, this division extends to a third-level AI-driven divide marked by the consequences of disadvantages and inequalities in using algorithmic systems (van Deursen & Helsper, 2015; Verständig et al., 2016). This entails challenges like unequal developmental contribution opportunities, algorithmic biases, and unfair practices in AI algorithms such as algorithmic colonialism, which denotes the tendency of algorithms to inequitably favor certain groups while disadvantaging or exploiting others (Mohamed et al., 2020).

In light of these considerations, the purpose of this chapter is twofold. First, it expounds on the emerging global AI-driven divide in journalism by drawing on theoretical conceptualizations of the digital divide and thematizing challenges as outlined in existing literature, thereby delineating constraints to the attempted cosmopolitanization. Second, I introduce a conceptual framework proposing potential pillars for advancing a truly cosmopolitan perspective to bridge the North–South divide and enhance our understanding of the inequalities and challenges associated with implementing AI tools in journalism, which is a worthwhile consideration for future empirical investigations. Accordingly, the original value of this contribution lies in providing a cosmopolitan approach to algorithm-driven journalism that cultivates global, collaborative, and interdisciplinary academic synergy, considers the AI-driven divide a global phenomenon, and incorporates cosmopolitan algorithms into AI ethical and governance frameworks. While these pillars are not exhaustive, they offer valuable starting points for journalism scholars and journalists to begin reflecting on the practices they may adopt to contribute to the cosmopolitanization of algorithm-driven journalism research.

After outlining the driving factors behind the cosmopolitanization of algorithm-driven journalism, this chapter provides exemplary insights into how knowledge gaps, structural shortcomings, language barriers, lack of journalistic competencies, and approaches to journalism ethics can contribute to the emergence of global inequalities when integrating AI technologies into newsrooms. Building upon these theoretical concepts and reflections on the cosmopolitanization of research, I present pillars of cosmopolitan algorithm-driven journalism to be considered in future journalism research and provide concluding remarks.

Driving factors behind the cosmopolitanization of algorithm-driven journalism

There is a growing impetus to cosmopolitanize algorithm-driven journalism research in response to the evolving global media landscape precipitated by the increasing adoption of AI technologies in newsrooms, which introduces many conceptual opportunities and potentials for journalistic practices (Newman, 2024). Algorithmic systems are steadily becoming an integral part of the journalism profession, and they are already employed throughout the entire journalistic news value chain. This trend belongs to the field of computational journalism (Anderson, 2013) and is referred to as algorithm-driven journalism, involving the utilization of AI methods such as machine learning and deep learning along with automation technologies to perform tasks that require human-like journalistic intelligence (Sarısakaloğlu, 2025). AI tools assist in automatically gathering and selecting information, identifying relevant topics, performing fact-checking, verifying sources, generating textual and visual news content in various languages, and disseminating news articles (e.g., news recommendation, personalization of news content, chatbots, etc.) with minimal or no human intervention (Sarısakaloğlu, 2024). Furthermore, algorithmic tools can facilitate collaborative editing and streamline editorial workflow, allowing multiple authors to work collaboratively on journalistic content from diverse locations around the globe. These advancements offer the possibility of enhancing the journalism profession by divesting human journalists of the burden of routine editorial tasks and transforming traditional newsrooms into sociotechnical entities, where human and anthropomorphized artificial journalists interact in a symbiotic relationship at all levels of journalistic content generation and distribution (Sarısakaloğlu, 2022).

The use of algorithmic systems not only provides journalists with novel avenues for communication processes but also heralds a change in the field of journalism. It urges us to redefine journalistic workflows, routine practices, job tasks, and responsibilities, thus ushering in new competence requirements for journalists and introducing fresh approaches to the profession (Sarısakaloğlu, 2022). Accordingly, the drive toward algorithmic adoption in journalism poses challenges to news organizations worldwide, irrespective of their geographical location, size, or available resources (Beckett & Yaseen, 2023). However, these challenges are accentuated when it comes to the newsrooms of the Global South, with the exception of China, one of the world's leading innovators in the development of AI systems, primarily due to knowledge gaps, structural limitations (e.g., resource constraints, infrastructural challenges, etc.), language barriers, and lack of necessary competencies (Beckett & Yaseen, 2023; Candelon et al., 2023; Gondwe, 2023a; Yu et al., 2023).

Despite the growing significance of these challenges, our knowledge remains limited. The results of my systematic literature review on algorithm-driven journalism (Sarsakaloğlu, 2025), analyzing 348 peer-reviewed journal articles retrieved across 10 databases up to December 2022, revealed an imbalance in the countries under investigation, with a predominant focus on the Global North, particularly the United States, the United Kingdom, and Germany. In contrast, studies on the use of algorithmic systems in newsrooms in the Global South have provided evidence of underrepresentation (Guanah et al., 2020; Kothari & Cruikshank, 2022; Munoriyarwa et al., 2023; Okiyi & Nsude, 2019). However, to incorporate the Global South into the envisioned "global village" (McLuhan, 1964), wherein technology enables active participation in the global media landscape, it is vital to nurture a profound sense of interconnectedness beyond geographical and cultural boundaries. It is paramount to transition to a truly cosmopolitan vision of algorithm-driven journalism to achieve a more equitable distribution of the benefits derived from algorithmic systems.

The use of AI tools raises concerns about the growing disparity between news organizations in the Global North, where implementing algorithmic systems is more feasible, and their counterparts in the Global South, where resource constraints are commonplace among many news organizations. Hence, our attention should extend beyond investigating the opportunities associated with AI technologies. To foster a cosmopolitan research perspective on algorithm-driven journalism, the disparities and challenges encountered by newsrooms in Global South countries need to be acknowledged to create a more inclusive approach and ensure the equitable use of AI technologies across the globe.

Unveiling the global divide in algorithmizing journalism

The study of media- and technology-related inequalities among different segments of the population has a long tradition in the social sciences. The first research endeavors date back to the early 1970s, when communication scholars Tichenor et al. (1970) introduced the knowledge gap hypothesis, which addresses phenomena related to the unequal utilization of mass media among diverse social groups. The hypothesis asserts that social groups with a higher socioeconomic status or formal education are more adept at acquiring knowledge and new competencies conveyed through mass media compared with their socioeconomically disadvantaged counterparts (Tichenor et al., 1970). This perspective implies that countries in the Global South are possibly being excluded from the opportunity of using media in general.

Building upon the knowledge gap hypothesis, conceptualizations of the digital divide emerged in the 1990s as Western Europe began to experience increased digitalization and Internet usage became widespread. Analytically, these conceptualizations can be grouped into three categories, providing a framework for the investigation and description of phenomena related to unequal access to technological opportunities among population groups, these being the aforementioned first-level, second-level, and third-level digital divides (DiMaggio et al., 2004; Norris, 2001; Warschauer, 2003). In general, the digital divide refers to global (e.g., between geographic areas), institutional (e.g., among organizations), or individual (e.g., based on personal demographics) disparities in access to information and communication technologies (Norris, 2001; Organization for Economic Co-operation and Development [OECD], 2001; Riggins & Dewan, 2005). With the increasing use of AI technologies, an AI-driven divide (or algorithmic divide) is emerging, which can be considered a subdimension of the digital divide (Carter et al., 2020). In the following section, the levels of the AI-driven divide are described in relation to global disparities.

First-level AI-driven divide

Research on the first-level digital divide addresses structural disparities that prevent certain population groups from having equal access to digital communication infrastructure, such as computers and the Internet (van Dijk, 2006). These inequalities may be driven by physical, technical, and social constraints on access to digital media and the Internet (van Dijk, 2018) and compounded by inadequate economic resources, among other factors. Building upon Norris's (2001) definition of the global digital divide, which focuses on delineating disparities in access to information and communication technologies between geographic areas, for the purposes of algorithm-driven journalism research, it can be inferred that the global first-level AI-driven divide is primarily rooted in the unequal availability of and access to data and algorithmic technologies in newsrooms, which is particularly pronounced between industrialized and peripheral countries (North–South divide). Such disparities can have substantial implications for journalism in the Global South, where journalists may not have access to the same digital toolbox and information resources as their counterparts in the Global North (Mellor, 2024).

Given that algorithmizing journalistic practices can enhance the efficiency of editorial processes, Western media organizations in particular are becoming increasingly reliant on and investing in AI systems, as highlighted in the *Digital News Project 2024* report by the Reuters Institute (Newman, 2024). Notably, media organizations in the United States are pioneers in integrating algorithm-driven tools into their newsrooms. The Los Angeles Times introduced Quakebot to generate earthquake news, and The Associated Press adopted the Wordsmith software to create automated news stories. Innovative tools, such as The Washington Post's text generator Heliograf, and more recently generative systems like Open AI's ChatGPT, have proven to be useful tools when employed in the execution of journalistic tasks (Gondwe, 2023b; Kunova, 2023). The DALL·E program, which can create images based on the text descriptions provided, further highlights the forward-thinking approach of US-American media organizations. In Europe, news agencies, including Agence France-Presse and the German Press Agency, and news media such as The Guardian and the Süddeutsche Zeitung, as well as many others, incorporate AI tools into their daily workflows. This trend is mirrored in Chinese newsrooms, where, for instance, Xinhua News (2018) introduced the world's first AI news anchor in collaboration with the technology company Sogou. In contrast, to compare this situation to numerous newsrooms across the Global South, the findings of a global survey on the adoption of AI technologies in newsrooms that garnered responses from 105 news and media organizations across 46 countries revealed that respondents, particularly those from sub-Saharan Africa, Middle East and North Africa (MENA), and the Asia-Pacific, emphasized infrastructure challenges due to low Internet penetration, which complicates efforts to facilitate the use of algorithmic systems (Beckett & Yaseen, 2023). Additionally, barriers to accessing specific AI tools like ChatGPT exist in some countries (e.g., Cuba, Russia, and Venezuela) due to a range of factors, such as sanctions by the United States, political reasons, or lack of support for the native languages of several countries (Beckett & Yaseen, 2023; OpenAI, n.d.). Furthermore, for example, in Nigerian newsrooms, a notable lag in adopting such advanced technologies is evident due to the unavailability of data and limited knowledge of AI technologies (Okiyi & Nsude, 2019).

From a cosmopolitan perspective, collaborative research efforts involving scholars from the Global North and the Global South are needed to tackle the underlying causes of the first-level AI-driven divide and to develop inclusive solutions. To achieve this, communication and technology scholars should shift their focus away from exploring the implementation of AI technologies solely in Western newsrooms. Instead, they should strive to consider diverse contexts beyond the West in scholarly analysis, such as scrutinizing how technology transfer, defined as "the process of conveying results stemming from scientific and technological research to the ... wider society" (European Commission, 2023), can be carried out by news organizations currently employing AI technologies to generate and distribute journalistic content for those lacking access. Such investigations would provide an opportunity to promote the effective transfer of technology and knowledge. By exchanging academic knowledge, expertise, resources, and best practices, new approaches to addressing the AI-driven divide can be developed to meet the specific needs and challenges of different regions.

Second-level AI-driven divide

The mere availability of and access to digitally networked media, however, does not necessarily imply competent usage (Gunkel, 2003). Individuals with digital skills and a deeper understanding of new technologies are better positioned to leverage the opportunities offered by digital applications. This gives rise to the exploration of the second-level digital divide, which focuses on gaps in technological competencies and the resulting differences in the usage of networked media (DiMaggio et al., 2004; Hargittai et al., 2019; van Dijk, 2006; Warschauer, 2003). Accordingly, research on the second level has shifted focus away from access discrepancies to analyzing inequalities so as to understand the reasons behind disparities in digital knowledge and skills and the subsequent differences in technology usage practices (Gurstein, 2011; Hargittai, 2002; van Dijk, 2005). In the realm of algorithm-driven journalism, such inequality manifests in differences in taking advantage of intelligent technologies and preventing effective and meaningful engagement with new tools in journalism practice. This limitation may be viewed as a constraint in providing innovative journalistic content or meeting the standards of emerging trends in journalism. To ensure the seamless integration of AI tools, overcoming the algorithmic competence gap is of utmost importance.

Notably, the bulk of research in this domain concentrates on Western countries (Graßl et al., 2022; Min & Fink, 2021; van Dalen, 2012), which can be attributed to the fact that Western countries, as previously outlined, are at the forefront of AI technology development and its integration into newsrooms, leading to a heightened interest in investigating journalistic competencies in handling AI. However, Western-centric investigations may be inadequate for elucidating journalistic practices across diverse cultural and regional contexts beyond the Western sphere. Studies have indicated that in newsrooms in the Global South, such as in African newsrooms, where proficiency in handling AI technologies is lacking (Kothari & Cruikshank, 2022; Munoriyarwa et al., 2023; Okiyi & Nsude, 2019), journalists often find themselves playing catch-up or imitating practices from developed countries, irrespective of whether these approaches align with their own journalistic cultures (Mellor, 2024). Hence, journalism research that seeks a cosmopolitan perspective should include, for instance, an examination of the challenges encountered by journalists in non-Western regions when using AI technologies to acknowledge and understand competence gaps in a global context, rather than solely from a Western viewpoint. In this way, non-Western evidence can be incorporated into existing approaches to describing journalistic competencies.

Third-level AI-driven divide

It is essential to explore the consequences of disadvantages and inequalities in users' ability to benefit from these technologies, referred to as the third-level or zero-level digital divide (Iske & Kutscher, 2020; Ragnedda & Muschert, 2018). This concept encompasses challenges that emerge, for instance, from the deployment of algorithmic mechanisms and advancements in technological infrastructure related to net neutrality and hardware design (Iske & Kutscher, 2020; van Deursen & Helsper, 2015; Verständig et al., 2016). In their analysis of the challenges arising from technological-structural changes to media education research, Iske and Kutscher (2020) emphasized that this divide leads to further digital marginalization, evident in phenomena such as the algorithmic personalization of information and the problem of achieving transparent, neutral, and fair technology use. Previous studies have underscored the importance of developing an adequate level of awareness of algorithmic tools (e.g., Gran et al., 2021) and acquiring algorithmic literacy as essential components in effectively navigating the impact of these systems (e.g., Eubanks, 2018; Mohamed et al., 2020). It is reasonable to infer that journalists with a sufficient understanding of algorithmic technologies, an awareness of ethical challenges, and an adept handling of AI tools will eventually benefit more from these systems and be less susceptible to potential harm.

Against this backdrop, additional research is needed to address disparities in journalists' awareness, perceptions, and attitudes regarding the strengths and risks of algorithmic systems in newsrooms in Global South countries. Such studies could not only enhance effective employment but also be mindful of differences in values and norms worldwide, thus advancing the development of ethical approaches to algorithm-driven journalism that are sensitive to diverse cultural contexts beyond the West.

Challenges faced by inclusive algorithm-driven journalism

To facilitate the cosmopolitanization of algorithm-driven journalism research, scholars need to thoroughly engage with the diversity of practices and challenges related to the global usage of AI tools, since the development of these technologies does not occur in a vacuum but relies on the data provided by individuals that inevitably reflect societal realities, including power imbalances and distortions (Couldry & Mejias, 2019b; Mhlambi, 2020; O'Neil, 2017; Steensen, 2019). Big tech companies in the Global North provide digital infrastructure and resources that enable the performance and operation of digital systems, while countries in the Global South are primarily viewed as valuable resources for creating raw data on account of the digital footprints they leave (Couldry & Mejias, 2019a, 2019b). News agencies can use this data to tailor journalistic content by analyzing user behavior, preferences, and trends, or it can be commodified and sold to entities such as advertising companies. When describing this phenomenon, Couldry and Mejias (2019a) alluded to the concept of data colonialism, defining it as a process that "combines the predatory extractive practices of historical colonialism with the abstract quantification methods of computing" (p. 337), wherein "the exploitation of human beings through data" becomes normalized (p. 336). In the context of AI technologies, Mohamed et al. (2020) argued that colonial power structures can become entrenched, potentially leading to the dominance of specific nations or regions and perpetuating or exacerbating existing power imbalances and inequalities reminiscent of historical colonial structures. They referred to this phenomenon as *algorithmic colonialism*, which posits that algorithms may disproportionately benefit specific groups while marginalizing or exploiting others (Mohamed et al., 2020). In academia, this has sparked a range of ethical questions, including concerns about the exclusion of the Global South in the design of algorithmic systems, algorithmic biases, and unfair practices leading to generating cultural stereotypes and discriminatory outcomes (e.g., Porlezza & Schapals, 2024). These challenges need to be critically reflected upon in future research to facilitate a comprehensive understanding and balanced portrayal of the ethical issues encountered in the field.

The root of the imbalance in the systems design process can be traced back to the exclusion of algorithm developers and journalistic actors from marginalized communities. When teams of developers working on algorithmic tools have a homogenous composition that predominantly comprises programmers and journalists from the Western world, there is a risk that specific ethical values and norms inherent to journalism in the

Global South might not be properly incorporated or may be disregarded in the software code. The lack of diversity in development teams regarding expertise in contexts from the Global South may have a disproportionately negative impact on those on the side of the AI-driven divide (Mhlambi, 2020). This imbalance is also reflected in the findings of a global survey on the implementation of AI technologies in newsrooms, which shows that newsrooms in the Global South face language barriers related to AI tools (Beckett & Yaseen, 2023). This challenge arises from the fact that many tools are predominantly designed for users in industrialized countries and for English speakers, overlooking the need for applications catering to underrepresented languages worldwide (Beckett & Yaseen, 2023). Furthermore, as algorithmic systems are predominantly developed in English-speaking contexts, data scientist Fu (2022) has contended that the absence of representative datasets for non-English languages results in a dearth of technologies capable of processing non-English data. This situation emphasizes the importance of diverse representation in technology development and accentuates the potential pitfalls of excluding certain voices, which contributes to inequitable impacts and perpetuates biases in algorithmic systems. To overcome the overrepresentation of English data in training algorithmic systems and to ensure a more balanced representation in datasets across different languages, efforts should be directed toward actively curating content in multiple languages to improve the accuracy of AI systems in non-English contexts.

Additionally, researchers revealed that a lack of diversity in the datasets used to train algorithms can lead to biases in AI systems, particularly if the training data is built on Western-centric settings, thereby contributing to a biased perspective that favors the Western context (Andersen, 2019; Fu, 2022). As algorithmic technologies are shaped by the cultural values and perspectives of their developers, there is a risk that distortions in data may also emerge due to conscious or unconscious prejudices on the part of algorithm developers, resulting in a systemic bias that adversely affects underrepresented groups (O'Neil, 2017). If a piece of information being classified as newsworthy or non-newsworthy hinges solely on news factors deemed important by Western software developers and journalists, algorithmic biases can emerge, given the diverse journalistic cultures of non-Western newsrooms (Sarısakaloğlu, 2024). An example of such algorithmic bias is when training data for algorithms is inclined toward negative narratives regarding countries in the Global South. Consequently, when reporting on these communities, algorithm-driven tools may inadvertently perpetuate harmful biases by generating news stories that excessively emphasize issues such as conflict, poverty, or disease. This skewed portrayal can lead to an oversight of positive aspects, such as economic growth, cultural richness, or technological advancements, and uphold a narrative of victimhood.

Moreover, scholars stress that using faulty, incomplete, or outdated datasets for input and algorithm training can create biases in the outcomes (Gondwe, 2023a; Mhlambi & Tiribelli, 2023; Noble, 2018; O'Neil, 2017; Zweig et al., 2018), which disproportionately affect individuals who are impoverished or disadvantaged (Eubanks, 2018; O'Neil, 2017). This concern is exacerbated by a lack of transparency in algorithmic processes (Diakopoulos, 2015), contributing to what is often referred to as the black box problem in AI (Fu, 2022; Russell & Norvig, 2021). To prevent unfair distortions in journalism, algorithm-driven journalistic processes should be transparent and disclose the underlying data and decision-making criteria of intelligent systems, thereby ensuring the accountability of algorithmic tools (Diakopoulos, 2015).

If biases or stereotypes are embedded in the database and are not critically examined or addressed, possibly due to a lack of diversity in development teams, they can even lead to cases of discrimination (Noble, 2018). For instance, discrimination becomes evident in algorithmic recommendation systems if news content is specifically presented to individuals of a particular ethnic background (Sarısakaloğlu, 2024) or when algorithms predominantly suggest news articles that align with a more Western-centric perspective to readers from the Global North, while infrequently recommending news stories primarily consumed by non-Western readers. Algorithmic discrimination not only confines readers' exposure to a narrow spectrum of information but also contributes to reinforcing ethnocentric viewpoints and facilitating a lack of cultural inclusivity in producing and disseminating algorithm-driven news content (see also Horz-Ishak in this book).

To address these challenges effectively, cosmopolitanism-oriented journalism research needs to develop an inclusive theoretical framework that respects and integrates the diversity of ethical and cultural values without favoring Western perspectives. Such a framework would enable scholars to critically reflect on the challenges posed by algorithmic systems and gain a deeper understanding of the implications of AI technologies in journalism on a global scale. Thus, a truly cosmopolitan approach to algorithm-driven journalism research would also facilitate the development of ethical guidelines that are applicable and relevant to different cultural contexts.

Navigating cosmopolitan algorithm-driven journalism research

The insights provided in the preceding sections strongly suggest that, as an analytical instrument, cosmopolitan algorithm-driven journalism aims to provide a deep understanding of deploying algorithmic systems in newsrooms across different global contexts and cultural values to transcend regional boundaries and acknowledge the multifaceted dynamics inherent in the contemporary media landscape. To achieve this, it is imperative to pay attention to phenomena that might evade the analytical scope of Western scholarship but hold significance in the context of the Global South. This entails considering three key pillars and reflections for cosmopolitan algorithm-driven journalism research, as illustrated in Figure 1.

In advancing cosmopolitan algorithm-driven journalism, cultivating inclusive academic knowledge production emerges as a crucial endeavor emphasizing the importance of scholarly exchange, diverse perspectives, and collaborative efforts across international borders and disciplinary boundaries. In my systematic literature review of algorithm-driven journalism research, I identified a notable geographical imbalance in authors' organizational affiliations, with a prevailing representation of institutions in Global North countries (Sarısakaloğlu, 2025). Within a corpus of 835 authors, 714 authors (85.5%) are from countries in the Global North, with the United States, the United Kingdom, the Netherlands, Spain, and Germany emerging as the five most frequently involved (Sarısakaloğlu, 2025). In contrast, only 121 authors (14.5%) are from the Global

South, with the most commonly represented five nations being China, Singapore, the United Arab Emirates, India and Brazil (Sarısakaloğlu, 2025).

Figure 1: Pillars for advancing cosmopolitan algorithm-driven journalism research (source: author's compilation)

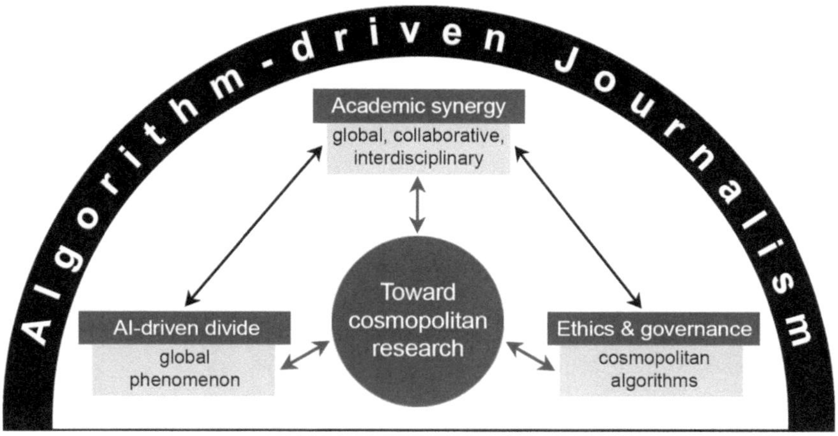

Cultivating global, collaborative, and interdisciplinary academic synergy

Notwithstanding the fact that algorithm-driven journalism studies are a generally nascent research subject across the world, the development of this emerging field in the Global South remains largely unchartered by scholarship compared with the Global North. In this context, particular attention should be paid to exploring how journalism scholars beyond the West can be empowered to actively contribute to the discourse on algorithm-driven journalism and fill the research gap in non-Western scholarship. Among the 348 articles in the aforementioned systematic literature review, over half are written collaboratively (234 articles) (Sarısakaloğlu, 2025). However, as observed in systematic literature reviews on comparative journalism studies (e.g., Hanusch & Vos, 2020), it is worth noting that collaborations in the research of algorithm-driven journalism are most prevalent among scholars affiliated with countries from the Global North (193 articles, 82.5%) (Sarısakaloğlu, 2025). In contrast, collaborations between Global North and Global South scholars account for only 16 articles (6.8%) (Sarısakaloğlu, 2025). These findings indicate power asymmetries in research and publication, which will continue to lead to the domination of Western institutions, resulting in a distortion of research outcomes and the marginalization of voices from beyond the Western world. Hence, ensuring inclusivity through encouraging global and collaborative research initiatives that engage scholars from both Western and non-Western backgrounds is of utmost importance. This could be further exemplified, for instance, by fostering partnerships between research institutions worldwide and bringing together diverse perspectives and expertise, regardless of their geographical location or cultural background, to facilitate equitable participation in academic knowledge production. Through North–South collaboration, researchers could collectively navigate the intricacies of algorithm-driven journalism, ensuring a global understanding of the use of algorithmic technologies in newsrooms and their impacts on journalism.

Furthermore, adopting an interdisciplinary participatory approach to algorithm-driven journalism (Sarısakaloğlu, 2024) is instrumental in integrating country-specific journalistic norms and values into algorithmic tools. This requires collaboration not only with computer scientists and journalism scholars in the Global North but also with their counterparts operating in the Global South. Such an approach would enrich academic discourse by facilitating the exchange of technical expertise in AI and journalism, a crucial step in overcoming the challenges posed by the global AI-driven divide. Additionally, it would have the potential to contribute to the development of ethical and culturally sensitive algorithmic systems, reinforcing cosmopolitan ideals in journalism research and practice.

AI-driven divide as a global phenomenon

Acknowledging the AI-driven divide as a global phenomenon can enhance efforts to describe its widespread impact on newsrooms globally and foster a more interconnected journalistic community. By doing so, researchers can develop approaches to address disparities triggered by algorithmic tools that are not only context-specific but also account for the diverse conditions and needs of newsrooms across different regions. This would ensure that the implications of AI technologies are understood and navigated on a global stage. One step in this direction would be to pay attention to the disparity in journalists' access to AI tools in newsrooms across the globe, their proficiency in using these systems, and the outcomes of their engagement with AI. To adequately investigate the AI-driven divide in newsrooms as a global phenomenon, however, it is essential to critically re-examine the dominant Western theoretical approaches when describing journalistic processes, along with the research designs employed in exploring algorithm-driven journalism, with the aim of assuring the applicability of theoretical frameworks and research evidence across diverse global contexts.

For example, adopting an approach to exploring the competencies of journalists in algorithm-driven newsrooms can benefit from leveraging conceptualizations of journalistic competencies proposed by Western scholars, such as the work of Nölleke-Przybylski et al. (2020). Drawing on previous German-language competence research, they identified components of journalistic competence that delineate skills and tasks essential in digital work environments, including technical competence (Nölleke-Przybylski et al., 2020). In addition to the technical skills and factual knowledge necessary for the operation and programming of software and tools (Nölleke-Przybylski et al., 2020), other scholars have emphasized the application of computational thinking skills to critically reflect on and analyze the abundance of data and generated information to avoid potential biases in journalistic content (Diakopoulos, 2019; Sarısakaloğlu, 2024). The accumulation of all the competencies required for effective interaction with AI tools can be defined as AI literacy (Long & Magerko, 2020) or as *algorithmic capital* in a Bourdieusian sense. The higher the level of algorithmic capital among journalists in the Global South, the more they will be able to harness the full potential of these technologies, and the lower the global second-level AI-driven divide will become.

Nevertheless, given the diverse journalistic cultures with distinct standards and professional views of journalists, these theoretical concepts may fall short of capturing journalistic competencies beyond the Western context. In response to this, a comparative research approach can be employed to investigate the extent to which the competency requirements of journalists from the Global North and the Global South differ in the context of algorithm-driven news production and distribution, thus expanding existing approaches to journalistic competencies. However, it is challenging to conduct empirical studies due to the limited adoption of AI technologies in many Global South newsrooms, as evidenced in African (Kothari & Cruikshank, 2022) and Pakistani (Jamil, 2023) newsrooms. In contrast, US-American tech companies, media organizations, and research institutions are leveraging AI significantly to enhance journalistic practices.

This further complicates the exploration of technological consequences for journalists and their work, hindering efforts to adapt educational programs for journalists and raise awareness regarding technological and ethical challenges to ensure responsible use and overcome imbalances in knowledge production between newsrooms in the Global North and the Global South (Kothari & Cruikshank, 2022).

Incorporating cosmopolitan algorithms into AI ethical and governance frameworks

As outlined in the previous section, the development of algorithms for AI tools is predominantly anchored in developed countries, raising concerns about their applicability in diverse global settings, particularly as they may not adequately reflect the unique contexts of developing countries (Hamann, 2018). This Western bias has manifested in the form of inadequate representations of non-Western values in algorithmic systems, which underscores the need to broaden the perspective of algorithm-driven journalism research and to adeptly navigate ethical constraints to ensure that AI tools align with and respect journalistic values and norms beyond the Global North to produce culturally sensitive news content. To promote balanced and unbiased representations of global events in news coverage, it is key to design *cosmopolitan algorithms*. These algorithms ought to be collaboratively designed by both Western and non-Western developers trained on diverse data sources and equipped with the capability to prevent or alleviate distortions, stereotypes, or discriminatory patterns in news reporting. By improving fair judgment in AI tools, such algorithms would contribute to the creation of less biased journalistic content. However, putting such a framework into practice would require a collaborative knowledge exchange between scholars from the Global North and the Global South, as well as cross-cultural comparisons, to establish a shared understanding of the concept of fairness, given that ethical values are far from universal and their definitions depend on various factors such as location, historical considerations, and cultural context (Fu, 2022).

Since AI technologies pose ethical challenges not only in journalism but also in all social domains, ensuring trustworthy and inclusive AI has become the focus of various governance approaches (e.g., Porlezza, 2023). Political decision-makers, data protection authorities, and ethics committees are endeavoring to develop guidelines and measures to ensure a fair and trustworthy deployment of AI technologies (e.g., European Commission, 2019; European Parliament, 2022; Institute of Electrical and Electronics Engineers [IEEE], 2019; United Nations Educational, Scientific and Cultural Organization [UNESCO], 2022). In 2024, EU member states adopted the AI Act, the world's first and most comprehensive law regulating AI that ensures a unified framework for the ethical and responsible design and use of AI systems in Europe, aiming to minimize risks to society (European Parliament, 2024). However, these regulations do not necessarily guarantee the trustworthy use of AI technologies in Global South countries. Nonetheless, the AI Act could establish a global standard for AI regulation.

Moreover, regulations were developed to guide the responsible integration of AI technologies in the field of journalism. In partnership with a commission comprising civil society organizations, AI experts, media representatives, and journalists, Reporters Without Borders (Reporters sans Frontières, 2023) established the first international ethical framework for the use of AI in journalism, known as the Paris Charter on AI and Journalism. The charter delineates a set of fundamental ethical guidelines for journalistic entities worldwide to protect the diversity and integrity of information and news. For instance, it underscores the imperative that "journalists, media outlets and journalism support groups (...) should ensure that AI governance respects democratic values, and that diversity of people and cultures is reflected in the development of AI" (Reporters sans Frontières, 2023, p. 2). These efforts highlight the importance of establishing effective AI governance policies to foster technological innovation while guaranteeing the safe and responsible use of AI systems in alignment with ethical values and legal regulations.

Overall, while the charter sets forth essential ethical guidelines for journalistic entities globally, future research should investigate its practical implementation and identify areas of improvement to enhance the application of ethical frameworks in different national journalistic cultures and to develop theoretical perspectives, without presuming that Western concepts are universally applicable, but rather by considering the ethical values of other regions and the social and cultural context in which AI technologies are embedded.

Conclusion

This chapter advocates for the cosmopolitanization of communication and media research in the study and practice of algorithm-driven journalism by delineating a path toward a more inclusive approach to understanding journalism in the era of AI. To successfully cosmopolitanize algorithm-driven journalism, the inequalities and challenges stemming from the global AI-driven divide must be taken into account, along with recognition of the diversity of global journalistic cultures. As such, scholars investigating the integration of AI tools in journalism not only need to uncover the impediments that restrict journalists from the Global South in generating journalistic content with the assistance of non-human journalists in sociotechnical newsrooms and acting as equals alongside their counterparts in the Western environment, but they must also work toward solutions to overcome inequalities and ensure equitable participation. This goal can be achieved by (1) fostering global, collaborative, and interdisciplinary research efforts,

(2) acknowledging the AI-driven divide as a global phenomenon, and (3) integrating cosmopolitan algorithms into ethical and governance frameworks for AI. These pillars offer avenues to bridge the North–South divide and pave the way for a more equitable future in algorithm-driven journalism.

Overcoming biases inherent in Western-centric perspectives would necessitate co-creating academic knowledge. Instead of emulating Western practices, adopting a cosmopolitan research approach that ensures adequate representation of non-Western voices would help avoid power imbalances in research and broaden frameworks to incorporate the Global South perspective which is essential to attaining a comprehensive understanding of sociotechnical phenomena beyond the mainstream West and discerning blind spots imposed by Western-centric approaches.

Simultaneously, bridging the global research divide in algorithm-driven journalism also requires filling the knowledge gap through the adaptation of curricula and training programs in journalism education that will enable students to become cosmopolitan scholars. They should be equipped with an in-depth theoretical understanding of AI technologies' global implications by recognizing cultural diversity in journalism and the methodological competencies for examining phenomena from a cosmopolitan lens, ensuring that diverse voices and perspectives are represented.

Additionally, acquiring algorithmic capital is vital to avoid adding to the educational disparities created by the use of algorithmic systems and to cultivate future *AI-savvy cosmopolitan journalists*. Otherwise, the disparity in journalists' proficiency in the usage of emerging technologies may lead to a sense of disconnectedness from the *global village journalism* and potentially result in professional inefficiency. Therefore, the aim is not to merely facilitate technical access to AI technologies but to nurture future journalists as sophisticated contributors to the news value chain across the globe, equipping them with the technical, critical, and reflexive skills necessary to use new technologies and build fair and inclusive worldviews.

As cosmopolitan communication and media scholars, our task is to steer algorithm-driven journalism research in a direction that considers the implications of technological advancements on a global scale and contributes to a global mindset of sociotechnical phenomena.

References

Andersen, L. (2019). Artificial intelligence in international development: Avoiding ethical pitfalls. *Journal of Public & International Affairs*. https://jpia.princeton.edu/news/artificial-intelligence-international-development-avoiding-ethical-pitfalls

Anderson, C. W. (2013). Towards a sociology of computational and algorithmic journalism. *New Media and Society*, *15*(7), 1005–1021. https://doi.org/10.1177/1461444812465137

Badr, H., & Ganter, S. A. (2021). Towards cosmopolitan media and communication studies: Bringing diverse epistemic perspectives into the field. *Global Media Journal – German Edition*, *11*(1). https://doi.org/10.22032/dbt.49164

Beck, U. (2006). *Cosmopolitan vision*. Polity Press.

Beckett, C., & Yaseen, M. (2023). *Generating change: A global survey of what news organisations are doing with AI*. JournalismAI. https://static1.squarespace.com/static/64d6052 7c01ae7106f2646e9/t/656e400a1c23e22da0681e46/1701724190867/Generating+Chang e+_+The+Journalism+AI+report+_+English.pdf

Brennen, J. S., Howard, P. N., & Nielsen, R. K. (2018). *An industry-led debate: How UK media cover artificial intelligence*. Reuters Institute for the Study of Journalism. https://reutersinstitute.politics.ox.ac.uk/sites/default/files/2018-12/Brennen_ UK_Media_Coverage_of_AI_FINAL.pdf

Candelon, F., Courtaux, M., & Jha, G. (2023, January 6). *To make the Global South more competitive, its leading companies must become A.I. national champions*. Fortune. https://fortu ne.com/2023/01/06/artificial-intelligence-ai-global-south-national-champions/

Carpentier, N., Ganter, S. A., Ortega, F., & Torrico, E. (2020). A debate on post-colonialism and de-coloniality: Latin American and European perspectives on change and hope. In F. O. Paulino, G. Kaplún, M. Vincente-Mariño, & L. Custódio (Eds.), *Research traditions in dialogue: Communication studies in Latin America and Europe* (pp. 275–293). Media XXI.

Carter, L., Liu, D., & Cantrell, C. (2020). Exploring the intersection of the digital divide and artificial intelligence: A hermeneutic literature review. *AIS Transactions on Human-Computer Interaction, 12*(4), 253–275. https://doi.org/10.17705/1thci.00138

Couldry, N., & Mejias U. A. (2019a). Data colonialism: Rethinking big data's relation to the contemporary subject. *Television & New Media, 20*(4), 336–349. https://doi.org/10. 1177/1527476418796632

Couldry, N., & Mejias, U. A. (2019b). *The costs of connection: How data is colonizing human life and appropriating it for capitalism*. Stanford University Press.

Curran, J., & Park, M.-J. (2000). Beyond globalization theory. In J. Curran & M.-J. Park (Eds.), *De-Westernizing media studies* (pp. 2–15). Routledge.

Diakopoulos, N. (2015). Algorithmic accountability: Journalistic investigation of computational power structures. *Digital Journalism, 3*(3), 398–415. https://doi.org/10.1080/2 1670811.2014.976411

Diakopoulos, N. (2019). *Automating the news: How algorithms are rewriting the media*. Harvard University Press.

DiMaggio, P., Hargittai, E., Celeste, C., & Shafer, S. (2004). Digital inequality: From unequal access to differentiated use. In K. M. Neckerman (Ed.), *Social inequality* (pp. 355–400). Russell Sage Foundation.

Eubanks, V. (2018). *Automating inequality: How high-tech tools profile, police, and punish the poor*. St. Martin's Press.

European Commission. (2019). *Ethics guidelines for trustworthy AI*. https://data.europa.eu /doi/10.2759/346720

European Commission. (2023, September 13). *What is technology transfer?* https://knowle dge4policy.ec.europa.eu/technology-transfer/what-technology-transfer_en

European Parliament. (2022). *Artificial intelligence in a digital age: European Parliament resolution of 3 May 2022 on artificial intelligence in a digital age*. https://www.europarl.europ a.eu/doceo/document/TA-9-2022-0140_EN.pdf

European Parliament. (2024). *EU Artificial Intelligence Act – AI Act* [Final draft]. https://ww w.europarl.europa.eu/doceo/document/TA-9-2024-0138-FNL-COR01_EN.pdf

Fu, G. (2022). *Toward ethical artificial intelligence in international development*. Center for Digital Acceleration. https://www.dai.com/uploads/ethical-ai.pdf

Ganter, S. A., & Ortega, F. (2019). The invisibility of Latin American scholarship in European media and communication studies: Challenges and opportunities of de-Westernization and academic cosmopolitanism. *International Journal of Communication, 13*, 68–91. https://ijoc.org/index.php/ijoc/article/view/8449

Gondwe, G. (2023a). ChatGPT and the Global South: How are journalists in sub-Saharan Africa engaging with generative AI? *Online Media and Global Communication, 2*(2), 228–249. https://doi.org/10.1515/omgc-2023-0023

Gondwe, G. (2023b). *Exploring the multifaceted nature of generative AI in journalism studies: A typology of scholarly definitions*. Harvard Institute for Rebooting Social Media. https://scholar.harvard.edu/sites/scholar.harvard.edu/files/ggondwezunda/files/exploring_the_multifaceted_nature_of_generative_ai_in_journalism_studies.pdf

Gran, A.-B., Booth, P., & Bucher, T. (2021). To be or not to be algorithm aware: A question of a new digital divide? *Information, Communication & Society, 24*(12), 1779–1796. https://doi.org/10.1080/1369118X.2020.1736124

Graßl, M., Schützeneder, J., & Meier, K. (2022). Artificial intelligence as a tool of assistance: A scientific and practical perspective on AI in journalism. *Journalism Research, 5*(1), 3–24. https://doi.org/10.1453/2569-152X-12022-12049-en

Guanah, J. S., Agbanu, V. N., & Obi, I. (2020). Artificial intelligence and journalism practice in Nigeria: Perception of journalists in Benin City, Edo State. *International Review of Humanities Studies, 5*(2), 698–715. https://scholarhub.ui.ac.id/irhs/vol5/iss2/16/

Gunaratne, S. A. (2010). De-Westernizing communication/social science research: Opportunities and limitations. *Media, Culture & Society, 32*(3), 473–500. https://doi.org/10.1177/0163443709361159

Gunkel, D. J. (2003). Second thoughts: Toward a critique of the digital divide. *New Media & Society, 5*(4), 499–522. https://doi.org/10.1177/146144480354003

Gurstein, M. B. (2011). Open data: Empowering the empowered or effective data use for everyone? *First Monday, 16*(2). https://doi.org/10.5210/fm.v16i2.3316

Hamann, R. (2018, January 4). *Developing countries need to wake up to the risks of new technologies*. The Conversation. https://theconversation.com/developingcountries-need-to-wake-up-to-the-risks-of-new-technologies-87213

Hanitzsch, T. (2019). Journalism studies still needs to fix Western bias. *Journalism, 20*(1), 214–217. https://doi.org/10.1177/1464884918807353

Hanitzsch, T., Ahva, L., Alonso, M. O., Arroyave, J., Hermans, L., Hovden, J. F., Hughes, S., Josephi, B., Ramaprasad, J., Shapiro, I., & Vos, T. P. (2019). Journalistic culture in a global context: A conceptual roadmap. In T. Hanitzsch, F. Hanusch, J. Ramaprasad & A. S. de Beer (Eds.), *Worlds of journalism: Journalistic cultures around the globe* (pp. 23–45). Columbia University Press. https://doi.org/10.7312/hani18642-003

Hanusch, F., & Vos, T. P. (2020). Charting the development of a field: A systematic review of comparative studies of journalism. *International Communication Gazette, 82*(4), 319–341. https://doi.org/10.1177/1748048518822606

Hargittai, E. (2002). Second-level digital divide: Differences in people's online skills. *First Monday, 7*(4). https://doi.org/10.5210/fm.v7i4.942

Hargittai, E., Piper, A. M., & Morris, M. R. (2019). From Internet access to Internet skills: Digital inequality among older adults. *Universal Access in the Information Society, 18*(4), 881–890. https://doi.org/10.1007/s10209-018-0617-5

Institute of Electrical and Electronics Engineers. (2019). *Ethically aligned design: A vision for prioritizing human well-being with autonomous and intelligent systems*. https://standards.ieee.org/wp-content/uploads/import/documents/other/ead_v2.pdf

Iske, S., & Kutscher, N. (2020). Digitale Ungleichheiten im Kontext Sozialer Arbeit. In N. Kutscher, T. Ley, U. Seelmeyer, F. Siller, A. Tillmann, & I. Zorn (Eds.), *Handbuch soziale Arbeit und Digitalisierung* (pp. 115–128). Beltz Juventa.

Jamil, S. (2023). Evolving newsrooms and the second level of digital divide: Implications for journalistic practice in Pakistan. *Journalism Practice, 17*(9), 1864–1881. https://doi.org/10.1080/17512786.2022.2026244

Katz, E., Peters, J. D., Liebes, T., & Orloff, A. (2003). *Canonic texts in media research: Are there any? Should there be? How about these?* Polity Press.

Kothari, A., & Cruikshank, S. A. (2022). Artificial intelligence and journalism: An agenda for journalism research in Africa. *African Journalism Studies, 43*(1), 17–33. https://doi.org/10.1080/23743670.2021.1999840

Kozman, C. (2021). Reconceptualizing Arab media research: Moving from centrism toward inclusiveness and balance. *Journalism & Mass Communication Quarterly, 98*(1), 241–262. https://doi.org/10.1177/1077699020942924

Kunova, M. (2023, January 31). *Eight tasks ChatGPT can do for journalists*. Journalism. https://www.journalism.co.uk/news/how-can-journalists-use-chatgpt-/s2/a1005273/

Löffelholz, M., & Sarısakaloğlu, A. (2024). Paradigmengeschichte der Journalismusforschung. In M. Löffelholz & L. Rothenberger (Eds.), *Handbuch Journalismustheorien* (pp. 21–54). Springer VS. https://doi.org/10.1007/978-3-658-32151-2_2

Long, D., & Magerko, B. (2020). What is AI literacy? Competencies and design considerations. In *CHI '20: Proceedings of the 2020 CHI Conference on Human Factors in Computing Systems* (pp. 1–16). Association for Computing Machinery. https://doi.org/10.1145/3313831.3376727

McLuhan, M. (1964). *Understanding media: The extensions of man*. McGraw-Hill.

Mellor, N. (2024). The digital divide in the journalism sector. *Convergence, 30*(3), 1120–1133. https://doi.org/10.1177/13548565231176186

Mhlambi, S. (2020). *From rationality to relationality: Ubuntu as an ethical and human rights framework for artificial intelligence governance*. Carr Center for Human Rights.

Mhlambi, S., & Tiribelli, S. (2023). Decolonizing AI ethics: Relational autonomy as a means to counter AI harms. *Topoi, 42*(3), 867–880. https://doi.org/10.1007/s11245-022-09874-2

Min, S. J., & Fink, K. (2021). Keeping up with the technologies: Distressed journalistic labor in the pursuit of "shiny" technologies. *Journalism Studies, 22*(14), 1987–2004. https://doi.org/10.1080/1461670X.2021.1979425

Mohamed, S., Png, M.-T., & Isaac, W. (2020). Decolonial AI: Decolonial theory as sociotechnical foresight in artificial intelligence. *Philosophy & Technology, 33*(4), 659–684. https://doi.org/10.1007/s13347-020-00405-8

Munoriyarwa, A., Chiumbu, S., & Motsaathebe, G. (2023). Artificial intelligence practices in everyday news production: The case of South Africa's mainstream newsrooms. *Journalism Practice*, *17*(7), 1374–1392. https://doi.org/10.1080/17512786.2021.1984976

Newman, N. (2024). *Digital news project: Journalism, media, and technology trends and predictions 2024*. Reuters Institute for the Study of Journalism. https://reutersinstitute.politics.ox.ac.uk/sites/default/files/2024-01/Newman%20-%20Trends%20and%20Predictions%202024%20FINAL.pdf

Noble, S. U. (2018). *Algorithms of oppression: How search engines reinforce racism*. New York University Press.

Nölleke-Przybylski, P., Evers, T., & Altmeppen, K.-D. (2020). Catch me, if you can – eine Kompetenzperspektive auf Journalismus als Berufsfeld und Forschungsgegenstand. In J. Schützeneder, K. Meier, & N. Springer (Eds.), *Neujustierung der Journalistik/Journalismusforschung in der digitalen Gesellschaft: Proceedings zur Jahrestagung der Fachgruppe Journalistik/Journalismusforschung der Deutschen Gesellschaft für Publizistik- und Kommunikationswissenschaft 2019* (pp. 140–166). https://doi.org/10.21241/ssoar.70830

Norris, P. (2001). *Digital divide: Civic engagement, information poverty, and the Internet worldwide*. Cambridge University Press. https://doi.org/10.1017/CBO9781139164887

O'Neil, C. (2017, April). *The era of blind faith in big data must end* [Video]. TED Conferences. https://www.ted.com/talks/cathy_o_neil_the_era_of_blind_faith_in_big_data_must_end

Okiyi, G. O., & Nsude, I. (2019). Adopting artificial intelligence to journalistic practices in Nigeria: Challenges and way forward. *International Journal of Communication*, *24*(1), 141–161.

OpenAI. (n.d.). *Supported countries and territories*. https://platform.openai.com/docs/supported-countries

Organization for Economic Co-operation and Development. (2001). *Understanding the digital divide*. https://web-archive.oecd.org/2012-06-15/168628-1888451.pdf

Pavlik, J. (2000). The impact of technology on journalism. *Journalism Studies*, *1*(2), 229–237. https://doi.org/10.1080/14616700050028226

Porlezza, C. (2023). Promoting responsible AI: A European perspective on the governance of artificial intelligence in media and journalism. *Communications*, *48*(3), 370–394. https://doi.org/10.1515/commun-2022-0091

Porlezza, C., & Schapals, A. K. (2024). AI ethics in journalism (studies): An evolving field between research and practice. *Emerging Media*, *2*(3), 356–370. https://doi.org/10.1177/27523543241288818

Ragnedda, M., & Muschert, G. W. (Eds.). (2018). *Theorizing digital divides*. Routledge.

Reporters sans Frontières. (2023). *Paris charter on AI and journalism*. https://rsf.org/sites/default/files/medias/file/2023/11/Paris%20charter%20on%20AI%20in%20Journalism.pdf

Riggins, F. J., & Dewan, S. (2005). The digital divide: Current and future research directions. *Journal of the Association for Information Systems*, *6*(12), 298–337. https://doi.org/10.17705/1jais.00074

Russell, S., & Norvig, P. (2021). *Artificial intelligence: A modern approach* (4th ed.). Pearson.

Sarısakaloğlu, A. (2022). Algorithmisierung des Journalismus: Chancen und Herausforderungen künstlicher Intelligenzsysteme in sozio-technischen Newsrooms. *Communicatio Socialis*, *55*(3), 308–319. https://doi.org/10.5771/0010-3497-2022-3-308

Sarısakaloğlu, A. (2024). Künstliche Intelligenz und Journalismus. In M. Löffelholz & L. Rothenberger (Eds.), *Handbuch Journalismustheorien* (pp. 437–449). Springer VS. https://doi.org/10.1007/978-3-658-32151-2_30

Sarısakaloğlu, A. (2025). *Navigating the research landscape of algorithm-driven journalism: A systematic literature review of authorship, research trends, and future research pathways*. *Journalism Studies*, 1–27. https://doi.org/10.1080/1461670X.2024.2446326

Simon, F. M. (2024). *Artificial intelligence in the news: How AI retools, rationalizes, and reshapes journalism and the public arena*. Tow Center for Digital Journalism; Columbia University. https://doi.org/10.7916/ncm5-3v06

Steensen, S. (2019). Journalism's epistemic crisis and its solution: Disinformation, datafication and source criticism. *Journalism*, *20*(1), 185–189. https://doi.org/10.1177/1464884918809271

Tichenor, P. J., Donohue, G. A., & Olien, C. N. (1970). Mass media flow and differential growth in knowledge. *Public Opinion Quarterly*, *34*(2), 159–170.

UNESCO. (2022). *Recommendation on the ethics of artificial intelligence*. https://unesdoc.unesco.org/ark:/48223/pf0000381137

van Dalen, A. (2012). The algorithms behind the headlines: How machine-written news redefines the core skills of human journalists. *Journalism Practice*, *6*(5–6), 648–658. https://doi.org/10.1080/17512786.2012.667268

van Deursen, A. J. A. M., & Helsper, E. J. (2015). The third-level digital divide: Who benefits most from being online? In L. Robinson, S. R. Cotten, J. Schulz, T. M. Hale, & A. Williams (Eds.), *Communication and information technologies annual* (pp. 29–52). Emerald. https://doi.org/10.1108/S2050-206020150000010002

van Dijk, J. A. G. M. (2005). *The deepening divide – inequality in the information society*. Sage.

van Dijk, J. A. G. M. (2006). Digital divide research, achievements and shortcomings. *Poetics*, *34*(4–5), 221–235. https://doi.org/10.1016/j.poetic.2006.05.004

van Dijk, J. A. G. M. (2018). Afterword: The state of digital divide theory. In M. Ragnedda & G. W. Muschert (Eds.), *Theorizing digital divides* (pp. 199–206). Routledge.

Verständig, D., Klein, A., & Iske, S. (2016). Zero-level digital divide: Neues Netz und neue Ungleichheiten. *Si:So*, *21*(1), 50–55. https://nbn-resolving.org/urn:nbn:de:hbz:467-11973

Wahl-Jorgensen, K., & Hanitzsch, T. (2009). Introduction: On why and how we should do journalism studies. In K. Wahl-Jorgensen & T. Hanitzsch (Eds.), *The handbook of journalism studies* (pp. 3–16). Routledge.

Waisbord, S. (2015). De-Westernization and cosmopolitan media studies. In C.-C. Lee (Ed.), *Internationalizing "International Communication"* (pp. 178–200). University of Michigan Press. https://doi.org/10.2307/j.ctv65sxh2.11

Waisbord, S., & Mellado, C. (2014). De-Westernizing communication studies: A reassessment. *Communication Theory*, *24*(4), 361–372. https://doi.org/10.1111/comt.12044

Wang, G. (2011). *De-Westernizing communication research: Altering questions and changing frameworks*. Routledge.

Warschauer, M. (2003). *Technology and social inclusion: Rethinking the digital divide*. MIT Press. https://doi.org/10.7551/mitpress/6699.001.0001

Wasserman, H., & de Beer, A. S. (2009). Towards de-Westernizing journalism studies. In K. Wahl-Jorgersen & T. Hanitzsch (Eds.), *The handbook of journalism studies* (pp. 448–458). Routledge.

Xinhua. (2018, November 8). *World's first AI news anchor makes "his" China debut*. http://www.xinhuanet.com/english/2018-11/08/c_137591813.htm

Yu, D., Rosenfeld, H., & Gupta, A. (2023, January 16–20). *The 'AI divide' between the Global North and Global South*. World Economic Forum Annual Meeting, Davos, Switzerland. https://www.weforum.org/agenda/2023/01/davos23-ai-divide-global-north-global-south/

Zelizer, B. (2019). Why journalism is about more than digital technology. *Digital Journalism*, 7(3), 343–350. https://doi.org/10.1080/21670811.2019.1571932

Zweig, K. A., Fischer, S., & Lischka, K. (2018). *Wo Maschinen irren können: Fehlerquellen und Verantwortlichkeiten in Prozessen algorithmischer Entscheidungsfindung* [Working Paper]. Bertelsmann Stiftung. https://www.bertelsmann-stiftung.de/fileadmin/files/BSt/Publikationen/GrauePublikationen/WoMaschinenIrrenKoennen.pdf

Fact-checking

Broadening the research horizon with Latin American perspectives on the fight against misinformation

Regina Cazzamatta

Introduction

The emergence of fact-checking organizations in the early 2000s in the USA marked a journalistic reformist initiative. These agencies were considered a democratic-building tool (Amazeen, 2020; Graves, 2018), challenging traditional concepts of objectivity, known as the "he said/she said" style of journalism (Graves, 2016; Lawrence & Schafer, 2012). Subsequently, a second wave of fact-checking organizations emerged in response to post-truth politics marked by disinformation and disrupted public spheres (Bennett & Livingston, 2018), triggered by events like the 2016 Brexit referendum and the election of far-right populists. This global disinformation phenomenon led to a significant increase in fact-checking organizations, from 44 in 2014 (Adair, 2014) to 417 in 2023 across 100 countries (Stencel et al., 2023).

These indicators are unequivocal evidence of the rise of a global movement (Graves, 2018, 2022; Lauer & Graves, 2024). Despite their reform-oriented, less radical nature and structural diversity, independent fact-checkers, following Diani's (1992) definition, can be considered "networks of informal interactions among various individuals, groups, or organizations engaged in political or cultural conflicts [such as disinformation and post-truth politics], based on shared collective identities" (p. 1).

As communication becomes increasingly globalized, Western-originated forms of journalism, such as the fact-checking movement, are disseminating to various regions worldwide (Mutsvairo et al., 2019). However, while fact-checking as a research topic holds relevance across borders and is inherently comparative, there is a notable absence of systematic and comprehensive comparative projects, with only a few exceptions (see e.g., Graves & Cherubini, 2016; Humprecht, 2019, 2020). The absence of crucial academic geopolitical diversity across several analytical levels has long been a topic of discussion within communication studies (Demeter et al., 2022). Scholars actively strive to internationalize the field and overcome geographic disparities in knowledge production, deemed a form of "epistemic violence" (Ekdale et al., 2022, p. 1944). Deep internationalization entails adopting more inclusive approaches to foster comprehensive knowledge that reflects global perspectives (Badr & Ganter, 2021). This cosmopolitan perspective involves moving beyond Anglo-American and Eurocentric viewpoints in both knowledge production and analytical tools.

Latin American countries, despite internal diversity, share a relatively homogeneous historical, cultural, and linguistic background (Amazeen, 2020; Graves, 2018; Lawrence & Schafer, 2012). Considered a noncore region in media and communication studies, scholars have emphasized the peripheral status of Latin America in research designs, underscoring its practical indivisibility from European media and communication studies (Ganter & Ortega, 2019). Thus, this chapter aims to broaden the scope of analysis in fact-checking research by exploring Latin American perspectives. The objective is to observe the influence of the Latin American context on the fact-checking subfield, contributing to alternative knowledge in this area. Latin America presents a compelling case for studying the work of fact-checkers due to a unique combination of factors (Cazzamatta et al., 2024): extremely low resilience to disinformation, high Internet penetration, a preference for social media over television as a primary news source (Newman et al., 2023), and widespread use of WhatsApp (Mello, 2023).

Regarding fact-checking practices, organizations positioned beyond Western regions have encountered enduring challenges, such as insufficient or inappropriate resources, coupled with political opposition that jeopardizes the continuity of their operations (Palau-Sampio, 2018; Vinhas & Bastos, 2025). Furthermore, fact-checkers are confronted with varying levels of national disinformation resilience (Humprecht et al., 2020; Rodríguez-Pérez & García-Vargas, 2021). Building upon the argument presented by Wasserman and Madrid-Morales (2022) regarding disinformation as a context-dependent concept, the fact-checking movement is similarly influenced by local specificities. It must navigate challenges inherent in power dynamics within national information ecosystems. Nonetheless, a recent study by Nieminen and Rapeli (2019) highlighted that the fact-checking literature had primarily focused on actors in the USA. Their article revealed that 77% of the analyzed studies centered on the US-American context. In recent times, a handful of studies have broadened their scope by examining fact-checking projects in Latin America (Cazzamatta & Santos, 2024; Damasceno, 2022; Lelo, 2022; Moreno-Gil et al., 2021; Palau-Sampio, 2018; Recuero et al., 2022; Rodríguez-Pérez & Seibt, 2022). However, it is worth noting that most studies on fact-checking and the epistemologies of digital journalism (see Ekström et al., 2020; Steensen et al., 2024) still remain heavily concentrated on the Global North.

Ganter and Ortega (2019) emphasized that discussing Latin American contexts, as proposed here, is more prevalent than integrating scholarly work from within the region into intellectual discourses. Latin American scholars have a history of exploring imported ideas by interpreting them through local lenses, adapting arguments generated elsewhere to fit local realities, and scrutinizing the relevance of Western theories to address local issues (Waisbord, 2022; see also Averbeck-Lietz et al. in this book). Nonetheless, fact-checking research is also an emerging field in Latin America. By looking at Scielo—a bibliographic database of open-access journals with a high focus on Latin American production (although Spain, Portugal, and South Africa are also included)—only 29 studies containing the word "fact-checking" are available. Most are related to disinformation or online political communication (de Albuquerque et al., 2023), not necessarily to fact-checking practices. Considering that "[d]e-westernization interrogates the provenance and the positionality of academic knowledge" (Waisbord, 2022, p. 26), it is crucial to acknowledge here that the Latin American voices and authors included in this chapter are primarily based in the Global North or published in major Western academic journals.

That said, this chapter will present the work and difficulties faced by fact-checkers operating within the Latin American context in an attempt to build a more comprehensive and global network of analysis. First, I will provide a general introduction, offering an overview of the main issues within fact-checking studies. Before delving into the Latin American fact-checking movement and its challenges, I will discuss the idiosyncrasies of Latin American media systems these organizations must navigate and other digital indicators that make the region a disinformation laboratory.

Establishing the context: The emergence of fact-checking practices and their diversity worldwide

Fact-checking comes into play when information, whether in the form of public statements or content on digital platforms, has already made a significant social impact, reverberated in public debates, or gained substantial traction in the online environment (Rodríguez-Pérez & Seibt, 2022).[1] Fact-checking organizations endeavor to identify, verify, correct, and diminish the visibility of misinformation, occasionally collaborating with governmental entities and platform companies (Bélair-Gagnon et al., 2022). However, since 2016, there has been a notable reorientation of fact-checking organizations worldwide. The shift has moved from verifying statements made by politicians and public figures to actively monitoring and addressing the spread of viral misinformation on social networks (Bélair-Gagnon et al., 2023; Cazzamatta, 2024; Cazzamatta & Santos, 2024; Graves et al., 2023).

Organizations are classified into newsroom (in-house) and independent non-governmental organization (NGO) models (Graves & Cherubini, 2016). The former is associated with conventional media, prevalent in Northwestern Europe and the USA. Despite having more significant resources, these fact-checking efforts are constrained by the editorial dependence of their outlets. The latter, those that are non-profit and independent, predominant in Eastern Europe and Latin America, often collaborate with traditional news outlets to increase their reach (Palau-Sampio, 2018; Vinhas & Bastos, 2025). They are free from corporate constraints and typically receive support from foundations and organizations dedicated to advancing democratic institutions. The fact-checking literature generally does not distinguish the types of media ties of the agencies. In the case of

1 In addition to their verification work, fact-checking organizations can also engage in various other activities, including media literacy projects (e.g., Chequeado, Correctiv), research, such as the development of AI tools (Chequeado, Full Fact, and Africa Check), and even participation in policymaking (Full Fact).

Latin America however, it is crucial to observe if these units are linked to corporate, established media or one of several independent digital-native media sites that have multiplied in the region, such as Animal Politico from Mexico (with the fact-checking unit El Sabueso) or La Silla Vacía (Detector de Mentiras) from Colombia. They are not nonprofit NGOs but still maintain an independent character.

Scholars have noted the diversity of actors involved in the fact-checking movement. Global fact-checking has emerged as a hybrid institution, encompassing not only news organizations but also academic, political, and civil society groups. This diverse assembly of individuals and organizations, which does not conform to a uniform structure according to the new institutionalist perspective focusing on balance, alignment, stability, and consistency (Reese, 2022), advocates for enhanced journalism and accountability reporting. It surpasses traditional professional jurisdictional conflicts, simultaneously broadening institutional boundaries (Reese, 2022).

Over the past decade, the USA-based Poynter Institute has organized an annual event known as the Global Fact Summit, bringing together professional actors from around the world. Additionally, the International Fact-Checking Network (IFCN) at Poynter has developed a code of principles to guide best practices in fact-checking globally. It is reasonable to assert that the IFCN has significantly contributed to institutionalizing and professionalizing the fact-checking movement (Graves & Lauer, 2020). Certainly, this process inevitably involves layering imported journalism values onto existing institutional arrangements within countries and regions.

Fact-checkers strive to establish trust in their roles and uphold their authority by being members of the IFCN and disclosing their methodologies. The IFCN guidelines establish boundaries defining organizations committed to truth-seeking and transparency. This certification is particularly crucial in contexts where state-operated fact-checkers align with domestic false narratives to pave the way for enacting restrictive legislation, as Schuldt (2021) identified in the cases of Singapore, Thailand, and Malaysia. Some countries, including those in Latin America, have also faced challenges from "fake" fact-checkers—pseudo-initiatives that adopt the strategy, aesthetics, and style of legitimate verified fact-checkers, leading their audience to perceive them as credible sources of serious journalism, even though they are disseminating misinformation (Montaña-Niño et al., 2024; Moshirnia, 2020).

The generally rapid proliferation of fact-checking practices can be attributed to their alignment with the fundamental principles of professional journalism. Within the US-American context, it has also been perceived as an "interpretative turn" (Graves, 2016), challenging the traditional and orthodox understanding of objectivity. In contrast to other journalistic practices, fact-checking provides verdicts involving interpretative evaluation and deviating from the conventional comprehension and operationalization of objectivity rituals (Maras, 2013; Tuchman, 1972). While this interpretation is plausible and relevant in the US-American context, Latin American countries have long been dominated by powerful elites and oligarchs (Harlow, 2023; Waisbord, 2000), and the region had been challenging the traditional notion of objectivity long before the establishment of their first fact-checking agency. Latin American journalists have consistently regarded "objectivity" as a myth, replacing this norm with "an obligation to inform honestly, without consideration of secondary interests and in obedience to the public interest" (Restrepo, 1999, p. 230), demonstrating how universal values can be adapted to local realities. By analyzing previous Latin American codes of ethics, Restrepo observed that "neutrality is [was] read as a subtle but effective commitment to those who have the power" (p. 229) since only transmitting claims is not enough to make sense of the world events—a criticism regarding the operationalization of objectivity which the US-American fact-checkers engaged much later.

Latin American media systems, their extremely low resilience to disinformation, and its impact on fact-checking

The increasing adoption of fact-checking initiatives by traditional media outlets, along with collaborative efforts between independent organizations and established media entities, underscores the importance of examining the media system in which these organizations operate. While sharing some traits with the Mediterranean model (Hallin & Mancini, 2004), Latin America is broadly classified as "liberal captured," characterized by deregulation, dominance of conglomerates, high market concentration, regulatory inefficiencies, limited professionalization, discretionary allocation of public funds for advertising, and political influence (Echeverria et al., 2024; Guerrero, 2014). Media outlets, especially in Mexico, Brazil, and Argentina, have thrived with minimal regulatory frameworks driven by tightly managed, family-run enterprises (Becerra & Mastrini, 2017; Fuchs, 2021). Despite the efforts of scholars to classify Latin American media systems, typologize Latin American media systems is challenging due to diverse journalistic professionalism within the same country, intranational differences resembling liberal and polarized pluralist models, and greater instability with frequent changes between media and political systems compared with Europe or the USA (Echeverria et al., 2024; Hallin, 2020).

Furthermore, in the current digital media landscape, tech companies have taken a central role in shaping media structures by overseeing the platforms in which digital content is curated (Hallin, 2020). Thus, several studies have reflected on the necessity of adaptation, reconceptualization, and inclusion of new indicators (Hallin, 2020; Humprecht et al., 2020; Mancini, 2020; Mattoni & Ceccobelli, 2018). A helpful framework for cross-national comparative research to assess disinformation resilience, intrinsically associated with challenges faced by fact-checkers, was proposed by Humprecht and colleagues (2020) based on seven macro-level indicators: polarization of society, populist communication, weak public service broadcasting, fragmented audiences, size of the digital advertisement market, and high social media use. These variables were initially tested across 18 Western democracies and subsequently replicated for Latin American countries (Rodríguez-Pérez & García-Vargas, 2021), thus enhancing and enriching this framework (Wang, 2011). The operationalization of these variables within Latin America has demonstrated that the continent is much less resilient to disinformation compared with Europe or the USA, significantly impacting the daily activities of fact-checkers.

As an example, fact-checking practices are unquestionably influenced by polarization. Several analyses have shown that individuals tend to prefer fact-checks that align with their preexisting beliefs and are more likely to avoid those that contradict their political views, indicating a confirmation bias in the selection of corrective messages (see e.g., Hameleers & van der Meer, 2020; Shin & Thorson, 2017). Furthermore, other studies have observed that partisans often distrust fact-checkers and question their impartiality (see e.g., Young et al., 2018). Although polarization is a global phenomenon, data from the Varieties of Democracy (V-Dem) project[2] have revealed a significant increase in polarization over the past two decades in Latin America, making it one of the world's most polarized regions, ranking only behind Eastern Europe and Central Asia (United Nations Development Program [UNDP], 2023).

Studies have additionally associated the escalation of disinformation with the emergence of far-right populism[3] (see e.g., Bennett & Livingston, 2018; Hallin, 2019; Hameleers, 2020; Hameleers & Minihold, 2022; Marwick & Lewis, 2017). This trend is also evident in Brazil (Recuero et al., 2022; Santini et al., 2021) and across Latin America (Lupu et al., 2020). Right-wing politicians employ discourses deviating from established institutional norms, neglecting the importance of facts (Frankfurt, 2005) and impeding information and data accessibility. Given that fact-checkers can only verify factual statements and not opinions, the core of their epistemic work is significantly impacted in the face of heightened levels of populism and the corresponding lack of public accountability.

High social media use for news also has consequences for fact-checking practices, influencing decisions on which platforms to closely monitor and how disinformation spreads and goes viral. Widespread reliance on social media, with its potential to amplify online falsehoods, heightens the likelihood of populist disinformation agents building networks of followers who share their partisan perspectives, further reinforcing fragmentation. When measuring the daily time spent on social media per country, Latin American nations occupy prominent positions on the global list. Brazil ranks second, followed by Colombia (6th), Chile (7th), Mexico (9th), and Argentina (11th). Social media use for news in Latin America, in contrast to Europe, has outpaced TV as a source of news (Newman et al., 2023), leading to the uncontrolled dissemination of disinformation.

The lack of media trust is also concerning, as it often prompts individuals to resort to alternative, often partisan news sources, fostering distrust in fact-checking messages and reinforcing social media use. Illustratively, during Chile's *estallido social* (social outbreak) in 2019, television faced unprecedented criticism, despite increased viewership. Widespread distrust of traditional journalism led many Chileans to shift their attention to social media and messaging apps (Bachmann et al., 2022). In 2020, Chile emerged as one of the few countries globally where individuals generally placed more trust in information from social media (34%) than traditional media outlets (30%) (Newman et al., 2020).

This intricate landscape marked by partisan media ownership and extremely low disinformation resilience underscores Latin America as a fertile ground for disinformation,

2 The V-Dem Project, organized by the University of Gothenburg in Sweden, is a research initiative that offers a comprehensive dataset on democracy worldwide, focusing on the diversity of democratic practices and experiences across various countries.

3 For a comprehensive definition of populism in several contexts, see Kaltwasser et al. (2017).

which fact-checkers strive to counteract. Thus, we will delve into the difficulties faced by Latin American fact-checkers in navigating the described informational ecosystems.

The Latin American fact-checking movement and its challenges

Landscape structure: Digital native start-ups x corporate media

Different from the USA or Western Europe, the significant majority of Latin American fact-checkers are associated with independent fact-checking projects or digital native media sites (Graves & Cherubini, 2016), except in Brazil and Chile, where traditional established media are increasingly embracing this evolving journalistic genre. Considering the region's media landscape, this structural pattern makes sense, as established media tend to be financially dependent on the state or economic elites (Hallin, 2020).

The pioneering platform in Latin America since 2010 has been Argentina's Chequeado, serving as a model for similar flagship projects in other countries. For instance, Brazil's Agência Lupa was founded in 2015, followed by Ecuador's Chequea and ColombiaCheck in 2016. Chile experienced a surge in fact-checking platforms in 2019, driven by massive demonstrations and increasing political polarization. In the same year, Agence France-Presse expanded its multilingual fact-checking service to include Uruguay, adding to its existing presence in Argentina, Brazil, Colombia, and Mexico. More recent projects include DeFacto from the Cuban news portal elTOQUE and Ecuador Verifica, both established in 2021. Latin America is now home to 48 fact-checking units across 13 countries, making up approximately 12.6% of the total number of such organizations worldwide (Stencel et al., 2023).

Chile boasts 12 fact-checking projects, while Brazil has nine, making them the countries with the most significant number of such initiatives (Stencel et al., 2023). The involvement of major press organizations in fact-checking efforts in these two nations has raised concerns. However, it is important to note some distinctions. In Brazil, certain media-affiliated fact-checking projects are still signatories of the IFCN. Only the two associated with the major media conglomerates in the country, Globo and Record, are not part of the IFCN. In an interview with the author, an IFCN external assessor noted:

> Concerning Brazil, yes, they are associated with the IFCN, but it is still difficult because it took a while. I am telling you about the transparency issue; it took a while for the outlets to manage to have some level of transparency suitable for the IFCN. Historically, the media in Brazil has not been very transparent. So, when it comes to them talking about revenue, staff, everything else ... Some may even reach a minimum threshold there, but it is still a historical problem. (M. Crispim, personal communication, August 28, 2023)

In Chile, however, at the time of writing, no organization with media ties has reached the minimum standard of transparency required to become an IFCN signatory. Only independent fact-checking initiatives such as FastCheckCl and Mala Espina Check have done so. This lack of transparency can be attributed, in part, to the unique press duopoly in Chile where two journalistic consortiums hold the most influence (Mellado et al., 2018; Sapiezynska & Lagos, 2016). Additionally, most fact-checking projects in the country are relatively new and gained momentum during the 2019 social unrest. It is possible that they need more time to adapt, similar to their counterparts in Brazil. Some scholars from Latin America have also observed that fact-checking has become a new business within journalism. Traditional newsrooms in the region have mobilized toward this new genre as an alternative means to combat disinformation campaigns (Lelo, 2022; Rodríguez-Pérez & Seibt, 2022). A fact-checker involved in a university project in Chile remarked:

> So, they [corporate media] are dealing with survival problems because they have many problems in the economic model here in Chile, and they are trying to prove something, que "les dé el éxito" [that will give them success]. Okay, today is fact-checking. Tomorrow is data journalism. Next week will be another special and pyrotechnic thing they must try to survive. Nevertheless, maybe they are not reflecting on this kind of exercise, discipline [transparency of fact-checking methods]. (C. R. F. Alarcon, personal communication, July 15, 2023)

However, Brazil and Chile are exceptions on the continent. In general, as I stated somewhere else, independent organizations—even in these two states—are more actively engaged in the general fact-checking discourse. Fact-checkers affiliated with media partners tend to adopt a more cautious and moderate approach when articulating their primary goals (e.g., combat the spread of falsehoods). In contrast, independent or NGO-associated agencies focus more on advancing specific causes and driving reforms. They expect to increase the consequences of spreading falsehoods, enhance the quality of public discourse, safeguard freedom of expression, empower citizens through information, bolster democratic participation and human rights, fortify democracy, scrutinize those in power, demand transparency from governments, contribute to investigative journalism, and amplify marginalized voices. Nonetheless, they face considerable problems in promoting such goals. Research evaluating the values and operations of fact-checking platforms in Latin America and Spain has identified three significant challenges: access to public data (due to availability and reliability), resource constraints, and limitations in reaching a broader audience (Lelo, 2022; Moreno-Gil et al., 2021).

Lack of information availability and factual accountability

As described earlier, fragmented media ecosystems with limited effective gatekeepers contribute to the proliferation of misperceptions driven by populist rhetoric. This trend is not limited to prominent examples like Bolsonaro's administration in Brazil, but is also evident in other countries, including Bolivia, Chile, Colombia, Costa Rica, Peru, and Uruguay. In these nations, there has been a rise in right-wing populist rhetoric characterized by conspiracy theories, rumors, science denialism, partisanship, and anti-media attitudes (Sanahuja et al., 2023; Siles et al., 2021). Given this context, it becomes crucial to assess the extent to which Latin American politicians and institutions still adhere to the norms of truth-telling.

Graves and Wells (2019) delved into the consequences of political discourse that has become disconnected from established institutional standards and practices within a media landscape driven by the attention economy. They drew a clear distinction between information accessibility (the ability to access facts and public records) and the responsibility for "factual accountability," which relates to the importance of facts and truth in society. They defined accountability as "a quality of public discourse grounded in a range of norms and institutional mechanisms" (p. 45). In an era characterized by post-truth dynamics, marked by populist communication and hostility toward the press, politicians and democratic institutions may no longer consistently adhere to truth-telling standards. Latin America is significantly affected by both dimensions of this problem. Natália Leal,[4] the content director at Agência Lupa, has shed light on this issue:

> We have some difficulties here in Brazil and Latin America.... We have a lack of transparency. A good part of the governments is hiding data. They are trying in some way to prevent access to the data that is necessary for our work and the work of all journalists. There is an attitude of these governments, which is very common here in Brazil. A position of creating their own facts, distorting reality to tell another story, and making a different narrative....

Because these professionals aim to verify the accuracy of political statements by relying on independent and generally trusted qualified sources (Graves, 2016), this erosion of factual accountability has significant implications for the work of fact-checkers. While the erosion of factual accountability is also a concern in the USA and other Western nations, these countries still maintain their high levels of "rational legal authority" (Mancini, 2013, p. 35). For instance, in the UK, there is a Statistic Authority that fact-checkers can use to compel politicians and members of parliament to correct their statements (Graves & Cherubini, 2016). In contrast, in Latin America, it appears that politicians may be learning in their media training how to circumvent fact-checkers rather than prioritizing accuracy and adherence to facts, an external IFCN assessor suspects (M. Crispim, personal communication, August 28, 2023). Additionally, the levels of investigation and research required to collect public information in Latin America differ significantly from those of countries with stable and well-established rules for overseeing public information and records.

Fact-checking, akin to data journalism, constitutes an intervention reliant on the transparency of public institutions in data provision. This prerequisite is not consistently met, even in nations with a longstanding tradition of legislation supporting freedom of information, such as Spain or Sweden (see e.g., Appelgren & Salaverría, 2018). Fact-checkers encounter a demanding and time-consuming process of requesting access from authorities and acquiring data in non-machine-readable formats in the most adverse situations.

4 She made this statement during an online fact-checking course organized by the IFCN in 2021.

Online harassment

Another significant challenge, in addition to the lack of data and the erosion of factual accountability, is the prevalence of massive social media violence. Studies indicate that online harassment extends across news production (see e.g., Harlow et al., 2023). New actors, such as social media agent provocateurs, often acting on behalf of governments and political parties, engage in coordinated harassment campaigns to influence the narrative on social media (Harlow et al., 2023). A fact-checker from Columbiacheck highlighted, during an online fact-checking course organized by the IFCN in 2021, the challenges they face: "We have seen a series of coordinated attacks, especially from far-right groups that seek to discredit our work by spreading lies about our funding, purposes, and interests." Furthermore, an external IFCN assessor stressed the need for localized assessments in the face of these difficulties. For example, in evaluating the Brazilian agency Aos Fatos, it was noted that they omitted employee biographies due to harassment concerns, despite it being an IFCN requirement: "I justified it in the assessment" (M. Crispim, personal communication, August 28, 2023).

The issue of social media attacks extends beyond individual readers expressing dissatisfaction. In fact, journalists perceive these threats as part of systematic and organized campaigns orchestrated by governments and supported by their partisan followers. This phenomenon has been particularly prevalent in El Salvador, Honduras, Nicaragua, and Brazil—countries with a history of journalistic harassment (Harlow et al., 2023). This problem also extends to the fact-checking community. In 2018, when several organizations in Brazil began collaborating with Facebook, a PDF file containing comprehensive details about 40 Brazilian journalists, including screenshots from their profiles, circulated as "evidence" of their supposed left-leaning orientations. This PDF file went viral among right-wing circles on WhatsApp, which has 120 million users in Brazil. The entire fact-checking teams of the Lupa and Aos Fatos agencies were featured in the document, leading to trolling and threats (Funke, 2018).

Resources and reach through platform partnerships?

In addition to the erosion of factual accountability, unreliable data, and online harassment, Latin American fact-checkers also have to secure financial resources and a broader readership: "I think the problem in Latin America . . . is how to get fact-checks to people who are not looking for them."[5] Some platform partnerships partially target these two main problems: financial sources and audience reach. However, various scholars are critical of these partnerships. Lelo (2022) observed, for instance, in the Brazilian context, that, due to a restricted number of staff and the growing financial dependence on tech companies, fact-checking organizations are modulating their verification to different platforms and prioritizing online debunking and neglecting political fact-checking. Although this type of collaboration between tech platforms and fact-checkers needs to be critically investigated (Bélair-Gagnon et al., 2023; Cazzamatta & Santos, 2024 Full Fact,

5 P. Uribe made this statement during an online fact-checking course organized by the IFCN in 2021.

2020; Graves et al., 2023), attributing the "debunking turn" (social media policing) solely to these collaborations seems too simplistic (Cazzamatta, 2024).

First, there is a monthly cap on revenue generated from paid debunking articles. Consequently, organizations cannot endlessly profit from debunking misinformation on Facebook (Bélair-Gagnon et al., 2023). Second, the observed "debunking turn" is a global trend that cuts across various types of organizations and countries. While it is plausible to hypothesize that Latin America (as well as other Global South countries) may produce fewer fact-checks (verification of public figures' statements), it is essential to recognize that the reasons should not be solely attributed to the platforms' partnerships. As demonstrated earlier, Latin America faces challenges such as weaker press development (Hallin & Papathanassopoulos, 2002) and extremely low resilience to disinformation (Rodríguez-Pérez & García-Vargas, 2021). The region experiences high social media usage, with some countries ranking among the world's top users. Social media has even surpassed television as a news source in several countries. Additionally, systematic online disinformation campaigns orchestrated by far-right movements and populist politicians are typically anonymous and challenging to detect (Lupu et al., 2020). Considering this dysfunctional online media environment underscores the importance of focusing on debunking to promote a healthier and more truthful online information environment in a broader sense (Palau-Sampio, 2018). Mark Zuckerberg's decision to replace Facebook's fact-checking partnerships with a community-based verification model, similar to Musk's approach with X, necessitates urgent scholarly analysis. Given the structural constraints discussed in this chapter, Latin American organizations are likely to be more affected than their European counterparts.

More problematic than supporting fact-checking organizations is the attempt of platforms to hinder Internet regulation laws in the region. Latin American states have less leverage to hold big tech companies accountable: "The European Union has the capacity to impose regulations on the platform that countries in Latin America or Africa do not" (Hallin, 2020, p. 5782). The most striking example is the platforms' reaction against the Brazilian bill PL2630, inspired by the German NetzDG and the European Digital Service Act (DSA). In the report titled "The Platforms' War against PL2630,"—produced by the independent think tank InternetLab, focused on digital technologies,—it is described how platforms attempted to influence public opinion days before the parliamentary vote, employing every possible resource to prevent the bill's approval (NetLab, 2023). At the core of this issue is the massive capital generated through digital advertising (huge in the region) in an industry currently lacking regulations and transparency (NetLab, 2023). Regardless of their partnerships, independent fact-checkers were actively involved in the discussions around the draft bill. They suggested improving the regulatory proposal (Aos Fatos & Lupa, 2023), taking a different stance from that of the platforms. In their manifesto on the matter, they made it clear that it is urgent to limit financial incentives for the dissemination of hatred and disinformation (Aos Fatos & Lupa, 2023). In the same way that collaborations with platforms seem problematic, the same can be said about partnerships with established media. It is more challenging to establish media alliances when prominent news organizations have ties with political affiliations and oligarchs (Graves & Wells, 2019).

Conclusion

In this chapter, I have underscored the challenges that Latin American fact-checkers encounter within the context of their media systems and various aspects of their online information environments. My aim is to contribute to global knowledge production by addressing geographic disparities in the field and to demonstrate that research on the fact-checking done in the Global South is equally relevant to that which occurs in the USA or Europe and necessitates no further specific justification. Although fact-checking practices originated in the USA, they have since spread globally, with fact-checkers emerging in diverse contexts and facing varying challenges that require academic attention. At the research level, discussing a more cosmopolitan approach to fact-checking practices in Latin America poses a significant challenge.

First, there needs to be more comparative analyses within the continent, as most literature still focuses on Western nations. Additionally, existing studies on the topic are produced mainly by Latin American scholars affiliated with Global North institutions, including the author herself, with only a few exceptions. Thus, the scarcity of voices in and within the continent (Mutsvairo et al., 2021) is much more pronounced. In respect to their specific fact-checking practices, Latin American fact-checkers face similar challenges to their global counterparts, including financial constraints, unreliable data, and limited audience reach. However, they have not consistently had the same access to infrastructure, financial resources, information, or capacity-building opportunities as organizations in Europe or the USA (Mutsvairo et al., 2019).

Second, as demonstrated throughout the chapter, universalizing approaches to media systems provide limited insights into fact-checking practices (see also Radue et al. on media systems in this book). Despite shared characteristics such as media concentration, absence of public service broadcasting, persistent patronage, and instrumentalization of journalism, there is substantial diversity within the Latin American media systems. For example, the media systems in Uruguay and Chile are quite different from those in Venezuela and Cuba. Considerable diversity in journalistic culture (Hanitzsch et al., 2019) is also evident within a single country and journalistic genre, such as fact-checking initiatives. Furthermore, understanding the daily challenges of fact-checking solely in light of media system traits disregards digital developments, such as high social media use for news. In Latin America, unlike Europe, social media have surpassed television as a primary news source (Newman et al., 2023), contributing to the uncontrollable spread of disinformation.

Transparency, impartiality, and fairness, as critical concepts transferred to the fact-checking culture, can be influenced by different macro-media system contexts and meso/micro perspectives from diverse fact-checking cultures. The criticism regarding the operationalization of objectivity understood as an "interpretative turn" in journalism with the emergence of US-American fact-checkers in the early 2000s has long been a tradition in Latin American journalism. Despite Latin American journalists embracing traditional norms rooted in a more liberal journalistic tradition, they are simultaneously largely supportive of a more active role in social change (Hanitzsch et al., 2019). The same can be applied to the fact-checking movement. Studies have demonstrated that fact-checkers in Latin America characterize their work as aligned with journalism's societal responsibility and in contrast to political polarization (Lelo, 2022; Moreno-Gil et al., 2021; Vinhas & Bastos, 2025).

Therefore, it is essential to recognize distinct fact-checking cultures that have emerged within traditional established media corporations, independent digital native media start-ups, or NGO-based organizations. As mentioned earlier, independent fact-checkers articulate their mission more engagingly, emphasizing goals such as improving democracy, media literacy, and the quality of public debate, in contrast to their counterparts in mainstream media, who adopt milder statements, focusing primarily on countering misinformation. This distinction may stem from issues such as media capture and the political instrumentalization of journalism. In Latin America, fact-checking is predominantly practiced by independent (media) organizations, with Chile and Brazil standing out as exceptions where corporate media have also embraced the genre.

Due to limited access to information and public data, which are fundamental requirements for fact-checking practices, political fact-checking in the region often takes on the character of investigative journalism. For instance, when data on unemployment, inflation, and other indicators are either unavailable or manipulated by governments, fact-checkers must develop their own indices or undertake independent investigations to gather accurate data. Without public data, fact-checkers must establish a database with relevant information gathered through bureaucratic mechanisms and investigative practices. Consequently, outside of Europe and North America, the fact-checking movement is primarily associated with independent media or investigative journalism initiatives. Organizations operating in countries with weaker rational-legal authority and data accountability must establish a professional relationship with the politicians they cover. They frequently encounter online threats (Harlow et al., 2023) and accusations of bias. Collaborations with established traditional media outlets also prove challenging given the close ties between prominent media and political elites. In this context, these fact-checkers lack a shared pool of public data for establishing institutionally recognized facts (Graves & Wells, 2019) and the media platforms to disseminate those facts effectively.

In summary, these organizations aim to establish a public service by prioritizing a civic agenda and holding political power accountable in a communication environment characterized by media concentration, commercial interests, patronage (Moreno-Gil et al., 2021), and extremely low resilience to disinformation (Rodríguez-Pérez & García-Vargas, 2021). Hence, research should adopt a more cosmopolitan approach, acknowledging the significance of macro contexts (media system traits influenced by digital indicators) and the high heterogeneity of fact-checking cultures across countries and organizations, especially by extending the research beyond the Western world. Despite the global nature of the fact-checking movement, one should critically consider global differences and seek contextual knowledge.

References

Adair, B. (2014, April 4). Duke study finds fact-checking growing around the world. *Duke Reporters' LAB*. https://reporterslab.org/duke-study-finds-fact-checking-growing-around-the-world/

Amazeen, M. A. (2020). Journalistic interventions: The structural factors affecting the global emergence of fact-checking. *Journalism, 21*(1), 95–111. https://doi.org/10.1177/1464884917730217

Aos Fatos, & Lupa. (2023, April 26). *Aos Fatos e Lupa apresentam propostaspara aperfeiçoar regulação das redes*. Aos Fatos. https://www.aosfatos.org/noticias/aos-fatos-e-lupa-apresentam-propostas-para-aperfeicoar-regulacao-das-redes/

Appelgren, E., & Salaverría, R. (2018). The promise of the transparency culture: A comparative study of access to public data in Spanish and Swedish newsrooms. *Journalism Practice, 12*(8), 986–996. https://doi.org/10.1080/17512786.2018.1511823

Bachmann, I., Grassau, D., & Labarca, C. (2022). Aliens, spies, and staged vandalism: Disinformation in the 2019 protests in Chile. In H. Wasserman & D. Madrid-Morales (Eds.), *Disinformation in the Global South* (pp. 74–87). Wiley Blackwell. https://doi.org/10.1002/9781119714491.ch6

Badr, H., & Ganter, S. A. (2021). Towards cosmopolitan media and communication studies: Bringing diverse epistemic perspectives into the field. *Global Media Journal – German Edition, 11*(1). https://doi.org/10.22032/dbt.49164

Becerra, M., & Mastrini, G. (2017). *La concentración infocomunicacional en América Latina (2000–2015): Nuevos medios y tecnologías, menos actores*. Universidad Nacional de Quilmes.

Bélair-Gagnon, V., Graves, L., Kalsnes, B., Steensen, S., & Westlund, O. (2022). Considering interinstitutional visibilities in combating misinformation. *Digital Journalism, 10*(5), 669–678. https://doi.org/10.1080/21670811.2022.2072923

Bélair-Gagnon, V., Larsen, R., Graves, L., & Westlund, O. (2023). Knowledge work in platform fact-checking partnerships. *International Journal of Communication, 17*, 1169–1189. https://ijoc.org/index.php/ijoc/article/view/19851

Bennett, W. L., & Livingston, S. (2018). The disinformation order: Disruptive communication and the decline of democratic institutions. *European Journal of Communication, 33*(2), 122–139. https://doi.org/10.1177/0267323118760317

Cazzamatta, R. (2024). The content homogenization of fact-checking through platform partnerships: A comparison between eight countries. *Journalism & Mass Communication Quarterly*. Advance online publication. https://doi.org/10.1177/10776990241261725

Cazzamatta, R., & Santos, A. (2024). Checking verifications during the 2022 Brazilian run-off election: How fact-checking organizations exposed falsehoods and contributed to the accuracy of the public debate. *Journalism, 25*(10), 2022–2043. https://doi.org/10.1177/14648849231196080

Cazzamatta, R., Santos, A., & Albuquerque, G. (2024). Unveiling disinformation: Mapping attacks on Brazil's electoral system and the response of the superior electoral court (2018–2023). *International Journal of Communication, 18*, 3551–3575. https://ijoc.org/index.php/ijoc/article/view/23008/

Damasceno, D. de R. (2022). *Marcas da mediatização no jornalismo fact-checking: Um estudo sobre a cobertura da pandemia de covid-19 realizada por Agência Lupa e Aos Fatos* [Doctoral dissertation, Universidade Federal da Bahia]. Repositorio. https://repositorio.ufba.br/handle/ri/36046

de Albuquerque, A., Recuero, R., & Alves Dos Santos Junior, M. (2023). Online communication studies in Brazil: Origins and state of the art. *Online Media and Global Communication*, 2(1), 100–121. https://doi.org/10.1515/omgc-2022-0068

Demeter, M., Goyanes, M., Navarro, F., Mihalik, J., & Mellado, C. (2022). Rethinking de-Westernization in communication studies: The Ibero-American movement in international publishing. *International Journal of Communication*, *16*, 3027–3046. https://ijoc.org/index.php/ijoc/article/view/18485

Diani, M. (1992). The concept of social movement. *The Sociological Review*, 40(1), 1–25. https://doi.org/10.1111/j.1467-954X.1992.tb02943.x

Echeverria, M., González, R. A., & Reyna, V. H. (2024). Bringing history back into media systems theory: Multiple modernities and institutional legacies in Latin America. *International Journal of Press/Politics*, 29(4), 940–959. https://doi.org/10.1177/19401612221141315

Ekdale, B., Biddle, K., Tully, M., Asuman, M., & Rinaldi, A. (2022). Global disparities in knowledge production within journalism studies: Are special issues the answer? *Journalism Studies*, 23(15), 1942–1961. https://doi.org/10.1080/1461670X.2022.2123846

Ekström, M., Lewis, S. C., & Westlund, O. (2020). Epistemologies of digital journalism and the study of misinformation. *New Media & Society*, 22(2), 205–212. https://doi.org/10.1177/1461444819856914

Frankfurt, H. G. (2005). *On bullshit*. Princeton University Press.

Fuchs, G. (2021). Who controls Latin America's media? In an evolving media ecosystem, concentrated ownership persists as conglomerates scramble to adapt to the digital age. *NACLA Report on the Americas*, 53(4), 349–358. https://doi.org/10.1080/10714839.2021.2000761

Full Fact. (2020). *The challenges of online fact checking: How technology can (and can't) help*. https://fullfact.org/media/uploads/coof-2020.pdf

Funke, D. (2018, September 10). *These fact-checkers were attacked online after partnering with Facebook*. Poynter. https://www.poynter.org/fact-checking/2018/these-fact-checkers-were-attacked-online-after-partnering-with-facebook/

Ganter, S. A., & Ortega, F. (2019). The invisibility of Latin American scholarship in European media and communication studies: Challenges and opportunities of de-Westernization and academic cosmopolitanism. *International Journal of Communication*, *13*, 68–91. https://ijoc.org/index.php/ijoc/article/view/8449

Graves, L. (2016). *Deciding what's true: The rise of political fact-checking in American journalism*. Columbia University Press.

Graves, L. (2018). Boundaries not drawn: Mapping the institutional roots of the global fact-checking movement. *Journalism Studies*, 19(5), 613–631. https://doi.org/10.1080/1461670X.2016.1196602

Graves, L. (2022). Fact-checking movement. In G. A. Borchard & C. H. Sterling (Eds.), *The Sage encyclopedia of journalism* (2nd ed., pp. 631–634). Sage.

Graves, L., & Cherubini, F. (2016). *The rise of fact-checking sites in Europe*. Reuters Institute for the Study of Journalism. https://reutersinstitute.politics.ox.ac.uk/sites/default/files/research/files/The%2520Rise%2520of%2520Fact-Checking%2520Sites%2520in%2520Europe.pdf

Graves, L., & Lauer, L. (2020). From movement to institution: The "global fact" summit as a field-configuring event. *Sociologica*, 14(2), 157–174. https://doi.org/10.6092/ISSN.1971-8853/11154

Graves, L., & Wells, C. (2019). From information availability to factual accountability: Reconsidering how truth matters for politicians, publics, and the news media. In J. E. Katz & K. K. Mays (Eds.), *Journalism and truth in an age of social media* (pp. 39–57). Oxford University Press. https://doi.org/10.1093/oso/9780190900250.003.0004

Graves, L., Bélair-Gagnon, V., & Larsen, R. (2023). From public reason to public health: Professional implications of the "debunking turn" in the global fact-checking field. *Digital Journalism, 12*(10), 1417–1436. https://doi.org/10.1080/21670811.2023.2218454

Guerrero, M. A. (2014). The 'captured liberal' model of media systems in Latin America. In M. A. Guerrero & M. Márquez-Ramírez (Eds.), *Media systems and communication policies in Latin America* (pp. 43–65). Palgrave Macmillan. https://doi.org/10.1057/9781137409058_3

Hallin, D. C. (2019). Mediatisation, neoliberalism and populisms: The case of Trump. *Contemporary Social Science*, 14(1), 14–25. https://doi.org/10.1080/21582041.2018.1426404

Hallin, D. C. (2020). Comparative research, system change, and the complexity of media systems. *International Journal of Communication*, 14, 5775–5786. https://ijoc.org/index.php/ijoc/article/view/14550

Hallin, D. C., & Mancini, P. (2004). *Comparing media systems: Three models of media and politics*. Cambridge University Press. https://doi.org/10.1017/CBO9780511790867

Hallin, D. C., & Papathanassopoulos, S. (2002). Political clientelism and the media: Southern Europe and Latin America in comparative perspective. *Media, Culture & Society*, 24(2), 175–195. https://doi.org/10.1177/016344370202400202

Hameleers, M. (2020). Populist disinformation: Exploring intersections between online populism and disinformation in the US and the Netherlands. *Politics and Governance*, 8(1), 146–157. https://doi.org/10.17645/pag.v8i1.2478

Hameleers, M., & Minihold, S. (2022). Constructing discourses on (un)truthfulness: Attributions of reality, misinformation, and disinformation by politicians in a comparative social media setting. *Communication Research*, 49(8), 1176–1199. https://doi.org/10.1177/0093650220982762

Hameleers, M., & van der Meer, T. G. L. A. (2020). Misinformation and polarization in a high-choice media environment: How effective are political fact-checkers? *Communication Research*, 47(2), 227–250. https://doi.org/10.1177/0093650218819671

Hanitzsch, T., Hanusch, F., Ramaprasad, J., & de Beer, A. S. (Eds.). (2019). *Worlds of journalism: Journalistic cultures around the globe*. Columbia University Press. https://doi.org/10.7312/hani18642

Harlow, S. (2023). *Digital-native news and the remaking of Latin American mainstream and alternative journalism*. Routledge.

Harlow, S., Wallace, R., & Chacón, L. C. (2023). Digital (in)security in Latin America: The dimensions of social media violence against the press and journalists' coping strategies. *Digital Journalism, 11*(10), 1829–1847. https://doi.org/10.1080/21670811.2022.2128390

Humprecht, E. (2019). Where 'fake news' flourishes: A comparison across four Western democracies. *Information, Communication & Society, 22*(13), 1973–1988. https://doi.org/10.1080/1369118X.2018.1474241

Humprecht, E. (2020). How do they debunk "fake news"? A cross-national comparison of transparency in fact checks. *Digital Journalism, 8*(3), 310–327. https://doi.org/10.1080/21670811.2019.1691031

Humprecht, E., Esser, F., & Van Aelst, P. (2020). Resilience to online disinformation: A framework for cross-national comparative research. *International Journal of Press/Politics, 25*(3), 493–516. https://doi.org/10.1177/1940161219900126

Kaltwasser, C. R., Taggart, P., Espejo, P. O., & Ostiguy, P. (2017). Populism: An overview of the concept and the state of the art. In C. R. Kaltwasser, P. Taggart, P. O. Espejo, & P. Ostiguy (Eds.), *The Oxford handbook of populism* (pp. 1–24). Oxford University Press. https://doi.org/10.1093/oxfordhb/9780198803560.013.34

Lauer, L., & Graves, L. (2024). How to grow a transnational field: A network analysis of the global fact-checking movement. *New Media & Society*. Advance online publication. https://doi.org/10.1177/14614448241227856

Lawrence, R. G., & Schafer, M. L. (2012). Debunking Sarah Palin: Mainstream news coverage of 'death panels.' *Journalism, 13*(6), 766–782. https://doi.org/10.1177/1464884911431389

Lelo, T. (2022). The rise of the Brazilian fact-checking movement: Between economic sustainability and editorial independence. *Journalism Studies, 23*(9), 1077–1095. https://doi.org/10.1080/1461670X.2022.2069588

Lupu, N., Bustamante, M. V. R., & Zechmeister, E. J. (2020). Social media disruption: Messaging mistrust in Latin America. *Journal of Democracy, 31*(3), 160–171. https://muse.jhu.edu/article/760080

Mancini, P. (2013). Rational legal authority, formal and informal rules in the news media. In M. Price, S. Verhulst, & L. Morgan (Eds.), *Routledge handbook of media law* (pp. 35–46). Routledge.

Mancini, P. (2020). Comparing media systems and the digital age. *International Journal of Communication, 14*, 5761–5774. https://ijoc.org/index.php/ijoc/article/view/14553

Maras, S. (2013). *Objectivity in journalism: Keys concepts in journalism*. Polity Press.

Marwick, A., & Lewis, R. (2017). *Media manipulation and disinformation online*. Data & Society. https://datasociety.net/wp-content/uploads/2017/05/DataAndSociety_MediaManipulationAndDisinformationOnline-1.pdf

Mattoni, A., & Ceccobelli, D. (2018). Comparing hybrid media systems in the digital age: A theoretical framework for analysis. *European Journal of Communication, 33*(5), 540–557. https://doi.org/10.1177/0267323118784831

Mellado, C., Humanes, M. L., & Márquez-Ramírez, M. (2018). The influence of journalistic role performance on objective reporting: A comparative study of Chilean, Mexican, and Spanish news. *International Communication Gazette, 80*(3), 250–272. https://doi.org/10.1177/1748048517711673

Mello, P. C. (2023, November 5). Brasil é o país do WhatsApp', diz presidente do aplicativo. *Folha de São Paulo*. https://www1.folha.uol.com.br/mercado/2023/11/brasil-e-o-pais-do-whatsapp-diz-presidente-do-aplicativo.shtml

Montaña-Niño, S. X., Vziatysheva, V., Dehghan, E., Badola, A., Zhu, G., Vinhas, O., Riedlinger, M., & Glazunova, S. (2024). Fact-checkers on the fringe: Investigating methods and practices 7 associated with contested areas of fact-checking beyond the borders. *Media and Communication*, *12*, Article 8688. https://doi.org/10.17645/mac.8688.

Moreno-Gil, V., Ramon, X., & Rodríguez-Martínez, R. (2021). Fact-checking interventions as counteroffensives to disinformation growth: Standards, values, and practices in Latin America and Spain. *Media and Communication*, *9*(1), 251–263. https://doi.org/10.17645/mac.v9i1.3443

Moshirnia, A. (2020). Who will check the checkers? False factcheckers and memetic misinformation. *Utah Law Review*, *4*(4), 1029–1073. https://doi.org/10.26054/0DDVTA6FPV

Mutsvairo, B., Bebawi, S., & Borges-Rey, E. (Eds.). (2019). *Data journalism in the Global South*. Palgrave Macmillan. https://doi.org/10.1007/978-3-030-25177-2

Mutsvairo, B., Borges-Rey, E., Bebawi, S., Márquez-Ramírez, M., Mellado, C., Mabweazara, H. M., Demeter, M., Głowacki, M., Badr, H., & Thussu, D. (2021). Ontologies of journalism in the Global South. *Journalism & Mass Communication Quarterly*, *98*(4), 996–1016. https://doi.org/10.1177/10776990211048883

NetLab. (2023). *The war of the platforms against PL 2630*. https://www.netlab.eco.br/en/post/the-war-from-the-platforms-against-pl-2630

Newman, N., Fletcher, R., Eddy, K., Robertson, C. T., & Nielsen, R. K. (2023). *Digital news report 2023*. Reuters Institute for the Study of Journalism. https://reutersinstitute.politics.ox.ac.uk/sites/default/files/2023-06/Digital_News_Report_2023.pdf

Newman, N., Fletcher, R., Schulz, A., Andi, S., & Nielsen, R. K. (2020). *Reuters institute digital news report 2020*. Reuters Institute for the Study of Journalism. https://reutersinstitute.politics.ox.ac.uk/sites/default/files/2020-06/DNR_2020_FINAL.pdf

Nieminen, S., & Rapeli, L. (2019). Fighting misperceptions and doubting journalists' objectivity: A review of fact-checking literature. *Political Studies Review*, *17*(3), 296–309. https://doi.org/10.1177/1478929918786852

Palau-Sampio, D. (2018). Fact-checking and scrutiny of power: Supervision of public discourses in new media platforms from Latin America. *Communication & Society*, *31*(3), 347–363. https://doi.org/10.15581/003.31.3.347-363

Recuero, R., Soares, F. B., Vinhas, O., Volcan, T., Hüttner, L. R. G., & Silva, V. (2022). Bolsonaro and the far right: How disinformation about COVID-19 circulates on Facebook in Brazil. *International Journal of Communication*, *16*, 148–171. https://ijoc.org/index.php/ijoc/article/view/17724

Reese, S. D. (2022). The institution of journalism: Conceptualizing the press in a hybrid media system. *Digital Journalism*, *10*(2), 253–266. https://doi.org/10.1080/21670811.2021.1977669

Restrepo, J. D. (1999). Latin America: Dead letter codes. In M. Kunczik (Ed.), *Ethics in journalism: A reader on their perception in the Third World* (pp. 225–244). Friedrich Ebert Stiftung.

Rodríguez-Pérez, C., & García-Vargas, G. R. (2021). Understanding which factors promote exposure to online disinformation. In G. López-García, D. Palau-Sampio, B. Palomo, E. Campos-Domínguez, & P. Masip (Eds.), *Politics of disinformation: The influence of fake news on the public sphere* (pp. 173–186). Wiley Blackwell. https://doi.org/10.1002/9781119743347.ch13

Rodríguez-Pérez, C., & Seibt, T. (2022). The Brazilian fact-checkers criteria: An analysis of the guiding purposes, principles, and routines of this journalistic practice. *Brazilian Journalism Research, 18*(2), 350–373. https://doi.org/10.25200/BJR.v18n2.2022.1510

Sanahuja, J. A., Burian, C. L., & Vitelli, M. (2023). The rise of the new far right in Latin America: Crisis of globalization, authoritarian path dependence and civilian-military relations. In R. Pinheiro-Machado & T. Vargas-Maia (Eds.), *The rise of the radical right in the Global South* (pp. 112–124). Routledge.

Santini, R. M., Tucci, G., Salles, D., & de Almeida, A. R. D. (2021). Do you believe in fake after all? WhatsApp disinformation campaign during the Brazilian 2018 presidential election. In G. López-García, D. Palau-Sampio, B. Palomo, E. Campos-Domínguez, & P. Masip (Eds.), *Politics of disinformation: The influence of fake news on the public sphere* (pp. 51–66). Wiley Blackwell. https://doi.org/10.1002/9781119743347.ch4

Sapiezynska, E., & Lagos, C. (2016). Media freedom indexes in democracies: A critical perspective through the cases of Poland and Chile. *International Journal of Communication, 10*, 549–570. https://ijoc.org/index.php/ijoc/article/view/3679

Schuldt, L. (2021). Official truths in a war on fake news: Governmental fact-checking in Malaysia, Singapore, and Thailand. *Journal of Current Southeast Asian Affairs, 40*(2), 340–371. https://doi.org/10.1177/18681034211008908

Shin, J., & Thorson, K. (2017). Partisan selective sharing: The biased diffusion of fact-checking messages on social media. *Journal of Communication, 67*(2), 233–255. https://doi.org/10.1111/jcom.12284

Siles, I., Tristán, L., & Carazo, C. (2021). Populism, media, and misinformation in Latin America. In H. Tumber & S. Waisbord (Eds.), *The Routledge companion to media disinformation and populism* (pp. 356–365). Routledge.

Steensen, S., Kalsnes, B., & Westlund, O. (2024). The limits of live fact-checking: Epistemological consequences of introducing a breaking news logic to political fact-checking. *New Media & Society, 26*(11), 6347–6365. https://doi.org/10.1177/14614448231151436

Stencel, M., Ryan, E., & Luther, J. (2023, June 21). Misinformation spreads, but fact-checking has leveled off. *Duke Reporters' LAB*. https://reporterslab.org/tag/fact-checking-census/

Tuchman, G. (1972). Objectivity as strategic ritual: An examination of newsmen's notions of objectivity. *American Journal of Sociology, 77*(4), 660–679. https://www.jstor.org/stable/2776752

United Nations Development Programme. (2023, February 28). *"With me, or against me": The intensification of political polarization in Latin America and the Caribbean*. https://www.undp.org/latin-america/blog/me-or-against-me-intensification-political-polarization-latin-america-and-caribbean

Vinhas, O., & Bastos, M. (2025). When fact-checking is not WEIRD: Negotiating consensus outside Western, educated, industrialized, rich, and democratic countries.

International Journal of Press/Politics, 30(1), 256–276. https://doi.org/10.1177/1940161223 1221801

Waisbord, S. (2000). *Watchdog journalism in South America: News, accountability, and democracy*. Columbia University Press.

Waisbord, S. (2022). What is next for de-Westernizing communication studies? *Journal of Multicultural Discourses, 17*(1), 26–33. https://doi.org/10.1080/17447143.2022.204164 5

Wang, G. (2011). Beyond de-Westernizing communication research: An introduction. In G. Wang (Ed.), *De-Westernizing communication research: Altering questions and changing frameworks* (pp. 1–18). Routledge.

Wasserman, H., & Madrid-Morales, D. (Eds.). (2022). *Disinformation in the Global South*. Wiley Blackwell.

Young, D. G., Jamieson, K. H., Poulsen, S., & Goldring, A. (2018). Fact-checking effectiveness as a function of format and tone: Evaluating FactCheck.org and FlackCheck.org. *Journalism & Mass Communication Quarterly, 95*(1), 49–75. https://doi.org/10.1177/10776 99017710453

The industry known as 'media development'

Analyzing media assistance from a cosmopolitan perspective in mass communication

Susanne Fengler, Ines Drefs, Mira Keßler, Michel Leroy, Johanna Mack, Fabíola Ortiz dos Santos, Viviane Schönbächler, Stefan Wollnik & Roja Zaitoonie

Introduction: Media development in a changing geopolitical context—Cosmopolitan research perspectives are a "must"

For cosmopolitan communication studies, media development[1] is a field of utmost relevance. Whenever organizations—traditionally from the Global North—intervene in the media systems and journalism cultures of transformation or (post-)conflict societies, different concepts and norms of news and news making collide. Mostly "Western" actors—state- or privately funded, from major global actors like the NGO Internews to small but highly visible foundations like the Swiss-based Fondation Hirondelle—address media practitioners and newsrooms that act in profoundly different political and economic contexts—even if they share normative concepts of journalism with donors. The latter seek to contribute to professionalism and freedom of expression in "target countries." However, good intentions cannot be realized effectively without understanding the restrictions and rationalities of local media actors involved in the cooperation, both on the micro and meso levels (Fengler & Jorch, 2012). For example, more participatory forms of content production in sub-Saharan Africa—transforming media users into content producers or co-producers, particularly of investigative content—also increasingly expose journalists to threats from armed or extremist groups; thus, promoting "Western" journalism concepts can have unintended detrimental effects on local media practitioners (Schönbächler, 2023).

1 This chapter will focus on media development, also defined as media assistance or media action (Leroy, 2025). "Media development cooperation" means the efforts of different actors, often internationally, to enable, build, and strengthen free and independent media ("media development") and to use media as a tool ("media for development") to work toward development goals (Manyozo, 2012; Scott, 2014). It includes but is not limited to journalism training, advocacy for press freedom and right to information, establishment of media outlets, community engagement, production of media contents, media literacy, and communication campaigns.

Media development activities often transport normative ideas of how media systems should function. However,

> Western models of media systems cannot be easily applied to new democracies. Instead, new hybrid forms of political communication are emerging that blend liberal ideals of a free press with the trajectories of the past, indigenous values and the constraints and experiences of transition. (Voltmer, 2013, p. 23)

Especially in fragile states, media development is part of the set of influences on the transformation of media systems. The cases of Bosnia and Herzegovina (Ahmetašević, 2024) and Afghanistan (Wollenberg & Bahar, 2023) are just two recent examples of how norms of free media that are introduced from the outside into a context where conflicts are still ongoing do not permeate the media system in-depth, so newly established institutions or achievements remain vulnerable. Misconceptions and shortcomings often become obvious only much later—the return of the Taliban to power in Afghanistan and the instant collapse of decades of development initiatives to strengthen civil society and freedom of expression is one striking example. Another is the fragmented, polarized, and "captured" broadcasting landscape in the Balkans, where many stations established by Western donors after the war in former Yugoslavia were "captured" by local media oligarchs once the donor community stopped funding and moved to other global "hotspots". A "cosmopolitan" approach to such spectacular failures also implies acknowledging particular conditions, such as imbalances and power asymmetries (Keßler, 2022), and the self-interest of local media practitioners (Fengler & Ruß-Mohl, 2008). The fact that a majority of studies analyzing the impact of media development have been commissioned and produced by the media donors themselves thus far limits a comprehensive assessment of the context-sensitive applicability of media development initiatives. All actors involved in such evaluations might have a strong interest in not questioning the effectiveness of funding (Noske-Turner, 2017).

At the time of this writing, the "era of innocence" in media development—that started with strong hopes for worldwide democratization processes after 1989/90—has come to an end. Already, the failure of the "Arab rebellion" made that obvious, and the geopolitical context for media development has changed profoundly with the Russian invasion of Ukraine and now the escalation in the Middle East. Reacting to these shifting dynamics, the USA has signaled its readiness to invest more in strengthening civil societies worldwide—while France has completely withdrawn from all (media) development activities in the Sahel region after the coups in Niger, Mali, and Burkina Faso. Given the increasing polarization of the world, it will also be more challenging for actors like Germany who have been acting for most of the post-WWII period without a geopolitical agenda to conceptualize and implement media development initiatives that local decision-makers will not perceive as Trojan horses for political interference. While the "autocratization" (Nord et al., 2024) of states across continents continues, explicit legal restrictions have also been implemented in countries such as India to discourage foreign media development projects. Awareness about media capture (Mungiu-Pippidi & Ghinea, 2012) through media development actors has also increased.

The cultural challenges of media interventions in Afghanistan and Bosnia and Herzegovina may also serve to illustrate the need for a cosmopolitan perspective in the analysis of media development, which helps to renew theoretical perspectives with critical and plural approaches: Who is theorizing media development cooperation, how, and for what? The legacy of colonial hierarchies and continuing postcolonial tensions are not only topics in practical media development cooperation and in the collaboration between partners from different countries. They are also at the foundation of theories and theory building. Who is doing what research about whom? Awareness and recognition of imbalances and power asymmetries are part of such an approach. "Local ownership," "trust," and "participation," for instance, are popular terms in the industry of international media cooperation that hardly anyone would be opposed to. However, research by Waisbord (2008) and Noske-Turner (2017) has suggested that participatory approaches to media development cooperation are often undercut by institutional imperatives and pressures resulting from fixed project cycles and bureaucratic systems. A less illusionist and idealistic—at least theoretically—but more critical and comprehensive, collaborative, and integrative, and thus cosmopolitan perspective may also serve to increase the sustainability of media development—which has been at the center of critical investigation within the Graduate School MEDAS 21, funded by VolkswagenStiftung from 2018 through 2023.[2] Eight research projects were initiated to provide independent academic analysis of the structures and processes of media development projects in sub-Saharan Africa and Asia. All of the studies using qualitative methods have placed the perspective of local actors and institutions at the core of the research, thus supporting the pledge made by Richter et al. (2023): "Nur mit Wissen über inter- und transnationale Zusammenhänge lässt sich auch an *Global Governance* teilhaben und lassen sich weltweite Interdependenzen mitgestalten" (*"Only with knowledge of inter- and transnational interrelationships is it possible to participate in global governance and help shape worldwide interdependencies,"* [p. 4]). However, due to the space restrictions of a book chapter, this chapter will provide only a highly condensed overview of MEDAS 21 key research results. In a kaleidoscopic manner, it will highlight the Graduate School MEDAS 21 findings on key issues of a cosmopolitan approach to media development: postcolonial Western but also new forms of authoritarian non-Western influences on structures in media development; cultural challenges of media interventions via media development; influences and sustainability of media development on the transformation of media systems in the Global South; and collaborative and integrative approaches in media development to counter power asymmetries and promote participatory approaches.

Media development: An "industry" between postcolonial and authoritarian influences

A "cosmopolitan" perspective on media development also demands a sober analysis of donors' motives. Foreign aid for media assistance is a global phenomenon with

2 https://brost.org/medas21/

historic links to public diplomacy, but it is also rooted in colonialism and market expansion (Lugo-Ocando, 2020). International organizations have often treated media development as something that is done to other countries—as a supply-driven service performed and funded by outsiders mainly operating internationally (Miller, 2009). A donor-driven approach can imply that "locals" may be required to abandon their historical methods to adapt to donor-mandated practices (McPhail, 2009). Within the media development domain, international NGOs enjoy a unique position of non-partisanship seen as anti-political, apolitical, and supra-political, despite their financial support coming from states or regional bodies. Media assistance is, by its very nature, an act of political-cultural intervention (Miller, 2009, pp. 26–27). The legacy of history is a key consideration for the so-called media development sector: most of the implementers (INGOs or semi-public bodies) were born at the time of the collapse of the Soviet world, the war in the Balkans, and the genocide of the Tutsis in Rwanda (Leroy, 2021). Their credo thus often stems from post-Cold War geopolitics. The existence of mechanisms that can be regarded as prefiguring media development in the pre-independence Global South remains under-researched, although they are increasingly being addressed by scholars in the British (Potter, 2012), Portuguese (Ribeiro, 2014), and French contexts (Asseraf, 2019).

This situated view is compounded by the fact that the forms taken by transnational state media influence strategies have undergone profound changes since the end of the twentieth century, in line with changes in the geopolitical context and the emergence of new dissemination formats, including digital ones (Mattelart et al., 2022, p. 106). This combination of public and private diplomacies in the service of dedicated agendas highlights the interweaving of different spheres of influence, which is reshaping mass communication. Thus, from the 1990s onward in Eurasia and even more so in the following decade in Africa (Frère, 2012, 2022), a standard of intervention made the promotion of "democracy" the main thrust of media development strategies.

At the same time, new actors, including autocratic states, have entered the "media development cooperation" stage in past years, bringing different value systems. China's increasing investments in media in Africa and beyond (e.g., Gagliardone, 2015; Kumar, 2022) and the engagement of other global players from the BRICS countries to Turkey (de Albuquerque & Lycarião, 2018) and the Gulf States make it even more relevant to reconsider which values and norms are actually supported in media development cooperation and for what purposes. Increasing and value-driven competition between traditional "Western" and emerging "non-Western" media development actors should also draw more attention to the agency of actors from countries that receive media development support (see e.g., Lugo-Ocando, 2020). Myers (2018), for instance, described how Nigerian newspaper editors "talk about being able to 'circumscribe,' 'define,' 'pick and choose,' and that they seem to be able to assert their own strategies in the face of donor power" (p. 38). This raises the question of the power relations (Berger, 2010; Harris, 2018; Keßler, 2022) between various media development actors as well as their respective room for maneuvering.

Media development and peacebuilding: Case studies on UN radios in peacekeeping operations

In the context of conflicts—whether they be armed, political, or social struggles—the media plays a pivotal role. When conflicts emerge, the media often become, intentionally or not, key actors as they disseminate, withhold, and manipulate information in ways that can influence the unfolding of the conflicts (Mano, 2021). Scholarly debate has more frequently supported the media's agency in escalating conflicts but also in contributing to peace by defusing tensions and enhancing the prospects of conflict transformation, particularly through radio with its large outreach in hard-to-reach places (Hoffmann & Hawkins, 2015; Legatis, 2015; Maweu & Mare, 2021; Orgeret, 2021; Rodríguez, 2015; Santos & Schönbächler, 2022; see also Schleicher & Sarısakaloğlu in this book).

In the 1990s, the UN started deploying so-called UN radios in some of its peacekeeping operations.[3] The first of these radios, Radio UNTAC in Cambodia, went on air in 1992. Since then, the UN has set up more than a dozen radio stations to support its peacekeeping operations (see Betz & Papper, 2015; Lehmann, 2015; Orme, 2010). Peacekeeping operations often take place in countries with poorly developed and disrupted media landscapes. In this context, UN radios are sometimes the only non-partisan media source available. They provide the local population with reliable information about peacekeeping operations, the peace process, and other relevant issues. Against this backdrop, it is not surprising that some UN radios became the largest and most popular broadcasters across the country. However, UN radios are not without controversy. What is the exact role of UN radios? What impact do UN radios have on the local media landscape? What happens to UN radios when a peacekeeping operation ends?

After and during peacekeeping operations, UN media (particularly radio) have often become the standard of accuracy and professionalism for local media (Loewenberg, 2006). On the other hand, UN media have also become a "controversial matter" (Oksamytna, 2018, p. 80), as the assumption that it is appropriate for external actors to "educate" the local population in human rights and democracy is disputable and problematic.

In general, research in media development has frequently relied on theories that specify the role of Western mainstream news media in armed, militarized conflicts (Bratic, 2015; Budka & Bräuchler, 2020; Stupart, 2021), neglecting perspective and lived experiences of "local" journalists who work and reside in conflict areas (Santos & Schönbächler, 2022). Therefore, it is essential to involve local practitioners and experts in cross-border research.

Zaitoonie's (2025) case studies of Liberia, Sierra Leone, and Côte d'Ivoire have focused on the short- and long-term impact of the UN's media strategies in peacekeeping operations and thus reflect on media sustainability as well. The three countries were hit by severe and protracted civil wars. In the course of these civil wars, the UN adopted

3 Peacekeeping operations are UN interventions consisting of military, police, and civilian components that seek to stabilize conflict-affected countries after a peace agreement has been concluded (for further detail, see Gowan, 2018).

peacekeeping operations with comprehensive media strategies that included the establishment of UN radios: Radio UNAMSIL in Sierra Leone, UNMIL Radio in Liberia, and ONUCI FM in Côte d'Ivoire. The preliminary results of Zaitoonie's case studies revealed that the UN's media strategies—the UN radios in particular—significantly contributed to domestic media development in the respective countries. All three radios became major broadcasters with large audiences and a high degree of professionalism, which had a considerable spillover effect on the local media landscape. In respect to this context, UN media development expert Bill Orme (2010) claimed that UN radios, by design or default, "contributed more to media development in certain post-conflict countries than any other concurrent media assistance programs" (p. 8). Notwithstanding, UN radios also face certain challenges: becoming major domestic broadcasters bears the risk of distorting the local media market. Various key informants pointed to the fact that UN radios poached the best media workers—and the audiences—from local media outlets, which could not compete with the UN's considerable resources. Another point of criticism is the ambiguous role of UN radios. While they formally represent a strategic communications tool of the UN, they are often perceived as de facto public broadcasters. Nevertheless, UN radios are bound to the UN's peacekeeping mandates and cannot act completely independently, even if their staff have a great deal of editorial freedom. Until the end of the 1990s, UN radios were usually packed up and removed at a peacekeeping operation's end, which could leave behind a sudden void in the domestic media landscapes (see Betz & Papper, 2015, p. 174; Orme, 2010, p. 9). Therefore, more and more voices have called for the continuation of UN radios beyond a peacekeeping operation's end. However, the case studies have indicated that a lack of financial resources and editorial independence can severely endanger the quality of successor radio stations.

Santos's study (2024) has confirmed these constraints faced by journalists working for UN missions. She has studied the situation in the Central African Republic (CAR), where attempts at state building have failed, and the state has been plunged into an eternal spiral of violence. Santos compared Guira, a radio station run by the UN mission in the CAR (MINUSCA), and radio Ndeke Luka, which is run by the Swiss-based NGO Fondation Hirondelle, with regard to the norms and agency of media professionals belonging to these two stations. Agency refers not only to the intentions people have in doing things but also to their "capability of doing those things" (Giddens, 1984, p. 9). Through the journalists' narrated experiences, it was possible to detect layers of differences in how both radios shaped local journalists' spaces of action and how the journalists from these radio stations exercised their agency. Central African journalists working for the UN radio tended to have more prestige but less agency. Landing a position at a UN radio station would be a significant professional step for the journalists toward a higher social status, and their improved image would have positive outcomes for their peers, family, surroundings, and ultimately the country. However, while they had more resources for their work, there were also more rules that local UN journalists had to abide by, as well as more restrictions of what they could publish and broadcast. For example, they were not allowed to reach out to representatives of armed groups to avoid diplomatic imbroglios. The contrasting degrees of agency of journalists working in the UN radio Guira and the NGO-run radio Ndeke Luka became evident in their coverage of the process of the peace talks and the event of the signature of the Khartoum peace accords in February 2019.[4] As narrative analysis revealed, Hirondelle's Ndeke Luka showed a more investigative approach by questioning the actors involved in the peace negotiations and attempting to delve into the specific details about the accords and to publicly unveil what was being secretly discussed. On the other hand, the narratives about the peace talks aired by the UN's Guira mainly portrayed the official top-down perspective of the ceremony of the peace accord signature without delving into the nuances of how these accords were negotiated. The peace deal that had been discussed and agreed upon behind closed doors in Khartoum appeared as a given fact and as taken for granted. Guira's approach neglected that it had been a construction process under intense negotiation. Additionally, the views of the society at large about such accords were not included in the UN broadcast, and when it did attempt to reach out to the people, the goal was mainly to "sensitize" the population and raise their awareness about that new piece of document.

Assessments of the context sensitive applicability of media development initiatives: Examples of gender-focused projects in a conflict state and of health journalism in Africa

Volatile conditions in conflict-affected areas require a particularly thoughtful media development approach, taking postcolonial structures, religion, gender, and related values into account (Geertsema-Sligh, 2019). However, Schönbächler's study (2023)[5] has shown how Western media development projects have contradictory effects on local female journalists. While women might gain access to media production, they are also more exposed to discrimination and risk. Due to persisting local social norms, women journalists continue to have less access to capacity building. Almost all female journalists interviewed by her in Burkina Faso claimed to be discriminated against when it came to being selected within their own newsrooms for attending training organized as part of media development initiatives.[6] However, many training organizers have noticed the lack of participation of women journalists and have thus adopted positive discrimination, explicitly asking for the participation of women journalists in their training. According to Schönbächler (2023), this practice has been highly appreciated by the interviewed Burkinabè women journalists, who have not only been able to develop their skills but also to earn additional income. Female journalists are also more often hired for media development projects, which can cause them to be increasingly exposed to

4 The signature of the *Accord politique pour la paix et la reconciliation* (APPR; Political agreement for peace and reconciliation) signed by the Central African government and 14 armed groups was brokered by the United Nations and the African Union.

5 Passages have already been published in Schönbächler (2023). Short text items have already appeared in onsite actors' agency within international media development (Drefs, 2022a).

6 In one case, the per diems were even used as an argument to change plans at the last minute: a male colleague was sent instead of the woman who was supposed to attend the training. The reason given was that the male journalist needed the money because his wife was in the hospital. This example illustrates very well how monetization of aid combined with local gender role expectations can lead to gender discrimination.

risks. In particular, when women journalists address topics related to gender, schooling, early/forced marriage, female genital mutilation, and family planning—often pushed by media development initiatives—they are threatened by transnational groups in Burkina Faso (Yaméogo, 2018). Almost all radio stations in the area covered by Schönbächler's study (2022) have experienced armed groups threatening them, their staff, and their families. As donors like to encourage journalistic formats that are more open to the voices of journalists and listeners, female radio journalists are also more vulnerable to abuse when covering these sensitive topics in interactive radio shows. Thus, her study has revealed a trade-off between giving voice and ensuring protection.

Promoting coverage of health issues in African countries is another example of cultural and normative challenges to media development, as socially prevailing patterns of interpretation in the project countries may stand in the way of professional health journalism Western donors support as agents of change. The divergence between cultural values on the one hand and generally accepted social knowledge on the other (Schimank, 2007, p. 126) can lead to cognitive dissonance concerning journalists' understanding of their professional role. In many sub-Saharan African countries, Western funders of media development train journalists to cover health issues comprehensively to make an impact on public health. However, according to Wollnik's study (2025), certain health topics are forgotten or neglected in African mass media. In this context, journalists from Uganda resort to "self-censorship" in their newsrooms after participating in training: they seek to avoid certain health topics that are undesirable in media houses due to prevailing social values. Examples are reporting on homosexuality, being considered a risk factor for infection with HIV, or deliberately induced abortions. According to a health journalist from Uganda in 2021 (Wollnik, 2022), covering these issues could require shifting the focus of reporting, for example, to (legal) access to post-abortion care, or applying the issue to certain social settings such as the special situation of refugees. Mental health disorders are another example. They have not only been on the rise in African countries since the COVID-19 pandemic (e.g. Chen et al., 2021; Sankoh et al., 2018) but are locally associated with religious superstition or witchcraft (Wollnik, 2022, p. 5). As a result, in some African countries, there is rarely any talk about these illnesses in the media (cf. Alemu et al., 2023).

Cultural challenges and power imbalances: Dealing with differences in media development between actors and countries

Media development donors may also be confronted with "media" that have little resemblance to Western concepts of newsrooms. Least developed countries (LDCs) are a particularly complicated terrain both for local media practitioners and international donors: in a context of permanent state fragility (instead of sudden conflict), they experience influencing by various interest groups, ambiguous constraints, and lack of resources, among other issues. Mack's study (2025)[7] investigated the media of Guinea-Bissau and their in-

7 Note: Some contents of this paragraph written by Mack have been published in the Central European Journal of Communication, Special Issue on Media Capture.

teractions with international actors. In the media system of such a small, politically, and economically fragile country, which is also marked by cultural diversity (Kohl, 2012, p. 643), a basic task of media development actors has been to train authorities on how to deal with media, whose stability is rather dependent on the will of current political leaders. The few donors from the Global North "compete" with new international actors, such as Christian and Muslim religious actors as well as the Chinese embassy, who do not prioritize media development but rather stress ideological loyalty. Journalists face restrictions and risks, and they often work without a stable payment. Due to the financial dependency of large parts of the media systems on either state/party or donor funding, journalists accept per diems and other benefits from politicians—and donors—despite being aware of the ethical compromise (Sampaio-Dias, 2019, p. 2352). Suitable partners for media development projects are difficult to select, as state media depend on the president and private media either belong to or are financed by politicians, so donors often have to compromise. Donor fatigue is one consequence of the continuous fragility (Walker & Gomes, 2021, p. 12), and "agenda setting [being] upside-down" (de Barros, 2012, p. 99) is another; that is, rather than the media setting the agenda, the "donor topics" determine the public debate.

Journalistic training projects can have unintended effects, as they may be characterized by an imbalance of agency between representatives of donors and trainees from the Global South. A study focusing on the negotiation of knowledge and positions between trainers and trainees (Keßler, 2024) hinted at enduring stereotypical role conceptions even in a phase of intense debates about postcolonialism. The author's examples point toward a lack of "cosmopolitanism" among the trainers tasked with addressing media actors from other journalism cultures and working environments. Keßler examined journalism training in order to detect if a "dominance of Western ideas" of development and journalism (Banda, 2013; Lugo-Ocando, 2020; Manyozo, 2012, pp. 200–207; Miller, 2009; Murphy & Scotton, 1987; Phiri & Fourie, 2011) may still be visible in concrete interactions between trainers from the so-called Global North and trainees from the Global South. Despite having been described in the literature as partially overcome (Communication Initiative et al., 2007, p. 44; Drefs & Thomass, 2019; Tufte & Mefalopolus, 2009), various practices of inequality emerged in Keßler's participant observation of training (see also Keßler, 2022 and Keßler at al. in this book). For example, trainers and organizational leaders tended to emphasize the superiority of their experience and theories coming mostly from the USA and Europe, and they relativized trainees' professionalism, thus stereotyping and infantilizing trainees (as known from discourse analyses of development actors, see Kogen, 2018). However, the trainees in turn emphasized their unique role as agents of change in their countries, their context-specific knowledge, and different understandings of journalism, social change, and how to address the audience. Religion, in particular, emerged as a key aspect of the discussions and negotiations between trainers and trainees;[8] it becomes clear here that religion is an important influencing factor that both media development researchers and practitioners should pay more atten-

8 For example, both sides negotiated stereotypes of a misanthropic Islam on various occasions and the importance of religiosity for campaigns of social change and the definition of credible sources in general. In addition, the impossibility of an organ donation campaign being shown was dis-

tion to (e.g., Deutsche Gesellschaft für Internationale Zusammenarbeit [GIZ], 2023; see also Bundesministerium für Wirtschaftliche Entwicklung und Zusammenarbeit [BMZ], 2016; Jahnel, 2018).

Drefs's (2022b) qualitative study of representatives of on-site actors and their agency comes to a somewhat ambivalent conclusion. Her exploratory interview study confirmed power imbalances but also pointed toward the considerable authoritative resources on-site actors possess in interactions with donors. So-called "authoritative" resources enable agents to have an influence on relevant aspects of social situations, such as "organisation of time and space, chances for self-development, organisation between people" (Best, 2003, p. 6). Overall, the study indicates that on-site media development actors display knowledge about the rules that structure the context in which they act. They know that international media development offers a field of action that allows them to strive for their idealistic goals. Motives like fighting media capture or enabling investigative reporting go well with the democratic agenda of international media development organizations and their funders, which also indicates that from the perspective of on-site actors "shared democratic values" (Drefs & Thomass, 2019, p. 267) are fundamental building blocks for media development partnerships. However, whether on-site actors' democratic values are genuine or displayed to serve as an advantage in the competition for media development resources is raised as an issue that requires scrutiny. After all, the interviewees' doubts about other local actors' sincerity to the cause confirm Berger's (2010) problematization of the media development sector as a market for resources. Yet, even for "genuine" actors, media development serves as a source of revenue. Thus, they are very much aware of the expectations and bureaucratic requirements attached to this field of action. The way they act upon them ranges from embracing these rules and aligning one's own organizations' structures accordingly to playing along while not actually subscribing to them or even opposing them. Especially when it comes to structures that have been installed by other actors for whom there is quite a big imbalance in allocative resources—that is, material resources such as financial means, products, or raw materials—on-site partners seemed to prefer to circumvent rules they find hard or annoying to fulfill, rather than questioning them openly. This finding implies an important realization for those actors who are rich in allocative resources, such as governmental organizations or private foundations who fund media development programs. In the quest for improved effectiveness and impact, many of them have introduced managerialist-inspired funding conditions and formalized procedures for monitoring and evaluation (Elbers et al., 2014). Yet, when on-site actors do not see a chance to question the appropriateness of such conditions and procedures in their local contexts, they might just work around them and—in doing so—reduce them to absurdity.

In other situations, especially when on-site actors' positions are strengthened by a good deal of authoritative resources, they make strategic use of their capabilities. On-site actors' local expertise and a network of long-term and diverse partners can work as authoritative resources that give them leverage in shaping projects and, in extreme cases,

---

cussed due to the religiously based sanctity of the body. Furthermore, participating media actors also acted as spiritual or Christian public actors in their countries.

even allow them to make a difference by ending a partnership. As far as allocative resources are concerned, the power imbalance between external organizations that give out funds and on-site organizations that receive them cannot be denied. While this, at times, prompts on-site actors to take on projects that not everybody on their team is particularly interested in, many seem to base such decisions on broader strategic considerations intended to benefit their organization overall and to enable other concrete endeavors they want to undertake. For Drefs (2022a), the question arose whether an open and argument-based discussion about the relevance and urgency of specific funding lines would not be more beneficial for all parties involved within this interplay of on-site actors' authoritative resources and international organizations' or donors' allocative resources. It would allow on-site actors to explain why certain issues are higher on their agendas than others. Likewise, it would help donors avoid funding projects that do not have local ownership.

Rethinking the transformative dimension of media development to foster its sustainability

One of the topical issues that regularly crops up in development discourse is sustainability (Leroy, 2021). The desire for impartiality on the side of Western media development actors illustrates, above all, the difficulty of thinking about media development from a "transnational" and not just an "international" perspective. The standards for evaluation in development assistance are still those defined by the principles for evaluation (1991) of the Organization for Economic Co-operation and Development's Development Assistance Committee, the successor to the Organization for European Economic Co-operation, which in 1948 was responsible for administering the Marshall Plan. This inherited divide perpetuates a kind of (at least discursive) power imbalance. This has, for example, resulted in Germany and, more marginally, France accounting for the broadcasting of their overseas media as Official Development Assistance to the Global South. The border between inside and outside—or, to put it another way, belonging—is no longer perceived in the same way.

Media development projects have long been built on the assumption that media growth is equated with media development (Berger, 2022). The emergence of a concern for sustainability and social impact shows that the growth = development equation is not always verified. In the development sector, a number of meta-evaluations had already been carried out to measure the effectiveness of "interventions" against the facts, with, to give just three examples, a sample of 34 evaluations from 2003 through 2005 (Forss et al., 2008), 340 randomly selected evaluations from 2009 through 2012 (Hageboeck et al., 2013), and 72 evaluations from 2018 through 2022 (African Development Bank [AfDB], 2023). They respectively showed that 38% of the evaluations were rated as poor or not satisfactory on their sustainability analysis, 34% had imprecise findings (simply "some," "many," or "most"), and shortcomings in institutional sustainability and strengthening of capacities were found in 42% of cases.[9] A meta-analysis focusing solely on media development projects and bringing together a corpus of 289 evaluations from 1999 through 2019 reached similar results (Leroy, 2025), as many as 45% of the ex post evaluations did not provide an opinion on the sustainability of their project, and when a judgement was made, it was mostly depreciatory (27%). Also, the study found that a concern for the economic dimension of sustainability was growing, with the ideal of a business-oriented transformation. With numerous references to "adaptation," the results also showed a reactivation of the old "modernization" pattern. For half of the study period, this universal goal of media sustainability excluded and essentialized Africa as a continent where sustainability was not an issue, or even where media projects were doomed to be unsustainable.

Following universalist roots (Hume, 2004), media development has entered a post-missionary era (Noske-Turner, 2017) but struggles to define its very purpose. It is probably no coincidence that information is not a Sustainable Development Goal as such among the 17 to be reached by 2030 adopted by the United Nations. The one that comes closest is the sixteenth, which is the goal to *"promote[s] peaceful and inclusive societies for sustainable development, provide[s] access to justice for all and build[s] effective, accountable and inclusive institutions at all levels"* (UN, 2015, p. 14, emphasis added). The media only appear explicitly in Target 10 with Indicators 16.10.1 and 16.10.2 linked to the imprisonment and death of journalists, that is, criteria that are undoubtedly less debated than the very aims of media interventions. For the rest, it seems that agreement on objectives is difficult to reach. As a normative view of democracy or democratic formalism (Bonnard et al., 2021) is sometimes adopted by autocratic regimes to legitimize themselves, the focus is now shifting to a consideration of a true "participation" and the conditions that enable it to emerge, as summarized in a discussion paper based on findings from a project in Niger: *"It is crucial to ensure* broad *participation and to have participants agree on ground rules for their exchanges to create a productive dialogical setting"* (Drefs & Souleymane, 2023, p. 10, emphasis added).

Lessons from the MEDAS 21 research: Suggestions for a more cosmopolitan media development practice

MEDAS 21 has published a set of policy recommendations aiming to improve research and practice of media development.[10] These recommendations (summarized below) describe a roadmap for how to "cosmopolitanize" theory-building, empirical research, and actual practices in media development. Systematically including the knowledge of on-site actors and scholars in the Global South will be a relevant first step to challenge (and thereby improve) an industry that still lacks the academic scrutiny needed in a time of uncertain global prospects. The MEDAS 21 recommendations are as follows:

---

9 On 29 November 2023, the institution organized a dedicated learning event on the theme of sustainability in development interventions based on these findings. This last figure was quoted by AfDB Evaluator General Karen Rot-Münsterman in her Opening Remarks.

10 https://www.medas21.net/resources/.

- It is of utmost importance to provide wider access to data sources in the sector, many of which are still confidential. This is particularly the case for baseline studies and ex post evaluations of projects.
- In order to broaden the perspective, it is necessary to put an emphasis on learning from on-site actors (Drefs & Thomass, 2019). A reflection on problematic points in interactions between international partnerships in media development cooperation is still to be promoted in this matter.
- In the media development cooperation sector, there seem to be few considerations for failure. Negative evaluation often comes with a negative connotation, synonymous with unfulfilled objectives, frustration, or even defeat. Benequista et al. (2019) insisted that "practitioners of media development have also been reluctant to acknowledge scholarly critique of the field" (p. 6).
- While the online and offline lived realities can no longer be separated, research has to increasingly take into account digital spaces and practices as well as their interaction with offline spaces and practices (Schmidt-Lux & Wohlrab-Sahr, 2020).
- The state of research on media development cooperation in the poorest countries is often extremely precarious. This makes much greater higher education cooperation indispensable between universities themselves and between universities and practitioners over the long term.

References

African Development Bank. (2023). *Annual report 2022: Adding value to the AfDB's work.* https://idev.afdb.org/sites/default/files/documents/files/AR%202022%20Layout-EN-29.06.2023.pdf

Ahmetašević, N. (2024). *The media as a tool of international intervention: House of cards.* Routledge.

Alemu, W. G., Due, C., Muir-Cochrane, E., Mwanri, L., & Ziersch, A. (2023). Internalised stigma among people with mental illness in Africa, pooled effect estimates and subgroup analysis on each domain: Systematic review and meta-analysis. *BMC Psychiatry, 23,* Article 480. https://doi.org/10.1186/s12888-023-04950-2

Asseraf, A. (2019). *Electric news in colonial Algeria.* Oxford University Press.

Banda, F. (Ed.). (2013). *Model curricula for journalism education: A compendium of new syllabi.* UNESCO. https://unesdoc.unesco.org/ark:/48223/pf0000221199

de Barros, M. (2012). Os Mídia e os Direitos Humanos na Guiné-Bissau: o caso dos jornais Nô Pintcha, Diário de Bissau e Gazeta de Notícias. *Revista Africana De Mídias,* 20(1–2), 83–100. https://doi.org/10.57054/amr.v20i1-2.5164

Benequista, N., Abbott, S., Rothman, P., & Mano, W. (Eds.). (2019). *Mass communication and journalism: Historical perspectives and new frontiers.* Peter Lang.

Berger, G. (2010). Problematizing 'media development' as a bandwagon gets rolling. *International Communication Gazette,* 72(7), 547–565. https://doi.org/10.1177/1748048510378143

Berger, G. (2022, February 9–10). Interview by Benequista N.. Third Global Conference for Media Freedom, Tallin, Estonia.

Best, S. (2003). *A beginner's guide to social theory*. Sage. https://doi.org/10.4135/9781446219621

Betz, M., & Papper, H. (2015). UN peacekeeping radio: The way forward. In J. Hoffmann & V. Hawkins (Eds.), *Communication and peace: Mapping an emerging field* (pp. 163–178). Routledge.

Bonnard, P., Dakowska, D., & Gobille, B. (Eds.). (2021). *Faire, défaire la démocratie: De Moscou, Bogota et Téhéran au Conseil de l'Europe*. Karthala.

Bratic, V. (2015). Beyond journalism: Expanding the use of media in peacebuilding. In J. Hoffmann & V. Hawkins (Eds.), *Communication and peace: Mapping an emerging field* (pp. 148–162). Routledge.

Budka, P., & Bräuchler, B. (Eds.). (2020). *Theorising media and conflict*. Berghahn Books.

Bundesministerium für Wirtschaftliche Entwicklung und Zusammenarbeit. (Ed.). (2016). *Partner für den Wandel: Religionen und nachhaltige Entwicklung*. https://www.partner-religion-development.org/wp-content/uploads/2023/10/BMZ_Religionen_Entwicklung.pdf

Chen, J., Farah, N., Dong, R. K., Chen, R. Z., Xu, W., Yin, J., Chen, B. Z., Delios, A. Y., Miller, S., Wan, X., Ye, W., & Zhang, S. X. (2021). Mental health during the COVID-19 crisis in Africa: A systematic review and meta-analysis. *International Journal of Environmental Research and Public Health*, *18*(20), Article 10604. https://doi.org/10.3390/ijerph182010604

Communication Initiative, Food and Agriculture Organization of the United Nations, & World Bank. (2007). *World congress on communication for development: Lessons, challenges, and the way forward*. http://documents.worldbank.org/curated/en/290941468138271611/World-congress-on-communication-for-development-lessons-challenges-and-the-way-forward

de Albuquerque, A., & Lycarião, D. (2018). Winds of change? BRICS as a perspective in international media research. *International Journal of Communication*, *12*, 2873–2892. https://ijoc.org/index.php/ijoc/article/view/8549

Deutsche Gesellschaft für Internationale Zusammenarbeit. (2023). *Religiöse Akteur*innen in die Entwicklungszusammenarbeit einbeziehen: Sektorvorhaben Religion für nachhaltige Entwicklung*. https://www.giz.de/de/weltweit/128395.html

Drefs, I. (2022a). Onsite actors' agency within international media development. *Global Media Journal – German Edition*, *12*(2). https://doi.org/10.22032/dbt.55516

Drefs, I. (2022b). Explaining contemporary media development practice: Building blocks for an analytical framework based on organizational institutionalism. *Journal of Applied Journalism & Media Studies*, *10*(2), 141–161.

Drefs, I., & Souleymane, L. (2023). *How to foster constructive dialogue as part of media development? Lessons from a case study on interactive radio formats in Niger*. Deutsche Welle Akademie. https://p.dw.com/p/4Xk4i

Drefs, I., & Thomass, B. (2019). The participation approach in media development cooperation. In K. Voltmer, C. Christensen, I. Neverla, N. Stremlau, B. Thomass, N. Vladisavljević, & H. Wasserman (Eds.), *Media, communication and the struggle for democratic change: Case studies on contested transitions* (pp. 257–279). Palgrave Macmillan. https://doi.org/10.1007/978-3-030-16748-6_11

Elbers, W., Knippenberg, L. and Schulpen, L. (2014). Trust or control? Private development cooperation at the crossroads. *Public Administration and Development, 34*(1), 1–13. https://doi.org/10.1002/pad.1667

Fengler, S., & Jorch, J. (2012). Medienentwicklungszusammenarbeit – Stiefkind der Wissenschaft, Liebling der Praxis: Empirische Befunde und Lösungsvorschläge für Probleme der Praxis aus kommunikationswissenschaftlicher Sicht. In S. Fengler, T. Eberwein, & J. Jorch (Eds.), *Theoretisch praktisch!? Anwendungsoptionen und gesellschaftliche Relevanz der Kommunikations- und Medienforschung* (pp. 219–234). UVK.

Fengler, S., & Ruß-Mohl, S. (2008). Journalists and information-attention markets. *Journalism, 9*(6), 667–690. https://doi.org/10.1177/1464884908096240

Forss, K., Vedung, E., Kruse, S. E., Mwaiselage, A., & Nilsdotter, A. (2008). *Are Sida evaluations good enough? An assessment of 34 evaluation reports*. Sida. https://cdn.sida.se/publications/files/sida45265en-are-sida-evaluations-good-enough---an-assessment-of-34-evaluation-reports.pdf

Frère, M.-S. (2012). Perspectives on the media in 'another Africa'. *Ecquid Novi: African Journalism Studies, 33*(3), 1–12. https://doi.org/10.1080/02560054.2012.732218

Frère, M.-S. (2022). *Politics and journalism in Francophone Africa: Systems, practices and identities*. Palgrave Macmillan. https://doi.org/10.1007/978-3-030-99399-3

Gagliardone, I. (2015). Media development with Chinese characteristics. *Global Media Journal – German Edition*, 4(2). https://globalmediajournal.de/index.php/gmj/article/view/70

Geertsema-Sligh, M. (2019). Internews: A case study of gender in media development. *Journal of International Communication, 25*(2), 206–229. https://doi.org/10.1080/13216597.2019.1588141

Giddens, A. (1984). *The constitution of society: Outline of the theory of structuration*. Polity Press.

Gowan, R. (2018). Peace operations. In T. G. Weiss & S. Daws (Eds.), *The Oxford handbook on the United Nations* (2nd ed., pp. 420–445). Oxford University Press. https://doi.org/10.1093/oxfordhb/9780198803164.013.23

Hageboeck, M., Frumkin, M., & Monschein, S. (2013). *Meta-evaluation of quality and coverage of USAID Evaluations 2009–2012*. United States Agency for International Development. https://www.usaid.gov/evaluation/meta-evaluation-quality-and-coverage

Harris, S. T. G. (2018). Questioning the role of foreign aid in media system research. In B. Mutsvairo (Ed.), *The Palgrave handbook of media and communication research in Africa* (pp. 401–412). Palgrave Macmillan. https://doi.org/10.1007/978-3-319-70443-2_22

Hoffmann, J., & Hawkins, V. (Eds.). (2015). *Communication and peace: Mapping an emerging field*. Routledge.

Hume, E. (2004). *The media missionaries: American support for journalism excellence and press freedom around the globe*. Knight Foundation.

Jahnel, C. (2018). "The future is not ours to see": Postkoloniale Perspektiven auf den religious turn in der (deutschen) Entwicklungszusammenarbeit. In S. Pittl (Ed.), *Weltkirche und Mission: Bd. 10. Theologie und Postkolonialismus: Ansätze – Herausforderungen – Perspektiven* (pp. 168–190). Verlag Friedrich Pustet.

Keßler, M. (2022). How "insiders and outsiders" perceive media development work: Reflections on their relationship and cooperation. *Global Media Journal – German Edition, 12*(1). https://doi.org/10.22032/dbt.53059

Keßler, M. (2024). *Lehren und Lernen mit Asymmetrien: Eine qualitative Studie zu situativen Differenzierungspraktiken von Trainees und TrainerInnen einer internationalen journalistischen Weiterbildung als Teil von Medientwicklungsarbeit* [Doctoral dissertation, publication in preparation]. Institute for Media Studies, Ruhr-Universität Bochum.

Kogen, L. (2018). For celebrity communication about development to do good: Reframing purpose and discourses. In F. Enghel & J. Noske-Turner (Eds.), *Communication in international development: Doing good or looking good?* (pp. 118–134). Routledge.

Kohl, C. (2012). Diverse unity: Creole contributions to interethnic integration in Guinea-Bissau. *Nations and Nationalism, 18*(4), 643–662. https://doi.org/10.1111/j.1469-8129.2012.00545.x

Kumar, R. (2022, July 24). *How China uses the news media as a weapon in its propaganda war against the West*. Reuters Institute. https://reutersinstitute.politics.ox.ac.uk/news/how-china-uses-news-media-weapon-its-propaganda-war-against-west

Legatis, R. (2015). *Media-related peacebuilding in processes of conflict transformation*. Berghof Foundation. https://berghof-foundation.org/library/media-related-peacebuilding-in-processes-of-conflict-transformation

Lehmann, I. A. (2015). Still caught in the crossfire? UN peace operations and their information capacities. In J. Hoffmann & V. Hawkins (Eds.), *Communication and peace: Mapping an emerging field* (pp. 135–147). Routledge.

Leroy, M. (2021). *Sustainability: Going beyond the buzz-word* (Research Reviews for Media Development Practitioners No. 3). Forum Media and Development & MEDAS 21. https://fome.info/wp-content/uploads/2021/11/Media-Dev-Research-Reviews-3-Sustainability_1-1.pdf

Leroy, M. (2025). *The sustainability imperative in media development: A critical analysis of a self-serving myth*. Palgrave Macmillan.

Loewenberg, S. (2006). *United Nations media strategy: Recommendations for improvement in peacekeeping operations. Case study: UN Interim Administration Mission in Kosovo*. United Nations Peacekeeping. https://citeseerx.ist.psu.edu/document?repid=rep1&type=pdf&doi=c38dd6a06acf5dd564d00200f540395e533ec59e

Lugo-Ocando, J. (2020). *Foreign aid and journalism in the Global South: A mouthpiece for truth*. Lexington.

Mack, J. (2025). *Media system transformation in a context of stable instability: Conceptualizing media development in Guinea-Bissau*. [Doctoral dissertation, publication in preparation]. Institute for Communication Studies, University of Duisburg-Essen.

Mano, W. (2021). Peace and conflict journalism: An African perspective. In K. S. Orgeret (Ed.), *Insights on peace and conflict reporting* (pp. 50–60). Routledge.

Manyozo, L. (2012). *Media, communication and development: Three approaches*. Sage.

Mattelart, T., Idelson, B., & Pélissier, N. (2022). Les mutations des stratégies transnationales d'influence médiatique des États. *Questions de Communication, 41*, 105–114. https://doi.org/10.4000/questionsdecommunication.28590

Maweu, J., & Mare, A. (Eds.). (2021). *Media, conflict and peacebuilding in Africa: Conceptual and empirical considerations*. Routledge. https://doi.org/10.4324/9780429344862

McPhail, T. L. (Ed.). (2009). *Development communication: Reframing the role of the media*. Wiley Blackwell. https://doi.org/10.1002/9781444310740

Miller, J. (2009). NGOs and 'modernization' and 'democratization' of media: Situating media assistance. *Global Media and Communication, 5*(1), 9–33. https://doi.org/10.1177/1742766508101312

Mungiu-Pippidi, A., & Ghinea, C. (2012). Struggling with media capture: Romania. In E. Psychogiopoulou (Ed.), *Understanding media policies: A European perspective* (pp. 166–181). Palgrave Macmillan. https://doi.org/10.1057/9781137035288_11

Murphy, S. M., & Scotton, J. F. (1987). Dependency and journalism education in Africa: Are there alternative models. *Africa Media Review, 1*(3), 11–35.

Myers, M. (2018). Nigerian newspapers: The attractions and drawbacks of foreign aid funding. *African Journalism Studies, 39*(2), 30–41. https://doi.org/10.1080/23743670.2018.1473273

Nord, M., Lundstedt, M., Altman, D., Angiolillo, F., Borella, C., Fernandes, T., Gastaldi, L., God, A., Natsika, N., & Lindberg, S. (2024). *Democracy report 2024: Democracy winning and losing at the ballot.* V-Dem Institute. https://www.v-dem.net/documents/43/v-dem_dr2024_lowres.pdf

Noske-Turner, J. (2017). *Rethinking media development through evaluation: Beyond freedom.* Palgrave Macmillan. https://doi.org/10.1007/978-3-319-58568-0

OECD. (1991). *Principles for evaluation of development assistance.* Development Assistance Committee, OECD Publishing. https://www.oecd.org/content/oecd/en/topics/sub-issues/development-co-operation-evaluation-and-effectiveness.html

Oksamytna, K. (2018). Policy entrepreneurship by international bureaucracies: The evolution of public information in UN Peacekeeping. *International Peacekeeping, 25*(1), 79–104. https://doi.org/10.1080/13533312.2017.1395286

Orgeret, K. S. (Ed.). (2021). *Insights on peace and conflict reporting.* Routledge. https://doi.org/10.4324/9781003015628

Orme, B. (2010). *Broadcasting in UN blue: The unexamined past and uncertain future of peacekeeping radio.* Center for International Media Assistance. https://www.cima.ned.org/wp-content/uploads/2015/02/CIMA-UN_Radio.pdf

Phiri, S., & Fourie, P. J. (2011). Media development aid and the Westernisation of Africa: The case of the Open Society Initiative for Southern Africa (OSISA). *Communicatio, 37*(1), 80–99. https://doi.org/10.1080/02500167.2010.556575

Potter, S. J. (2012). *Broadcasting empire: The BBC and the British world, 1922–1970.* Oxford University Press. https://doi.org/10.1093/acprof:oso/9780199568963.001.0001

Ribeiro, N. (2014). Broadcasting to the Portuguese empire in Africa: Salazar's singular broadcasting policy. *Critical Arts, 28*(6), 920–937. https://doi.org/10.1080/02560046.2014.990630

Richter, C., Grüne, A., Hafez, K., Fiedler, A., Behmer, M., Horz-Ishak, C., Badr, H., Litvinenko, A., Hahn, O., Radue, M., Sarısakaloğlu, A., Löffelholz, M., Fengler, S., Illg, B., Hamidi, K., Hanitzsch, T., & Thomaß, B. (2023). Die "tiefe Internationalisierung" der deutschen Kommunikationswissenschaft? Eine Evaluation der Personal- und Forschungsstrukturen sowie der Lehrprogramme deutscher Hochschulen. *Global Media Journal – German Edition, 13*(1). https://doi.org/10.22032/dbt.57904

Rodríguez, C. (2015). Community media as performers of peace. In J. Hoffmann & V. Hawkins (Eds.), *Communication and peace: Mapping an emerging field* (pp. 289–302). Routledge.

Sampaio-Dias, S. (2019). Per diem payments as a form of censorship and control: The case of Guinea-Bissau's journalism. *Journalism Studies*, *20*(16), 2349–2365. https://doi.org/10.1080/1461670X.2019.1593883

Sankoh, O., Sevalie, S., & Weston, M. (2018). Mental health in Africa. *Lancet Global Health*, *6*(9), e954-e955. https://doi.org/10.1016/S2214-109X(18)30303-6

Santos, F. O. dos (2024). *Journalism and peacebuilding: Challenges & approaches to media development – a case study of Radio Guira and Ndeke Luka in the Central African Republic*. [Doctoral dissertation, publication in preparation]. Institute for Communication Studies, University of Duisburg-Essen.

Santos, F. O. dos, & Schönbächler, V. (2022). Influencing factors of 'local' conflict journalism and implications for media development: A critical appraisal. *Journal of Applied Journalism & Media Studies*, *11*(2), 189–209. https://doi.org/10.1386/ajms_00090_1

Schimank, U. (2007). Handeln in Konstellationen: Die reflexive Konstitution von handelndem Zusammenwirken und sozialen Strukturen. In K.-D. Altmeppen, T. Hanitzsch, & C. Schlüter (Eds.), *Journalismustheorie: Next Generation. Soziologische Grundlegung und theoretische Innovation* (pp. 121–137). VS Verlag für Sozialwissenschaften. https://doi.org/10.1007/978-3-531-90401-6_6

Schmidt-Lux, T, & Wohlrab-Sahr, M. (2020). Qualitative Online-Forschung. Methodische und methodologische Herausforderungen. *Zeitschrift für Qualitative Forschung (ZQF)*, *21*(1), 3–11. https://doi.org/10.3224/zqf.v21i1.01

Schönbächler, V. (2023). Intersectionality matters: Studying the contradictory effects of media development through women radio journalists in Burkina Faso. *Communication and Culture Review*, 4(1&2), 1–18. http://www.communicationandculturereview.in/index.php/communication_culture_review/article/view/63

Schönbächler, V. (2022). *Women journalists in proximity radio: Access, interaction, participation in conflict resolution and transformation processes in Burkina Faso* [Unpublished doctoral dissertation]. Ruhr-University Bochum.

Scott, M. (2014). *Media and development: Development matters*. Zed Books.

Stupart, R. (2021). Precarity, technology, identity: The sociology of conflict reporting in South Sudan. In J. Maweu & A. Mare (Eds.), *Media, conflict and peacebuilding in Africa: Conceptual and empirical considerations* (pp. 197–209). Routledge.

Tufte, T., & Mefalopolus, P. (2009). *Participatory communication: A practical guide* (World Bank working paper no. 170). http://documents.worldbank.org/curated/en/682081468166154717/Participatory-communication-a-practical-guide

United Nations. (2015). *Resolution adopted by the General Assembly on 25 September 2015. Transforming our world: the 2030 agenda for sustainable development* (A/RES/70/1). https://www.un.org/en/development/desa/population/migration/generalassembly/docs/globalcompact/A_RES_70_1_E.pdf

Voltmer, K. (2013). *The media in transitional democracies*. Polity Press.

Waisbord, S. (2008). The institutional challenges of participatory communication in international aid. *Social Identities*, *14*(4), 505–522. https://doi.org/10.1080/13504630802212009

Walker, S., & Gomes, A. (2021). *Mission not accomplished? Uniogbis closes amid uncertainty in Guinea-Bissau*. Global Initiative. https://globalinitiative.net/analysis/uniogbis-guinea-bissau

Wollenberg, A., & Bahar, H. (2023, October 5–6). *Critical review of media development in Afghanistan before Taliban takeover* [Paper presentation]. FoME Symposium 2023, Bonn, Germany.

Wollnik, S. (2022). *KAS health journalism conference 2022 – Reporting in COVID times and beyond.* KAS Media Africa. https://www.kas.de/documents/285576/0/KAS+Conference+on+health+journalism+%E2%80%93+Reporting+in+COVID+times+and+beyond.pdf/d6684ef6-77c4-4938-ec40-36c4311a619e?version=1.1&t=1645188665487

Wollnik, S. (2025). *Gesundheitsjournalismus und seine gesellschaftliche Bedeutung im Kontext globaler Medienentwicklungszusammenarbeit: Eine empirische Untersuchung am Beispiel des subsaharischen Afrikas.* [Doctoral dissertation, publication in preparation]. Institute for Communication Studies, University of Duisburg-Essen.

Yaméogo, L. (2018). Radiodiffusions et extrémisme violent: autopsie d'un journalisme assiégé. In CNP-NZ (Ed.), *Rapport sur l'état de la liberté de la presse 2017: Suivi de l'Etude sur "Radiodiffusions et extrémisme violent: autopsie d'un journalisme assiégé"* (pp. 106–155). Ouagadougou.

Zaitoonie, R. (2025). *The United Nations' media and public communications strategies in peacekeeping operations. A general policy analysis and case studies on Sierra Leone, Liberia, and Côte d'Ivoire.* [Doctoral dissertation, publication in preparation]. Faculty of Cultural Studies, TU Dortmund University.

Challenges and possibilities of context-specific differences in international journalism training

Mira Keßler with contributions from Kefa Hamidi & Beate Illg

Introduction

A key aspect of the deep internationalization of communication studies is a critical reflection about journalism training as part of media development. This chapter is based on a literature review and the research results of my doctoral thesis, and it offers insights into how people coming from different contexts and on different hierarchical levels (being a so-called donor vs. a receiver, a trainer vs. a training participant) negotiate their differences. The aim is raising awareness in journalism research as well as journalism training practices of context-specific differences at times of global interconnections, their challenges, and possibilities for a more participatory approach. In the following, communication and journalism studies, as well as the practice of journalism training, are two threads that are interwoven in a mutual learning process.

As part of media development work (see also Fengler et al. in this book), journalistic training and education are offered worldwide and across national borders. Most of the funding is provided by countries such as the USA, Germany, Japan, the UK and other European countries (Cauhapé-Cazaux & Kalathil, 2015, p. 9; Myers & Juma, 2018, pp. 20–21). The "recipients" of these fundings (Cauhapé-Cazaux & Kalathil, 2015, p. 3) are often located in countries of the so-called "Global South" (p. 16), which are, for example, referred to as "non-democratic countries" (p. 5) or "developing countries" (p. 16), with a focus on Africa and Asia (p. 12). For journalistic education and training, governmental, or non-governmental organizations (NGOs) either invite media professionals to their editorial offices or courses or travel to different countries to offer curricula or training for local media practitioners and students on-site. Against this background, this chapter assumes that it is essential to be prepared for the requirements of different work and life contexts in different media systems and regions. Regarding international training curricula and training offers, the lack of consideration of precisely these differences has already been discussed critically. In journalism studies, for example, the avoidance of these differences amounts to a "whiteness in journalism pedagogy," like a homogeneous White perspective on news selection or journalistic values and inequalities, and the necessity to decolonize knowledge and journalism education (Alemán, 2014; Glasser, 1992; Moyo, 2022).

Here, the key question of how we can achieve a deep internationalization of journalistic training is answered by delving into three parts. First, a look at the past shows that the awareness of context-specific differences and forms of inequality was lacking but has always been demanded. Second, an examination of the present based on empirical studies and the analysis of discourses by academics and practitioners shows a slow realization when challenges persist, and potentials for a "deeply international" journalistic education. Third, a glimpse into the future can show us how challenges and possibilities of context-specific differences can be better addressed in terms of participation and deeper internationalization.

Where do we come from?

Answering the question of where we come from shows why and how international journalistic education has developed under Western normative influences. After the end of World War II, media development efforts primarily in the USA and the UK were aimed at restoring a functioning news system in the defeated countries. Journalists were seen as essential here. In 1946, for example, the USA invited journalists from Germany, Japan, and Austria to come and receive further training and "professionalization," according to Wrenn (2011, p. 85–92). Moreover, the motives for these training courses were primarily based on an anti-fascist, anti-communist, and anti-totalitarian ideology and were therefore normatively oriented (Wrenn, 2011, p. 89). The predetermined ideal of the press was that it should be independent of the government and be self-financing. However, its possible lack of freedom due to its subjugation to free-market logic was ignored (Wrenn, 2011, p. 93). After the end of the Cold War and the German reunification in 1989, optimism grew even more that the media and their development could lead to democratization worldwide (Higgins, 2014; Myers et al., 2017, p. 8; Schiffrin & Behrman, 2011, p. 340). With the "victory" of liberal democracy over communism, media development efforts became a global endeavor, "at least from the perspective of Western donors" (Miller, 2009, p. 12). The USA and Great Britain once again dominated these efforts with their ideas of a "Western" democracy and a correspondingly free press.[1] On a global level, the United Nations Educational, Scientific and Cultural Organization (UNESCO) has had a funding program for the International Programme for the Development of Communication (IPDC) since 1980 (UNESCO, 2008, p. 12). It promoted an independent and pluralistic press and journalists through the Global Initiative for Excellence in Journalism Education program (UNESCO – International Programme for the Development of Communi-

1 Here, it is unfortunately not possible to go any further into the media development work and journalistic education and training offered by Russia and China, which is certainly relevant for a discussion of asymmetries and dominance in this field, too. Either way, the studies on financial support for media development show that the main donors are the USA and European countries; see Cauhapé-Cazaux and Kalathil (2015), Myers and Gilberds (2024), and Myers and Juma (2018).

cation, 2014). Journalists were seen as key players in development, democracy, and good governance.

As early as the 1980s, Asian countries formulated the inappropriateness of "Western" news reporting and called for a corresponding "de-Westernization," and this was discussed, for example, at a symposium in Bangkok in 1985 among different media scholars and practitioners (Xu, 2009, p. 360). Another early critical examination of this dominance and the "Western" influence on media systems is the MacBride Report, which was commissioned by UNESCO. It emphasized the importance of national and independent media being independent from Western and foreign influences (International Commission for the Study of Communication Problems, 1980). "Western aid —described in the report as "a 'Marshall Plan' for the development of third world communications"—was inappropriate as it would tend to reproduce Western values and transnational corporate interests in third world societies" (Raboy, 2019, p. 16).

Overall, various dependencies and influencing factors on journalistic ideas and practices in the so-called recipient countries were and are still discussed. Regarding the perception of editorial independence, according to Schiffrin (2017, p. 13), surveys of private foundations, intermediary organizations, and recipient organizations showed that grant recipients align themselves with the priorities of the donors. The author further noted that donors also exert influence through their demands to refrain from certain reports (such as not mentioning other NGOs) or to report on certain topics in a partially predetermined manner and another form of influence is journalists' engagement in reporting on selected "development projects" (pp. 18–19). Organizations in countries that primarily receive funding and are considered to be countries of the "Global South" are more dependent (Keßler, 2022). "In poor countries, media outlets may be more likely to accept requests from donors that they would decline if they could afford to say no" (Schiffrin, 2017, p. 21). In addition, funding from international organizations, such as the UN, can influence editorial decisions and lead to less critical reports about them or their member states (Scott et al., 2017, p. 173). The ideological understanding of what journalism is or should be is discussed as "Western" or Anglo-American and as normative (Josephi, 2005, pp. 575–576; see also Radue et al. on "prefix journalisms" in this book). For Asia, for example, the term "development journalism" was used to describe its own ideas of journalism that were distinct from the "West" (Kalyango et al., 2017; Xu, 2009).[2] Historically, its roots and approaches are seen in the 1960s in Asia, such as in the Philippines and India, and especially at the beginning as a differentiation from "Western news agencies" (Xu, 2009, pp. 358–361). Here, journalists engage in goal-oriented communication in the context of specific development projects (Grossenbacher, 1988, pp. 65–66; Kalyango et al., 2017).[3] Another key aspect of this understanding of the role of development journalism is the promotion of tolerance and cultural diversity (Kalyango et al., 2017, p. 584).

2 It should be noted here that there is no one form of "development journalism," even if it is described as a special form of journalism for "developing countries" and in the service of their national development and independence; see Grossenbacher (1988, pp. 59–62) and Xu (2009).

3 The motives of "development journalism," such as positive social change, are not necessarily only to be found among journalists from "developing countries" or in the service of "development efforts." A similar form is, for example, the "public journalism" of the 1990s in the USA; see Shafer (1998). Other common features of "development journalism" described by Musa und Domatob (2007) are

This historical development described here clarifies what possibilities and challenges can be associated with journalistic education and training, especially those offered and developed by external actors. In particular, global structural inequalities and context-specific differences between countries and regions need to be reflected upon. After looking back at the past, the question of where journalistic education and training stand today needs to be explored.

Where do we stand?

The dominance of "Western" ideas of development and journalism, which has been criticized by academics and practitioners alike (Banda, 2013; Lugo-Ocando, 2020; Manyozo, 2012, pp. 200–207; Miller, 2009; Murphy & Scotton, 1987; Phiri & Fourie, 2011), has long been addressed in media development work and research through participatory approaches. Furthermore, for some time, they have been described as partially overcome[4] (Communication Initiative et al., 2007, p. 44; Drefs & Thomass, 2019; Tufte & Mefalopolus, 2009). With these participatory approaches, an "external expert stance" of stakeholders, mostly from donor countries, should be avoided (World Bank, 1996, p. 3). Here, participation is not only associated with equality and self-determination but also with the effectiveness and sustainability of "development efforts," as well as with their success, relevance, and quality (World Bank, 1996, p. ix). Overall, the objectives and standards of further journalistic education and training are described in various ways, such as "good practice" (Banda, 2013, p. 9; UNESCO, 2013, p. 5), "professionalization," or "capacity-building" (Scott, 2014, p. 80). "Professionalization" is also one of five indicators for measuring the sustainability of media in respective countries developed by the Washington/USA-based NGO International Research & Exchanges Board (IREX, 2016). Despite these approaches and standards, the following subchapter will show that dominance and the resulting challenges have not yet been overcome.

In addition to the various objectives and motives, the content of journalistic training is diverse (in line with different understandings of media development work, categorized and discussed by Manyozo, 2012; Scott, 2014). Depending on the understanding of what media development is and what it is for, and how media is used for this purpose, there are journalistic courses on topics such as reporting on health or economic issues (Schiffrin & Behrman, 2011), elections (Myers et al., 2017, p. 19), the development of business models for the self-financing of media organizations (Myers et al., 2017, p. 21; Scott, 2014, pp.

---

discussed as so-called universal ethics, such as the claim to truth and objectivity and social responsibility (pp. 318–319).

4 However, recent studies show that these asymmetrical relationships still exist. For example, the study on the unequal negotiation processes of the 2030 Development Agenda within the UN, see Denk (2023), and for the study on the one-sided dominance of development cooperation between actors from Germany on the one hand and Peru and Bolivia on the other, see Nguyen (2016). Also, the many voices of thinkers from the "Global South" on the Western dominance of journalistic education and their calls for African perspectives illustrate this, for example Banda (2009), Ezumah (2019), Kasoma (1996), Nyamnjoh (2010), Phiri & Fourie (2011), Rodny-Gumede (2018), and Ugangu (2019).

85–86), science journalism (Lugo-Ocando, 2020, pp. 128–130), how to use new or digital technologies (Myers et al., 2017, p. 46; Scott, 2014, p. 98), and methods of investigative journalism (Lublinski et al., 2016).

Journalistic skills and quality are measured, among other things, by whether journalism schools or university programs exist in the respective countries (Center for International Media Assistance [CIMA], 2007, pp. 7, 9; Nelson & Susman-Peña, 2012, p. 23; UNESCO, 2013, pp. 6, 22). In the Model Curricula for Journalism Education, the UNESCO member states declare: "We do believe, however, that study in university disciplines should continue to be seen as basic to professional training in journalism" (UNESCO, 2007, p. 7). This orientation can be seen in terms of institutionalization and formalization (Schmidt & Martens-Edwards, 2015, p. 15) and follows a trend from the USA and Great Britain begun in the 1970s and 80s (Martens-Edwards, 2023, pp. 68–71; Murphy & Scotton, 1987, pp. 17–18). However, the preference for either university or practical training was not uniform between the UK and the USA, or even in European countries, which is clearly shown in the international comparison of journalistic education and training sponsored by UNESCO (Gaunt, 1992). This also applies to the question of who leads such courses: either practitioners or academics. Either way, when we talk about preferences, the question of who designs them is relevant. To answer this question, we use the example of the UNESCO training curricula to illustrate why and how they are discussed as a normative Western dominance. Overall, UNESCO's journalistic standards emphasize practical skills, reflections on the role of journalists and their function in society, and the importance of including other disciplines from arts and sciences (UNESCO, 2007, pp. 7–8). In 2007, UNESCO developed indicators for the quality of journalism education (Berger & Matras, 2007), based on European countries and the USA. They were then adapted to the African continent to define excellence centers of journalism here.[5] Besides the UNESCO curricula, standards for journalism education were also defined at the World Journalism Education Congress (WJEC) (MacKinnon, 2007). The WJEC is an alliance of academic associations worldwide that deal with journalism and mass communication and work together with UNESCO – International Programme for the Development of Communication (2020, p. 1). Their standards for journalistic education require, among other things, that educators are a mix of academics and practitioners (principle 3), that they must be proficient in a variety of computer-based technologies (9), that they are required to collaborate internationally (11), and, finally, that they have an interest in context-specific differences (10) (MacKinnon, 2007).

In 2020, the WJEC and UNESCO reaffirmed the importance of journalism education and training for the 2030 Development Agenda with the Paris Declaration on Freedom of Journalism Education at the World Journalism Education Congress: "We believe that journalism education has a fundamental role to play towards more inclusive societies and the United Nations' 2030 development agenda" (UNESCO – International Programme for the Development of Communication, 2020, p. 8). UNESCO has a publication series on journalism education and offers various curricula and training courses (Banda, 2013;

5 The authors noted here that "by no means should UNESCO ignore the institutions that are not in the potential Centres of Excellence list, but rather see its work with the latter as a way of leveraging impact on such Centres of Reference (and others)"; see Berger und Matras (2007, p. 28).

UNESCO, 2007).[6] The universally conceived model curricula are still discussed critically as in the past by several participants at two meetings, particularly regarding local differences and "Western dominance" (UNESCO, 2013). On the one hand, the curricula homogenize context-specific differences with their specific and different journalistic practices and enable the continuation and expansion of postcolonial privileges of former colonizers (Harris et al., 2023, pp. 155–156), as the above examples of factors influencing journalistic training have shown. On the other hand, they link journalistic standards with the promotion of democracy and has a corresponding focus (Harris et al., 2023, p. 159). However, "In this vein, some scholars have legitimately and rightly questioned whether the UNESCO model curriculum design can be 'universal'" (UNESCO, 2013, p. 9). In addition, literature, and material references that are relevant for contexts such as Asia, for example, are indicated as missing (UNESCO, 2013, pp. 28–29).[7] Furthermore, the curriculum content does not always correspond to local realities, as discussed for Africa (UNESCO, 2013).[8] Here, references to the history of African journalism must inform journalism education, contrary to the dominant neoliberal model (UNESCO, 2013, pp. 53–57). Above all, the connection between ideas of development and economic standards subjects journalistic education and training to neoliberal constraints. Even in the "developed" countries themselves,[9] journalism education and training curricula are based on "journalism experiences grounded in western nations" (Aujla-Sidhu, 2022, p. 1638). This is shown, for example, by an analysis of the reading lists and module descriptions of journalism courses accredited in the UK, which neglect non-White perspectives and marginalize knowledge of the so-called Global South (Aujla-Sidhu, 2022). This shows that what is practiced locally is also transferred to international journalistic training and education and vice versa.[10] In addition, these dominances lead to an equally dominant teaching method "that is centered on the idea of teachers (as subjects) 'depositing' knowledge into students (as objects), which they have to mechanically memorize" (Harris et al., 2023,

---

6 The first UNESCO model curricula were published in 2007, and a second version was published in 2013.

7 Asia is a large context but is addressed as such here; see Sundeep R. Muppidi on "Developing an 'Asian' framework for UNESCO's model curricula" in UNESCO (2013, pp. 28–30). In some cases, a South Asian experience is described and experts and researchers from India, Sri Lanka, Thailand, Philippines, Laos, and Indonesia are mentioned.

8 The paper refers to Africa as the bigger context, mostly referring to sub-Saharan African countries. In contrast to the West, Africa is described here as a single entity. "Training programs from the West are often out of tune with the realities on the ground in Africa." See Ibrahim Seaga Shaw on "UNESCO model curricula in Africa: Call for a more bottom-up approach" in UNESCO (2013, p. 53)

9 The UN declarations on the 2030 Agenda for Sustainable Development and the World Development Report for example distinguish between "developed" and "developing countries" (United Nations General Assembly, 2015, p. 3; United Nations Development Programme, 2001). The so-called developed countries are located primarily in the Northern hemisphere and primarily refer to the USA and European countries.

10 See also the literature review and the study of two journalism training courses in the USA by Alemán (2014). It shows that a so-called objectivity in journalism is equated with Whiteness or the perspective of White trainers and journalists, Alemán (2014, p. 75).

p. 157).[11] Additionally, journalism programs and model curricula are limited by poor theoretical foundations, as well as legal, technical, and economic constraints. They can also be limited because of ideological restrictions, which make a genuine "deep" internationalization of journalism training and the trainees' subsequent professional practice difficult to implement.

In this respect, various studies based on interviews have shown the difficulties of "Western" journalistic training in transferring their content and ideas to foreign local contexts. In Indonesia, for example, journalists could not adopt the practices presented by invited external trainers. The implementation of the content was made difficult by their different journalistic values, normative foundations, practical conditions, and contexts (Muchtar & Hanitzsch, 2013). However, trainers also encounter difficulties when they try to train journalists from other contexts. These are based, for example, on their lack of knowledge of local circumstances, practices, and a possible rejection of their training content by local journalists (Becker et al., 2002; Ognianova, 1995). Other challenges emerge from sponsoring government organizations or NGOs that follow their specific ideas of journalism and try to transfer them to other contexts (Noske-Turner, 2017, pp. 8–9; Schiffrin, 2010, p. 406; Thomaß, 2012, p. 397). Thus, the addressed or receiving countries are subject to various external influences. These can even devalue local journalism ideas and practices (Nyamnjoh, 2010). Furthermore, the dominance of colonial languages at universities and training institutions, as well as of their learning materials and university models, makes it difficult for journalistic education and training to be practiced independently and locally (Murphy & Scotton, 1987). In his analysis of various curricula for education and training in Bangladesh, Ullah (2022) also concluded that there existed the dominance of one-sided guidelines from sponsors. The Chair of Mass Communication and Journalism at Rajshahi University, Al Mamun, explained:

> Everything comes from the top; from DW Akademie, we have no evidence for what we will do. The DW Akademie people come and ask to launch something that many faculty members do not support, except junior-level faculty members, to show their enthusiasm to participate in those project activities. [Interview 15.03.2021]. (Ullah, 2022, p. 154)

It becomes clear here that recognizing power inequalities and country-specific agendas along with local differences in journalistic practices and perceptions are crucial for the efficiency and legitimacy of journalistic training—both for training in host countries and for external training offered in recipient countries. With regard to internationalization and thus the recognition of context-specific differences and the need for participation in creating and providing journalistic training programs, there are various possibilities:

> In participatory approaches to development there are no universal models to be followed or adopted. All societies are different and require different methods for real and endogenous social change to occur. . . . (Phiri & Fourie, 2011, p. 92)

---

11 In his critical examination of pedagogy of the oppressed, Freire (2021) also described this as "banking concept of education" (pp. 28–29).

In order to counter unequal power relations and therefore also the one-sided control and dominance of media development work, representatives of participatory approaches demand that the needs of the recipients and their own relevance are decisive (Manyozo, 2012, pp. 152–195; Nelson, 2019; Thomaß, 2012, p. 403). In addition to the benefits of greater effectiveness and sustainability, participation is also a fundamental human right (Tufte & Mefalopolus, 2009, p. 18). For journalistic training courses, this type of participation would mean that the participating journalists use their knowledge to determine and expand the content and design of the training courses. For trainers and external curriculum designers, it would mean that "their role is to act as a catalyst in the empowerment of others who are different from themselves, without controlling the process and while maintaining a genuine respect for local knowledge" (Scott, 2014, p. 73). Thus, practical knowledge and the experiences of the journalists and their socialization play a major role here (Glasser, 1992, p. 135). A so-called professionalism with universal and homogenizing assumptions tends to ignore their diversity.

> What ends up in our lectures and texts are not the vividly cultural and historical accounts of journalism that would honor the role of experience in the production of knowledge. What we proffer instead is either a sanitized and scientized version of the practice of journalism. . . . (Glasser, 1992, p. 138)

All these aspects of participation, such as the orientation toward local and practical knowledge, a relationship of trust at eye level (Ashman, 2001), and the inclusion of various interest groups (Drefs & Thomass, 2019), are also reflected in the UN's declaration of its 9th roundtable of communication for development:

> Communication for Development is about people, who are the drivers of their own development. . . . Communication for Development is a two-way process—it is about people coming together to identify problems, agree on visions for desirable futures, create solutions and empower the poorest. Participatory Communication for Development does not only apply to work with communities. It is an approach of equal importance to all stakeholders. Communication for Development is about the co-creation and sharing of knowledge. Communication for Development respects Indigenous knowledge and culture; local context is key. (Food and Agriculture Organization [FAO] et al., 2005, p. 8)

In addition to power inequalities and corresponding asymmetries in journalistic education and training, the different journalistic concepts and practices (also described as "journalism cultures"; see Hanitzsch et al., 2019) also play a decisive role in the question of context-specific and suitable offers. In sub-Saharan Africa, for example, Skjerdal (2012) has identified numerous normative journalism models, such as ujamaa journalism, Indigenous journalism, and azmari- and dagu-based journalism, to name a few. Moreover, the role of journalists in "crisis and transition countries," for instance, varies considerably. Accordingly, there is not only conflict-sensitive journalism. In their interviews with 100 journalists from Egypt, Kenya, Serbia, and South Africa, Lohner et al. (2019) found that the understanding of roles differs depending on the context and is sometimes mixed, such as that of objective, investigative, or mediating journalism. They concluded that journalistic values and practices need to be contextualized. According to Voltmer and Wasserman (2014), the understanding of key terms and concepts in the journalistic field can vary, as cultural traditions, historical experiences, and political values influence them. For example, the meaning of "freedom of the press" can therefore be understood differently (Voltmer & Wasserman, 2014). Language can also be a hybrid of local languages and "national languages" established by colonial countries and thus contradict a uniform journalistic form of language and use of terms (this is also referred to as the creolization of language; see Nyamnjoh, 2010, p. 18). The journalism and communication scientist Bebawi (2023), for example, described the unique aspects of an "Arab culture of journalism" and its role in journalistic training. Accordingly, Arabic is an emotional language that significantly influences reporting (p. 196). Stylistic devices here are exaggeration, overemphasis, and repetition. Furthermore, hypotheses are not tested as much as articles are researched and written according to an anticipated point of view (Bebawi, 2023, p. 197). Working on these stylistic devices and perspectives in further training courses takes time and can also be difficult for individual journalists to implement after returning to their newsrooms. Overall, it is important to design training courses in such a way that journalists learn "how 'to do' journalism within their own environments" (Bebawi & Onilov, 2023, p. 1). As an Australian trainer, Downey (2000) had similar experiences in Laos, Cambodia, and Vietnam. The local journalists could hardly apply the training content because the editors of their media organizations and their superiors in the ministries were not convinced. Other challenges were that Downey did not speak the local languages and that the participating journalists came from different media sectors. She had given the training herself with the help of local translators. She also emphasized the need for sufficient preparation time for training sessions in order to understand local conditions. Even though Downey (2000, pp. 110–111) described external trainers and especially short training (of 1 to 2 weeks) as a problem, she also saw advantages in their ability to share their experience and knowledge. The British trainers Gaber and Goldsmith (2023), who have worked in sub-Saharan Africa for over 20 years, have agreed with this view. Their conclusion was that the presence of international trainers can encourage local journalists to change (p. 188). Also, Duran, who himself is from France and has given media training mainly in Turkey, came to the following conclusion: "Journalistic training abroad is not impossible, nor is it completely useless" (2005, n. p.) Nevertheless, he emphasized that it is important to reflect on various differences. These include language, different ways of thinking and various ideologies, and the cultural and educational background of the participants. Furthermore, the relationship between the country of origin of the trainers and the country of origin of the course participants also plays a role (like between the USA and Iraq, when a US-American trainer is invited there), for example, when the countries are in a current conflict (Duran, 2005). In addition, there are "cultural differences" that can "complicate" the work of trainers, such as "gender differences (men and women sitting separately, for example), differences around levels of informality, acknowledging hierarchies" (Gaber & Goldsmith, 2023, p. 177).

Norwegian media scientist Skjerdal (2009) described his approach to address differences in journalistic values and practices. He co-developed the curriculum for a master's program at Addis Ababa University in Ethiopia, which was launched in 2004 and financed by the Norwegian government as part of its media development work through the Norwegian Agency for Development Cooperation (NORAD) (p. 24). Skjerdal (2009) described what he saw as an interplay between "universal" and locally specific journalistic values. For him, one example was the communitarian orientation of "public or civic journalism" in the USA in the 1990s. Similar orientations can be found in Ethiopia, even if this cannot simply be transferred to this context. According to this, the author noted, it is a matter of not simply adopting the communitarian orientation, "but of recovering it in a culture where it is already deeply rooted in history and communities" (p. 30). Overall, it is striking how context-specific differences are ignored, especially in the field of journalistic education and training, but still play a major role in areas such as intercultural pedagogy (Auernheimer, 2002; Prengel, 2019), intercultural and transcultural communication (Thomas et al., 2005), and media reception research (Stehling, 2015):

> It is curious that questions of cultural dependence are legitimate topics for research when applied to the consumption of media programs and the like but largely a non sequitur when applied to the consumption of media education. (Glasser, 1992, p. 135)

Another context, another journalism? The example of Afghanistan (Kefa Hamidi)

Another example of addressing different journalistic norms and practices is the proposal for a dual training program for journalists in Afghanistan developed in 2016 and 2017 (Hamidi & Brüchner, 2020; Kutsch & Hamidi, 2020).

This case study focused on the design and implementation of a dual training program for journalists in Afghanistan as well as the integration of mediation skills as a critical component of journalistic practice. The duality of the training program refers to combining theoretical education with simulated practical experience. The program was tailored to the specific needs and cultural context of Afghanistan, particularly its multiethnic and Islamic identity. Through a thorough needs assessment, including interviews with various stakeholders, such as Afghan university journalism faculty, media representatives, and university policymakers, a dual training program for professional journalism was designed. The findings of this analysis were used to develop the program, which aimed to combine theory with simulated practical experience. In addition, emphasis was placed on teaching mediation skills, which are recognized as integral to journalistic practice in Afghanistan. The case study not only shed light on the genesis of the program but also highlighted the challenges and progress made in working with Afghan universities and the Ministry of Education.

A dual training program for professional journalism, which has long been practiced in various European and non-European countries, did not exist in Afghanistan. The proposed program aimed to combine theory and simulated practice. It was based on recent international research and a needs assessment.

Religion and tradition shape the self-image of journalists in Afghanistan and other Islamic societies. Hence, the connection between journalistic values and the teachings of Islam should be taken into account in training programs. Accordingly, the development of the program followed a participatory approach. From summer 2016 to May 2017, a team from the University of Leipzig conducted a needs and feasibility analysis in the form of guided interviews. The central topics were educational goals, areas, and contents, as well as university teaching. We also asked about the structural framework of Afghan universities and colleges (such as admission restrictions for students, the academic profile of teaching staff, and physical and technical investments).

Mediation skills emerged as a new key journalistic skill in Afghanistan. On the one hand, the interviewees showed similar attitudes toward journalism training compared to international models, such as an orientation toward professional competence, specialization, presentation, and cross-media competence, organizational competence, and social orientation. In addition, the participants in the study expressed that a further component of journalism training should be added, which can be described as "mediation skills." This involves promoting the common good and national development through journalistic work. It includes techniques such as creating a "forum," seeking "consensus," and acting as a "teacher." The interviewees stated that this type of reporting can contribute to the positive development of the country. They emphasized the importance of mediation in a variety of conflict situations and as a contribution to national development and social harmony, which can be described under the concept of "constructive national identity." The concept of constructive national identity emphasizes the collective sense of belonging and shared values within a nation, leading to unity, social cohesion, and progress (Smith, 1991). Unlike traditional understandings of national identity, which are often based solely on historical or cultural factors, constructive national identity emphasizes the active construction and promotion of positive narratives about the nation (Anderson, 1983). This approach encourages citizens to participate in shaping and defining their national identities, thereby promoting inclusiveness and a sense of ownership among different social groups (Brubaker, 1996). By promoting a constructive national identity, societies aim to build resilience and adaptability in the face of challenges while fostering a shared vision for the future (Smith, 1991).

So far, however, these mediation skills have not been taught in academic journalism training in Afghanistan. The needs assessment showed that mediation skills should be systematically integrated into journalism training. This can help promote a constructive national identity. It also develops cultural sensitivity in a multicultural and multiethnic society, as well as conflict management skills. This is particularly important in a country characterized by long-standing conflicts and social challenges.

The study was groundbreaking, perhaps unique, in following the system over the years to understand its status quo, strengths, and weaknesses over the long term. Over a period of 22 years, the system has evolved rapidly and has faced many challenges. The study identified these challenges and shortcomings and suggested ways to improve the education system. Following the completion of the study, negotiations between the University of Leipzig and the Afghan Ministry of Higher Education made significant progress. It was agreed that the proposals developed in the study would be phased into the system, with the agreement to be signed and implemented over a five-year period. This was due to take place in 2021, but toward the end of that year, Afghanistan's state structures collapsed and power passed to the Taliban. The new regime reversed all the developments of the previous two decades and brought about a complete overhaul of the system.

Despite the setback in Afghanistan, there is now hope that the lessons learned and the proposals put forward can be applied to other countries going through a similar process. Many countries face similar challenges and could benefit from the lessons learned.

A different context, a different journalism? The examples of Bhutan and Nepal (Beate Illg)

There is no doubt about the "Western" influence on journalism in Asia—not only but also due to media training and support by so-called "highly developed" countries. This is clearly shown in qualitative interviews conducted in Nepal and Bhutan with journalists about their self-perceptions within the democratic process in their country[a] (see Illg, 2018, 2019). Though both countries changed in completely different ways, they both became democratic structures in 2008. Nepal became a Federal Democratic Republic after a 10-year guerrilla war initiated by the Maoists (1996–2006). The Kingdom of Bhutan became a constitutional monarchy due to the King's voluntary decision.

Journalists in both countries are highly intrinsically motivated—they strive to give "a voice to the voiceless," want to help to improve peoples' lives, and see media as the fourth estate. Advocacy journalism is the label that characterizes this—according to a "Western" perspective.

This motivation is especially remarkable because journalists in both countries are not well paid; in fact, sometimes they even work for a month without receiving salary. Most Nepali journalists need a second job to secure survival. Investigative journalists in Nepal often face threats and intimidations. In a small society of about 760,000 inhabitants like Bhutan, this is less common, but journalists often feel social pressure that leads to self-censorship. In addition, the Bhutanese Gross National Happiness sometimes causes reports about crime or other negative topics to be skipped.

Although it is obvious that the interviewed journalists refer to objectivity as a common standard of journalism, they often claim that it is needed "to find their own way" in Bhutan as well as in Nepal. Both countries are quite young democracies, and other democracies within their neighborhood are not considered a blueprint because they do not seem to work very well. The role of the media as a watchdog is probably even more important in young democracies with problems such as corruption (especially Nepal) or social control (particularly Bhutan). As a consequence, it is perhaps more necessary that journalists focus on the position of the citizens, the disadvantaged, and those who cannot stand up for their rights and interests. Maybe fairness is more relevant than objectivity and neutrality—as Kunda Dixit (personal communication, February, 2024), an internationally well-known Nepali editor, journalist, and lecturer, mentioned in an interview. Asked about his perspective on what is needed for a deep internalization of journalism education, Dixit mentioned that teaching journalism on a national level is obsolete because all relevant topics like economics, climate, and migration are internationally linked. He argued vehemently for highly specialized journalists, like scientists, who have to be able to speak to authorities: "We really need this global minded specialized journalist." In his view, "journalists need to know what the problem is—of their community, their country, the world; this becomes more and more important." This theme-centered perspective is his key aspect because "**why** to do journalism, **why** it is important—it becomes critical." Implicitly, the interviewee referred to the concept of advocacy journalism without little regard for that kind of classification.

Media literacy was mentioned in this context as well, and the problem that it is often the most modern technology has the highest credibility. But techniques can be learned by training on the job—it is not necessary to go to university for that. Dixit stated that there is probably still a lack of equipment in several countries, such as Nepal, where support can be useful—"but content is much more important." Working as a lecturer for several international institutions like New York University (at the Abu Dhabi campus), he argued for field trips and cross-cultural experiences because this would open young peoples' eyes, widen their horizon, and create empathy. In his view, field trips are extremely important "because students of journalism really need exposure to the outside world and to see it for themselves." This enables them to "write with empathy about some other place."

As previously mentioned, Dixit considered a global perspective to be more important than ever before. As a consequence, "a credible global media is needed." Therefore, student and faculty exchanges are required as well as joint media training to "make them work together and learn from each other." This could be a good start for "deep internationalization."

Where do we want to go?

After critically reflecting on the past and current issues of journalistic education and training, as well as differences in journalistic roles and practices, questions arise as to what we can learn from this and where the field needs to develop further. Altogether, these questions have various practical and scientific implications. As the literature review and empirical studies presented above have shown, it is essential to be prepared for the requirements of different work and life contexts in different media systems and regions. Overall, the universality (often a frequent synonym for objectivity) of knowledge needs to be questioned, as it is a legitimization of dominance relations (Aujla-Sidhu, 2022, p. 1642). It would be important for journalistic training and education actors to be sensitized to different life and work contexts, as well as to practices of inequality, while reflecting their own teaching content and ideas. In this way, journalistic training can then be designed in a dialogical and participatory way. Teachers become learners, and learners become teachers (Freire, 2022, p. 80). Methodological approaches, for example, can be where the addressed participants name and develop their own topics, which can be supplemented by the instructors through a dialogical exchange (Freire, 2022, p. 120). Furthermore, the references of knowledge and its embedding in historical and current

---

a In Nepal, Illg conducted 24 qualitative interviews with journalists in 2013 and 17 interviews in 2017; in Bhutan, she conducted 17 interviews with journalists in 2017.

geopolitical conditions are reflected upon. Overall, the most important prerequisite is a critical and reflective attitude on the part of course providers and trainers. These reflections can be described as an integral part of pedagogical professionalism (Heinemann & Sarabi, 2020, p. 310). In this context, it helps to abandon the concept of "development" and its implicit constructions (Ziai, 2014, p. 426). Another tool is cooperation with actors from civil, teaching, and media organizations from countries of the so-called Global South, where there are also thinkers and publications that deal with the decolonization of journalistic education and training that goes beyond internationalization and de-Westernization (Moyo, 2022, pp. 1570–1571). Accordingly, the de-Westernization approaches have ignored asymmetrical power relations and the associated production of knowledge between the Global North and South. The internationalization approaches, on the other hand, have been motivated by the market economy and served more to spread Western models (Moyo, 2022). For this reason, it is important to explicitly decolonize journalism studies and training:

> While there have been calls to make journalism education more international in its outlook, the terminology decolonizing is specifically used to denote the undoing of colonial legacies in creating ethnically and racially divided societies. (Aujla-Sidhu, 2022)

This includes the inclusion of postcolonial theories and discourses on inequality and racism, which also reflect current structures: "No person can escape . . . geographical, and racial hierarchies set by this power structure of global coloniality" (Moyo, 2022, p. 1569). In addition, current structures of media development work, including journalism training and education, should be seen in the context of their historical development. Furthermore, different journalistic orientations, such as the "constructivist models of journalism, advocate or campaign journalism" should be included in the curricula (Aujla-Sidhu, 2022, p. 1647). Other options include regular workshops for organization and teaching-learning staff to expand bibliographies and case studies, inviting guest speakers who respond to the diversity of course participants, and representing alternative or community news organizations (Aujla-Sidhu, 2022, p. 1649). "At the most basic level, the locus of enunciation is about the fact that journalism practice and academic training are culturally and socially shaped" (Moyo, 2022, p. 1569). Thus, so-called informal or practical knowledge plays a major role (Glasser, 1992, p. 138). Intercultural communication and education training can also sensitize educators and media development actors to contextual differences and corresponding challenges (as suggested by Armbruster, 2014; Ognianova, 1995). Previous stays of trainers abroad alone are not enough here (Brinkmann & van Weerdenburg, 2014, pp. 95–134). Also, for journalistic training and education donors, participatory measures are crucial to overcome inequalities and recognize the diversity of journalistic practices and needs. Despite structural and institutional limitations, there are various ways to do this (Keßler, 2022; Waisbord, 2008). However, to not remain just lip service, participatory approaches and journalistic education need to be reviewed and evaluated accordingly, especially by the recipients themselves (Carpentier, 2011, p. 52; Chambers, 2017, pp. 134–135; Noske-Turner, 2017; Tufte & Mefalopolus, 2009, p. 6). With regard to a participatory design and reflection of training courses (Drefs & Thomass, 2019), we can also learn from the findings of development cooperation in general (Chambers, 2002, 2017). Furthermore, local actors and their knowledge must guide decisions, and trustful cooperation on an equal footing must be established accordingly (Keßler, 2022). Finally, scientific results can be transferred even more into training materials and approaches.

For journalism research, it is advisable to investigate practices of inequality further and their effects on journalistic training, going beyond the numerous existing interview studies (Keßler, 2024). In particular, ethnographic approaches (Breidenstein et al., 2015; Knoblauch, 2005; Lilienthal, 2017) and inductive data analyses (Corbin & Strauss, 2015), which work without fixed categories and theories, can examine actual and everyday conditions of journalism training and interactions between different actors even more closely and avoid one's own (possibly "Western") assumptions. Moreover, for international and comparative journalist research, it is "vital that researchers hold to the consensus that they must continue to increase the number of non-Western case studies and incorporate a greater number of scholarly perspectives from the Global South" (Martens-Edwards, 2023, p. 436). Finally, educational research findings and approaches are also particularly insightful and applicable to the teaching and learning setting of journalism training courses (Akbaba et al., 2022; Malik, 2022; Mecheril & Plößer, 2009; Mills & Morton, 2013; Tervooren et al., 2014).

References

Akbaba, Y., Buchner, T., Heinemann, A. M. B., Pokitsch, D., & Thoma, N. (Eds.). (2022). *Lehren und Lernen in Differenzverhältnissen: Interdisziplinäre und Intersektionale Betrachtungen*. Springer VS. https://doi.org/10.1007/978-3-658-37328-3

Alemán, S. M. (2014). Locating whiteness in journalism pedagogy. *Critical Studies in Media Communication, 31*(1), 72–88. https://doi.org/10.1080/15295036.2013.808355

Anderson, B. (1983). *Imagined communities: Reflections on the origin and spread of nationalism*. Verso.

Armbruster, L. (2014). *Cooperation across cultures: An analysis of intercultural competence in development organizations* [IEE Working Paper]. Ruhr-Universität Bochum. https://www.econstor.eu/bitstream/10419/183559/1/wp-205.pdf

Ashman, D. (2001). Strengthening North-South partnerships for sustainable development. *Nonprofit and Voluntary Sector Quarterly, 30*(1), 74–98. https://doi.org/10.1177/0899764001301004

Auernheimer, G. (Ed.). (2002). *Interkulturelle Kompetenz und pädagogische Professionalität*. Leske & Budrich.

Aujla-Sidhu, G. (2022). Decolonizing journalism education to create civic and responsible journalists in the West. *Journalism Studies, 23*(13), 1638–1653. https://doi.org/10.1080/1461670X.2022.2096666

Banda, F. (2009). Explorando la educación en medios como práctica cívica en África [Exploring media education as civic praxis in Africa]. *Comunicar, 16*(32), 167–180. https://doi.org/10.3916/c32-2009-02-015

Banda, F. (Ed.). (2013). *Model curricula for journalism education: A compendium of new syllabi.* UNESCO. https://unesdoc.unesco.org/ark:/48223/pf0000221199

Bebawi, S. (2023). Understanding different journalisms. In S. Bebawi & O. Onilov (Eds.), *Different global journalisms: Cultures and contexts* (pp. 193–199). Palgrave Macmillan. https://doi.org/10.1007/978-3-031-18992-0_10

Bebawi, S. & Onilov, O. (Eds.). (2023). *Different global journalisms: Cultures and contexts*. Palgrave Macmillan. https://doi.org/10.1007/978-3-031-18992-0

Becker, L. B., McConnell, P. J., & Punathambekar, A. (2002, July 21–26). *The impact of journalism training on the trainers: Reflections of U.S. international fellows* [Paper presentation to the Professional Education Section of the International Association for Media and Communication Research]. IAMCR, Singapore.

Berger, G., & Matras, C. (2007). *Criteria and indicators for quality journalism training institutions & identifying potential centres of excellence in journalism in Africa*. UNESCO. https://unesdoc.unesco.org/ark:/48223/pf0000151496

Breidenstein, G., Hirschauer, S., Kalthoff, H., & Nieswand, B. (2015). *Ethnografie: Die Praxis der Feldforschung* (2nd ed.). UVK. https://elibrary.utb.de/doi/book/10.36198/9783838544977

Brinkmann, U., & van Weerdenburg, O. (2014). *Intercultural readiness: Four competences for working across cultures*. Palgrave Macmillan. https://doi.org/10.1057/9781137346988

Brubaker, R. (1996). *Nationalism reframed: Nationhood and the national question in the New Europe*. Cambridge University Press. https://doi.org/10.1017/CBO9780511558764

Carpentier, N. (2011). *Media and participation: A site of ideological-democratic struggle*. Intellect. https://doi.org/10.26530/OAPEN_606390

Cauhapé-Cazaux, E. G., & Kalathil, S. (2015). *Official development assistance for media: Figures and findings – A report by CIMA and the OECD.* Center for International Media Assistance; Organisation for Economic Co-operation and Development. https://www.cima.ned.org/publication/official-development-assistance-for-media-figures-and-findings/

Center for International Media Assistance. (2007). *Media assistance: Challenges and opportunities for the professional development of journalists* [Working group report]. https://www.cima.ned.org/wp-content/uploads/2015/02/CIMA-Professional_Development-Working_Group_Report.pdf

Chambers, R. (2002). *Participatory workshops: A sourcebook of 21 sets of ideas and activities.* Earthscan.

Chambers, R. (2017). *Can we know better? Reflections for development.* Practical Action.

Communication Initiative, Food and Agriculture Organization of the United Nations, & World Bank. (2007). *World congress on communication for development: Lessons, challenges, and the way forward.* http://documents.worldbank.org/curated/en/290941468138271611/World-congress-on-communication-for-development-lessons-challenges-and-the-way-forward

Corbin, J., & Strauss, A. (2015). *Basics of qualitative research: Techniques and procedures for developing grounded theory* (4th ed.). Sage.

Denk, A. (2023). *Nachhaltige Entwicklung und globale Ungleichheit: Eine wissenspolitologische Studie über die Entwicklungsagenda der Vereinten Nationen*. Nomos.

Downey, S. (2000). Journalism training in Laos, Cambodia and Vietnam. *Asia Pacific Media Educator, 8*, 99–112. https://ro.uow.edu.au/apme/vol1/iss8/9/

Drefs, I., & Thomass, B. (2019). The participation approach in media development cooperation. In K. Voltmer, C. Christensen, I. Neverla, N. Stremlau, B. Thomass, N. Vladisavljević, & H. Wasserman (Eds.), *Media, communication and the struggle for democratic change: Case studies on contested transitions* (pp. 257–279). Palgrave Macmillan. https://doi.org/10.1007/978-3-030-16748-6_11

Duran, R. (2005). *Trainers can remain foreign to local journalists*. Nieman Reports. https://niemanreports.org/articles/trainers-can-remain-foreign-to-local-journalists/

Ezumah, B. (2019). De-Westernizing African journalism curriculum through glocalization and hybridization. *Journalism & Mass Communication Educator*, 74(4), 452–467. https://doi.org/10.1177/1077695819849360

Food and Agriculture Organization, Governo Italiano, UNESCO, World Bank, International Development Research Centre & Centre, & Centre for Agricultural and Rural Cooperation. (2005). *Communication for development roundtable report: Focus on sustainable development*. https://www.fao.org/3/y5983e/y5983e05.htm

Freire, P. (2021). *Pedagogy of hope: Reliving pedagogy of the oppressed* (R. R. Barr, Trans.). Bloomsbury Academic. https://doi.org/10.5040/9781350190238

Freire, P. (2022). *Pedagogy of the oppressed* (50th anniversary ed.). Bloomsbury.

Gaber, I., & Goldsmith, N. (2023). Confessions of two well-meaning 'Mzungu' journalism trainers. In S. Bebawi & O. Onilov (Eds.), *Different global journalisms: Cultures and contexts* (pp. 173–191). Palgrave Macmillan. https://doi.org/10.1007/978-3-031-18992-0_9

Gaunt, P. (1992). *Making the newsmakers: International handbook on journalism training*. Greenwood Press.

Glasser, T. L. (1992). Professionalism and the derision of diversity: The case of the education of journalists. *Journal of Communication*, 42(2), 131–140. https://doi.org/10.1111/j.1460-2466.1992.tb00785.x

Grossenbacher, R. (1988). *Journalismus in Entwicklungsländern: Medien als Träger des sozialen Wandels?* Böhlau.

Hamidi, K., & Brüchner, A. (2020). Reforming journalism education on a tertiary level in Afghanistan: Recommendations for a dual education model. *Global Media Journal – German Edition*, 10(2). https://doi.org/10.22032/dbt.47741

Hanitzsch, T., Hanusch, F., Ramaprasad, J., & de Beer, A. S. (Eds.). (2019). *Worlds of journalism: Journalistic cultures around the globe*. Columbia University Press. https://doi.org/10.7312/hani18642

Harris, S. T. G., Mfula, C., & Manthalu, C. H. (2023). Piecemeal pedagogies: Reflecting on the nature, context, and impact of journalism training and education in Malawi and Zambia. In B. Mutsvairo, S. Bebawi, & E. Borges-Rey (Eds.), *The Routledge companion to journalism in the Global South* (pp. 154–165). Routledge. https://doi.org/10.4324/9781003298144-16

Heinemann, A. M. B., & Sarabi, S. A. (2020). Paternalistic benevolence – enabling violence: Teaching the hegemonic language in a double bind. *European Journal for Research on the Education and Learning of Adults*, 11(3), 309–320. https://doi.org/10.25656/01:21089

Higgins, D. (2014). The Western way? Democracy and the media assistance model. *Global Media Journal – German Edition*, *4*(2). https://www.db-thueringen.de/servlets/MCRFileNodeServlet/dbt_derivate_00031117/GMJ8_Higgins_final.pdf

Illg, B. (2018). Freiheit durch Demokratie – Freiheit in der Demokratie? In A. Czepek, M. Hellwig, B. Illg, & E. Nowak (Eds.), *Freiheit und Journalismus* (pp. 173–188). Nomos.

Illg, B. (2019). Self-perception of Nepalese journalists within the democratization process of the country. In B. Dernbach & B. Illg (Eds.), *Journalism and journalism education in developing countries* (pp. 116–129). Manipal Universal Press.

International Commission for the Study of Communication Problems. (1980). *Many voices, one world: Towards a new more just and more efficient world information and communication order*. UNESCO; Kogan Page. https://waccglobal.org/wp-content/uploads/2020/07/MacBride-Report-English.pdf

IREX. (2016). *Media sustainability index 2016: Development of sustainable independent media in Europe and Eurasia*. https://www.irex.org/sites/default/files/pdf/media-sustainability-index-europe-eurasia-2016-full.pdf.pdf

Josephi, B. (2005). Journalism in the global age: Between normative and empirical. *Gazette: The International Journal for Communication Studies*, *67*(6), 575–590. https://doi.org/10.1177/0016549205057564

Kalyango, Y., Hanusch, F., Ramaprasad, J., Skjerdal, T., Hasim, M. S., Muchtar, N., Ullah, M. S., Manda, L. Z., & Kamara, S. B. (2017). Journalists' development journalism role perceptions. *Journalism Studies*, *18*(5), 576–594. https://doi.org/10.1080/1461670X.2016.1254060

Kasoma, F. (1996). The foundations of African ethics (Afriethics) and the professional practice of journalism: The case for society-centred media morality. *Africa Media Review*, *10*(3), 93–116. https://d.lib.msu.edu/jamr/22

Keßler, M. (2022). How "insiders and outsiders" perceive media development work. Reflections on their relationship and cooperation. *Global Media Journal – German Edition*, *12*(1). https://doi.org/10.22032/dbt.53059

Keßler, M. (2024). *Lehren und Lernen mit Asymmetrien: Eine qualitative Studie zu situativen Differenzierungspraktiken von Trainees und TrainerInnen einer internationalen journalistischen Weiterbildung als Teil von Medienentwicklungsarbeit* [Doctoral dissertation, publication in preparation]. Institute for Media Studies, Ruhr-Universität Bochum.

Knoblauch, H. (2005). Focused ethnography. *Forum Qualitative Social Research*, *6*(3). https://doi.org/10.17169/fqs-6.3.20

Kutsch, A., & Hamidi, K. (2020). *Dual journalism education: Recommendations for the reform of journalism education at state universities in Afghanistan – in particular for the Faculty of Journalism and Public Communication at Nangarhar University*. University of Leipzig. https://journalism-education-afghanistan.org/

Lilienthal, V. (2017). Recherchejournalismus für das Gemeinwohl. Correctiv – eine Journalismusorganisation neuen Typs in der Entwicklung. *Medien & Kommunikationswissenschaft (M&K)*, *65*(4), 659–681. https://doi.org/10.5771/1615-634X-2017-4-659

Lohner, J., Neverla, I., & Banjac, S. (2019). Conflict-sensitive journalism? Journalistic role perceptions and practices in democratization conflicts. In K. Voltmer, C. Christensen, I. Neverla, N. Stremlau, B. Thomass, N. Vladisavljević, & H. Wasserman (Eds.), *Media, communication and the struggle for democratic change: Case studies on con-*

tested transitions (pp. 59–81). Palgrave Macmillan. https://doi.org/10.1007/978-3-030-16748-6_3

Lublinski, J., Spurk, C., Fleury, J.-M., Labassi, O., Mbarga, G., Nicolas, M. L., & Rizk, T. A. (2016). Triggering change – How investigative journalists in Sub-Saharan Africa contribute to solving problems in society. *Journalism, 17*(8), 1074–1094. https://doi.org/10.1177/1464884915597159

Lugo-Ocando, J. (2020). *Foreign aid and journalism in the Global South: A mouthpiece for truth*. Lexington Books.

MacKinnon, R. (2007, June). WJEC declaration of principles of journalism education. *RConversation*. https://rconversation.blogs.com/rconversation/wjec_declaration_of_princ.html

Malik, M. (2022). Wer lernt (was) auf wessen Kosten? In Y. Akbaba, T. Buchner, A. M. Heinemann, D. Pokitsch, & N. Thoma (Eds.), *Pädagogische Professionalität und Migrationsdiskurse: Lehren und Lernen in Differenzverhältnissen* (pp. 25–44). Springer VS. https://doi.org/10.1007/978-3-658-37328-3_2

Manyozo, L. (2012). *Media, communication and development: Three approaches*. Sage.

Martens-Edwards, E. (2023). *Journalism education in the context of development and digital transformation: A cross-national comparative analysis of academic journalism degree programs in Cambodia and Vietnam* [Doctoral dissertation, University of Bamberg]. University of Bamberg Press. https://doi.org/10.20378/irb-54140

Mecheril, P., & Plößer, M. (2009). Differenz und Pädagogik. In S. Andresen, R. Casale, T. Gabriel, R. Horlacher, S. L. Klee, & J. Oelkers (Eds.), *Handwörterbuch Erziehungswissenschaft* (pp. 194–208). Beltz.

Miller, J. (2009). NGOs and 'modernization' and 'democratization' of media: Situating media assistance. *Global Media and Communication, 5*(1), 9–33. https://doi.org/10.1177/1742766508101312

Mills, D., & Morton, M. (2013). *Ethnography in education: Research methods in education*. Sage. https://doi.org/10.4135/9781446251201

Moyo, L. (2022). Introduction: Journalism studies and the Global South – Theory, practice and pedagogy. *Journalism Studies, 23*(13), 1567–1577. https://doi.org/10.1080/1461670X.2022.2125896

Muchtar, N., & Hanitzsch, T. (2013). CULTURE CLASH: International media training and the difficult adoption of Western journalism practices among Indonesian radio journalists. *Journalism Practice, 7*(2), 184–198. https://doi.org/10.1080/17512786.2012.753242

Murphy, S. M., & Scotton, J. F. (1987). Dependency and journalism education in Africa: Are there alternative models. *Africa Media Review, 1*(3), 11–35.

Musa, B. A., & Domatob, J. K. (2007). Who is a development journalist? Perspectives on media ethics and professionalism in post-colonial societies. *Journal of Mass Media Ethics, 22*(4), 315–331. https://doi.org/10.1080/08900520701583602

Myers, M., & Gilberds, H. (2024). *Are donors taking the journalism crisis seriously? An analysis of official aid to media 2010–2019*. Center for International Media Assistance. https://www.cima.ned.org/publication/are-donors-taking-the-journalism-crisis-seriously/

Myers, M., & Juma, L. A. (2018). *Defending independent media: A comprehensive analysis of aid flows.* Center for International Media Assistance. https://www.cima.ned.org/publication/comprehensive-analysis-media-aid-flows/

Myers, M., Harford, N., & Bartholomew, K. (2017). *Media assistance: Review of the recent literature and other donors' approaches* [A capitalisation exercise (CapEx) for the Swiss Agency for Development and Cooperation (SDC)]. iMedia Associates.

Nelson, M. M. (2019). Redefining media development: A demand-driven approach. In N. Benequista, S. Abbott, P. Rothman, & W. Mano (Eds.), *International media development: Historical perspectives and new frontiers* (pp. 30–41). Peter Lang.

Nelson, M., & Susman-Peña, T. (2012). *Rethinking media development: A report on the media map project.* Internews; World Bank Institute.

Nguyen, L. K. (2016). *Interkulturelle Aspekte der Entwicklungszusammenarbeit: Eine ethnographische Studie zu deutschen Organisationen in Peru und Bolivien.* Waxmann Verlag.

Noske-Turner, J. (2017). *Rethinking media development through evaluation: Beyond freedom.* Palgrave Macmillan. https://doi.org/10.1007/978-3-319-58568-0

Nyamnjoh, F. B. (2010). *Africa's media: Between professional ethics and cultural belonging.* Friedrich Ebert Stiftung. https://library.fes.de/pdf-files/bueros/africa-media/07366.pdf

Ognianova, E. (1995). Farewell to parachute professors in East-Central Europe. *Journalism & Mass Communication Educator, 50*(1), 35–47. https://doi.org/10.1177/107769589505000105

Phiri, S., & Fourie, P. J. (2011). Media development aid and the Westernisation of Africa: The case of the Open Society Initiative for Southern Africa (OSISA). *Communicatio, 37*(1), 80–99. https://doi.org/10.1080/02500167.2010.556575

Prengel, A. (2019). *Pädagogik der Vielfalt: Verschiedenheit und Gleichberechtigung in Interkultureller, Feministischer und Integrativer Pädagogik* (4th ed.). Springer VS. https://doi.org/10.1007/978-3-658-21947-5

Raboy, M. (2019). Global media: New issues, old story. In N. Benequista, S. Abbott, P. Rothman, & W. Mano (Eds.), *International media development: Historical perspectives and new frontiers* (pp. 13–19). Peter Lang.

Rodny-Gumede, Y. (2018). A teaching philosophy of journalism education in the Global South: A South African case study. *Journalism, 19*(6), 747–761. https://doi.org/10.1177/1464884916657516

Schiffrin, A. (2010). NOT REALLY ENOUGH: Foreign donors and journalism training in Ghana, Nigeria and Uganda. *Journalism Practice, 4*(3), 405–416. https://doi.org/10.1080/17512781003643244

Schiffrin, A. (2017). *Same beds, different dreams? Charitable foundations and newsroom independence in the Global South.* Center for International Media Assistance. https://www.cima.ned.org/resource/beds-different-dreams-charitable-foundations-newsroom-independence-global-south/

Schiffrin, A., & Behrman, M. (2011). Does training make a difference? Evaluating journalism training programs in Sub-Saharan Africa. *Journalism & Mass Communication Educator, 66*(4), 340–360. https://doi.org/10.1177/107769581106600404

Schmidt, C., & Martens-Edwards, E. (2015). Introduction. In C. Schmidt (Ed.), *Standards of journalism education: An international comparative study in the context of media and development* (pp. 10–32). Vistas.

Scott, M. (2014). *Media and development: Development matters*. Zed Books.

Scott, M., Bunce, M., & Wright, K. (2017). Donor power and the news: The influence of foundation funding on international public service journalism. *International Journal of Press/Politics*, 22(2), 163–184. https://doi.org/10.1177/1940161217693394

Shafer, R. (1998). Comparing development journalism and public journalism as interventionist press models. *Asian Journal of Communication*, 8(1), 31–52. https://doi.org/10.1080/01292989809364752

Skjerdal, T. S. (2009). Between journalism universals and cultural particulars: Challenges facing the development of a journalism programme in an East African context. *Journal of African Media Studies*, 1(1), 23–34. https://doi.org/10.1386/jams.1.1.23/1

Skjerdal, T. S. (2012). The three alternative journalisms of Africa. *International Communication Gazette*, 74(7), 636–654. https://doi.org/10.1177/1748048512458559

Smith, A. D. (1991). *National identity*. Penguin Books.

Stehling, M. (2015). *Die Aneignung von Fernsehformaten im transkulturellen Vergleich: Eine Studie am Beispiel des Topmodel-Formats*. Springer VS. https://doi.org/10.1007/978-3-658-09351-8

Tervooren, A., Engel, N., Göhlich, M., Miethe, I., & Reh, S. (Eds.). (2014). *Ethnographie und Differenz in pädagogischen Feldern: Internationale Entwicklungen erziehungswissenschaftlicher Forschung*. transcript.

Thomas, A., Kinast, E.-U., & Schroll-Machl, S. (Eds.). (2005). *Handbuch Interkulturelle Kommunikation und Kooperation. Band 1: Grundlagen und Praxisfelder* (2nd ed.). Vandenhoeck & Ruprecht.

Thomaß, B. (2012). Die Ermöglichung des Diskurses: Journalistische Ethik als Gegenstand transkultureller Seminare. In B. Dernbach & W. Loosen (Eds.), *Didaktik der Journalistik: Konzepte, Methoden und Beispiele aus der Journalistenausbildung* (pp. 393–404). VS Verlag für Sozialwissenschaften. https://doi.org/10.1007/978-3-531-93447-1_31

Tufte, T., & Mefalopolus, P. (2009). *Participatory communication: A practical guide* (World Bank Working Paper No. 170). http://documents.worldbank.org/curated/en/682081468166154717/Participatory-communication-a-practical-guide

Ugangu, W. (2019). Journalism training in a changing African society: Case of Kenya. In B. Dernbach & B. Illg (Eds.), *Journalism and journalism education in developing countries* (pp. 202–216). Manipal Universal Press.

Ullah, M. S. (2022). Under the wheel of decolonization and recolonization: The crossroads of journalism education in South Asia. In D. Garrisi & X. Kuang (Eds.), *Journalism Pedagogy in Transitional Countries* (pp. 139–163). Palgrave Macmillan. https://doi.org/10.1007/978-3-031-13749-5_7

UNESCO. (2007). *Model curricula for journalism education*. https://unesdoc.unesco.org/ark:/48223/pf0000151209

UNESCO. (2008). *Media development indicators: A framework for assessing media development*. https://unesdoc.unesco.org/ark:/48223/pf0000163102

UNESCO. (2013). *Teaching journalism in developing countries and emerging democracies: The case of UNESCO's model curricula*.

UNESCO – International Programme for the Development of Communication. (2014). *Global initiative for excellence in journalism education*. https://unesdoc.unesco.org/ark:/48223/pf0000230044

UNESCO – International Programme for the Development of Communication. (2020). *Journalism education: Paris declaration on freedom of journalism education*. https://unesdoc.unesco.org/ark:/48223/pf0000374531

United Nations Development Program (2001). *Human Development Report 2001: Making New Technologies Work for Human Development*. New York.

United Nations General Assembly. (2015). *Transforming our world: the 2030 Agenda for Sustainable Development*. Resolution adopted by the General Assembly on 25 September 2015 (A/RES/70/1). https://sdgs.un.org/2030agenda

Voltmer, K., & Wasserman, H. (2014). Journalistic norms between universality and domestication: Journalists' interpretations of press freedom in six new democracies. *Global Media and Communication, 10*(2), 177–192. https://doi.org/10.1177/1742766514540073

Waisbord, S. (2008). The institutional challenges of participatory communication in international aid. *Social Identities, 14*(4), 505–522. https://doi.org/10.1080/13504630802212009

World Bank. (1996). *The World Bank participation sourcebook*. https://documents1.worldbank.org/curated/en/289471468741587739/pdf/multi-page.pdf

Wrenn, M. C. (2011). Making the world safe for autonomy? The US initiative to reorient "foreign journalists" 1945–1970. In B. Franklin & D. Mensing (Eds.), *Journalism education, training and employment* (pp. 83–96). Routledge.

Xu, X. (2009). Development journalism. In K. Wahl-Jorgensen & T. Hanitzsch (Eds.), *The handbook of journalism studies* (pp. 357–370). Routledge.

Ziai, A. (2014). Post-Development-Ansätze: Konsequenzen für die Entwicklungstheorie. In F. Müller, E. Sondermann, I. Wehr, C. Jakobeit, & A. Ziai (Eds.), *Entwicklungstheorien: weltgesellschaftliche Transformationen, entwicklungspolitische Herausforderungen, theoretische Innovationen* (pp. 405–434). Nomos.

International research collaborations and networks

Challenges and solutions to assembling and working in international research teams

Sophia C. Volk

Introduction

International research collaboration has increased in almost all scientific disciplines in recent years (Wagner, 2018), including communication and media studies. Collaborative forms of knowledge production across borders can bring many advantages: When researchers from different countries work together and bring relevant contextual knowledge about the languages, cultures, and media of the countries being studied, common problems of comparative research, such as ethnocentrism, paternalistic views, parochialism, or safari research, can be reduced (Hantrais, 2009). Collaboration across borders enables researchers to gain access to and collect data about different populations, enabling larger-scale comparative studies. More fundamentally, collaboration is considered a highly effective means to facilitate innovative research through the pooling of resources, leading to increased productivity and disciplinary progress (e.g., Akkerman et al., 2012; Beaver, 2004; Katz & Martin, 1997).

This chapter traces the current composition of international research teams in communication and media studies, building on findings from systematic reviews of English-language journals, anecdotal reflections, and a qualitative study about comparative communication scholarship (Volk, 2021). It shows that international research teams are often dominated by Western scholars and are not truly international, arguing that more diverse and inclusive research teams are needed to produce more meaningful and context sensitive research. Therefore, the chapter explores typical challenges to assembling and collaborating in international research teams across geographic distances and cultures, focusing on the obstacles that emerge at the level of the team, the research process, the project management, and the contextual environment. Moreover, the chapter identifies typical solutions to mitigate or overcome such challenges and to create conditions for fruitful collaborations. Lastly, it calls for the composition of more diverse research teams and formulates demands for researchers, associations, and funding agencies.

International research collaborations in communication and media studies

Empirical evidence from scientometric analyses of journal publishing shows that co-authorships—a common proxy for research collaboration—have grown steadily and substantially over the past decade across virtually all scientific disciplines (e.g., Gazni et al., 2012; Henriksen, 2016; Wuchty et al., 2007). Some scholars have described this development as a "collaborative turn" of science (Olechnicka et al., 2019, p. 176) or as a "team science revolution" (Bozeman & Youtie, 2017, p. 2). Over time, co-authorships have not only increased in size, but have also become more internationally diverse (Kwiek, 2020; Wagner et al., 2015). Studies indicate that more than 25% of articles are published by international co-author teams where authors are spread across multiple countries (Hu et al., 2020).

This trend has arguably also affected the field of communication and media studies: From the 1980s to 2013, co-authorship in communication studies has increased from less than 30% to nearly 60%, according to a scientometric analysis of 4.5 million articles in the Web of Science (Henriksen, 2016). Supporting this development, a systematic review of 441 comparative communication studies published from 2000 through 2015 by Hanusch and Vos (2020) found that the proportion of co-authorships was 56%, and international co-authorships accounted for 26%. A more recent systematic review of 335 comparative communication studies published from 2015 through 2019 revealed that co-authorships rose to 76% and that international co-authorships even accounted for 36% (Volk, 2021).

Despite trends toward greater diversity in recent years, on closer inspection, co-authoring teams in communication and media studies are often not truly international. A look at the authors' geographical affiliations reveals that most international co-authorships consist of researchers from the West who often collaborate with other researchers from the West and less often with researchers from non-Western countries. The proportion of comparative studies produced without any "Western" participation is consequently low and ranges below 15% (Hanusch & Vos, 2020; Volk, 2021). Even in larger international co-authoring teams—composited of scholars from across the world—researchers from resource-rich Western countries are often overrepresented in numbers and occupy the "driver seat" by serving as project leaders (e.g., Hanitzsch, 2008; Kraidy, 2018; Stevenson, 2003; Wilke & Heimprecht, 2012). The dominance of Western scholars is also evident in co-citation analyses; for example, an analysis of 147 comparative studies published from 1979 through 2014 showed that the 50 most-cited scholars were exclusively from Europe or the United States (So, 2017)—indicating a powerful publishing and citation circle. This imbalance also has consequences for the objects of investigation: both the countries (e.g., Lind et al., 2025) and languages (e.g., Lind & Volk, 2025) studied in comparative research are disproportionately often Western and rich countries, resulting in a Western-centric bias and blind spots in many of communication and media studies.

Of course, the dominance of Western scholars is not limited to comparative research but reflects a deeper stratification of communication and media studies. Across the discipline, the underrepresentation of scholars from the Southern Hemisphere is evident in co-authorships and citation patterns, editorial positions in communication journals, or positions in professional associations (e.g., Chakravartty et al., 2018; Demeter, 2018; Goyanes & de Marcos, 2020; Waisbord, 2019). Possible reasons for this disparity are the more favorable opportunity structures in Anglophone and (Western) European countries, where there are better funding opportunities and researchers have privileged access to travel funds to build and maintain international networks with colleagues (e.g., Esser, 2019; Mutsvairo et al., 2021). But even in Western countries, particularly funding for cross-national research endeavors remains limited, as grants are typically awarded at the national level (Esser & Hanitzsch, 2012). This often means that only those who can acquire their own national funding participate in collaborative projects (Lauerer & Hanitzsch, 2019), which in turn is more likely to be the case in Western countries and for well-networked researchers. Although scholars have problematized such structural global inequalities and called for "de-Westernization" or "de-centering" communication and media studies (e.g., Curran & Park, 2000; Waisbord & Mellado, 2014), a "deep internationalization" is far from being achieved (e.g., Badr et al., 2020).

The lack of truly international research teams can be problematic for the further development of communication and media studies for several reasons. The underrepresentation of diverse epistemological perspectives from the Global South (e.g., Africa, Latin America, Eastern Europe, East Asia, or the Middle East) can lead to a certain type of research that mainly produces knowledge about Western countries and ignores other world regions or theoretical perspectives (e.g., Demeter et al., 2023; Richter, 2016; Suzina, 2021; Takahashi, 2023; see also Radue et al. on media systems research in this book). Furthermore, it can lead to research that takes mostly US-American or Western-connoted concepts as universal and applies them to non-Western contexts without much critical reflection (e.g., Averbeck-Lietz & Löblich, 2017; Mutsvairo et al., 2021). And even if research teams are at first sight internationally diverse, collaborations may be "lopsided" (Kraidy, 2018), for example, when Western scholars take the lead and impose their research perspectives on scholars from non-Western regions who merely become producers of datasets instead of being able to contribute their deep contextual expertise (Hanitzsch, 2008; Lauerer, 2023). A one-sided team composition can thus be a hindrance to the further development of concepts and contextualized approaches in comparative studies. Against this backdrop, it appears essential that the composition of international research teams becomes more diverse and inclusive—especially with regard to the inclusion of more scholars from underrepresented countries in the Global South (e.g., Hanitzsch et al., 2019).

Types, levels, and phases of international research collaborations

Research collaboration can occur at the level of individuals, universities, funding agencies, or nation-states and can take various forms: from inter- or transdisciplinary to international collaborations, from temporary to permanent collaborations, from informal to formal collaborations, and from small-scale to large-scale collaborations (Bozeman et al., 2013; Chompalov et al., 2002; Hara et al., 2003; Wang & Hicks, 2015). Research collaboration is often defined as a temporary social process in which at least two individual scientists divide labor and pool their expertise to jointly produce knowledge (Katz & Martin, 1997; Laudel, 2002). When at least three individual scientists work interdependently on the joint production of knowledge, they can be referred to as a research team (e.g., Salas et al., 2017; Ulnicane, 2015).

Even though research collaborations bring many advantages, any type of teamwork is also associated with challenges, which may constrain or prevent fruitful collaboration (Bozeman et al., 2013; Bozeman & Youtie, 2017; Sonnenwald, 2007). International research teams, which often work on temporary projects across geographic locations and time zones, arguably face more complications than national teams due to cultural, epistemological, or contextual differences between team members (Kosmuetzky, 2018; Mante-Meijer & Haddon, 2005; Walsh & Maloney, 2007). Although science studies, the sociology of science, the science of team science, and organizational psychology have been empirically researching collaborations for some time, very little is known to date about collaborations, especially in the social sciences and in an international setting (e.g., Chen et al., 2019).

This also applies to international research teams in communication and media studies. Such research teams can be organized very differently (Esser & Hanitzsch, 2012): some follow a truly collaborative model of teamwork where all collaborators are equal (e.g., Lauerer & Hanitzsch, 2019), whereas others are organized in a centralized way and directed by one country with leadership authority (e.g., Volk et al., 2025), while others are very independent and work rather individually than collaboratively. However, the various challenges that arise in international research teams in communication and media studies have hardly been researched to this point (Powers & Vera-Zambrano, 2018). While several studies have attempted to identify which solutions can help to mitigate typical challenges in the natural sciences, that is, which factors can facilitate or enable fruitful collaboration (e.g., Bozeman & Youtie, 2017; Meissner et al., 2022; Olson et al., 2008; Stokols et al., 2008), there are no empirical studies that have focused specifically on communication and media studies, except for a qualitative study by Volk (2021).

While it can be assumed that some of the challenges of international collaboration are common to all disciplines, the field of communication and media studies may face particular hurdles due to its fragmentation, diversification, and hyper specialization and its lack of a "theoretical or analytical center that could lend intellectual coherence to a vast academic field" (Waisbord, 2019, p. 121). The fragmentation of communication and media studies arguably makes it more difficult for international teams to find a common theoretical and methodological basis. In contrast to researchers in the natural sciences, communication researchers cannot simply assume, for example, that theories about freedom of the press, journalistic ideals, or hate speech "travel" across national borders and have the same meaning in different contexts (Hasebrink & Herzog, 2002; Hasebrink et al., 2009; Wirth & Kolb, 2004). Rather, they must take into account that many theories carry with them a certain "cultural baggage" (e.g., liberal-democratic bias due to their development in Western countries) and that non-Western contexts require different theories. This, in turn, requires theoretical flexibility and openness to the development of new theories on the part of all participants. This specificity of the discipline can thus be a particular obstacle to collaboration and may even be exacerbated by the different national academic cultures of communication scholars, who may bring different epistemologies (e.g., positivist, empiricist, constructivist, or hermeneutic) to a team (e.g., Averbeck-Lietz & Löblich, 2017).

Following the conceptualization by Volk (2021), which has built on the literature on team science (e.g., Engelbrecht et al., 2016; Sonnenwald 2007) and theories of temporary organizations (e.g., Kosmuetzky, 2018; Wöhlert, 2020), both the challenges and solutions of international research collaboration can be differentiated according to the "level" and the "phase" in which they occur during a collaboration process (Figure 1).

Figure 1: Levels and phases of research collaborations (source: adapted from Volk, 2021, p. 361)

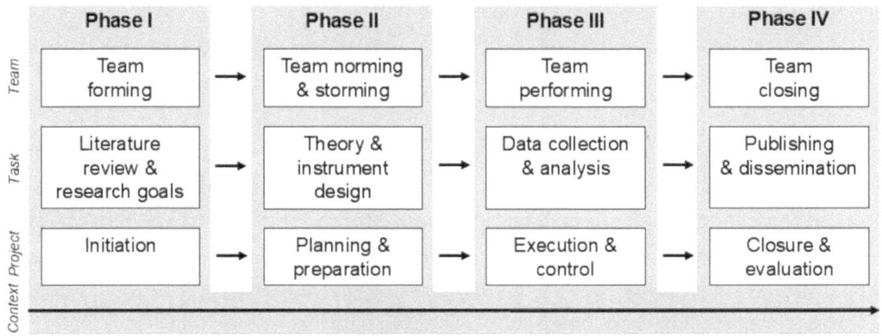

First, a conceptual distinction can be made between four levels at which challenges and solutions may arise (Volk, 2021):

- At the *team level*, which refers to the social dynamics of a collaboration, challenges and solutions can occur, for example, with regard to team diversity, geographic distance, trust among or commitment of team members.
- At the *task level*, which refers to the actual research of a collaboration, challenges and solutions can occur, for example, with regard to the choice of methods, labor division, or coordination.
- At the *project level*, which refers to the structure and organization of a collaboration, challenges and solutions can occur, for example, with regard to the acquisition of project funding, project management, leadership, or communication.
- At the *context level*, which refers to the environmental conditions of a collaboration, challenges and solutions can occur, for example, with respect to academic or national cultures, or surrounding geopolitical and societal contexts.

Second, a conceptual distinction can be made according to the phase during which particular *team-*, *task-*, and *project-*related challenges emerge and solutions are implemented; empirical research suggests that *contextual* challenges and solutions do not fluctuate over time (Volk, 2021).

I. In the *initiation phase*, challenges can arise and solutions may be implemented, especially when it comes to putting together a diverse team and agreeing on research questions and the organizational structure of a project.

II. In the *planning phase*, challenges may arise and solutions may be implemented, particularly in the search for a common basis in the teams, the consensus on theoretical frameworks and instruments, and the preparation and planning of data collection.
III. In the *execution phase*, challenges may arise and solutions may be implemented, particularly with regard to the ongoing commitment and performance of team members, overcoming obstacles to data collection and analysis, and monitoring progress and budget.
IV. In the *closing phase*, challenges may arise, and solutions may be implemented, especially when it comes to maintaining the motivation of team members, finalizing and disseminating publications, and evaluating and reporting on the success of the project.

Based on these distinctions, in what follows, the chapter provides an overview of typical challenges and typical solutions of international research collaborations in communication and media studies. It relies both on anecdotal reflections of research projects and the qualitative study by Volk (2021), which was based on expert interviews with 15 communication scholars from 10 countries who served as project leaders of international research teams and reported on their experiences with collaborative research projects.

Typical challenges of international research collaborations

International research collaborations in communication and media studies may face various challenges at the level of the team, task, project, and context, which often overlap and fluctuate over time. As there is no quantitative research on the scope and frequency of these challenges, only approximations of typical challenges can be given based on the qualitative study mentioned above.

Team-related challenges

Team-related challenges are manifold and likely more pronounced than challenges at other levels. A frequently mentioned problem in the initiation phase concerns the difficult team building and time-consuming search for potential collaborators, particularly from the Global South. Since personal networks or friendships seem to prominently influence team composition (e.g., see Cohen et al., 2013; Wilke & Heimprecht, 2012; Wirth & Kolb, 2004) and researchers from non-Western countries are often less represented at international conferences where networks are built due to limited travel funds, they are potentially excluded from such networks (e.g., see Takahashi, 2023). In addition, due to the comparatively small size of the discipline, there are only a limited number of colleagues from the Southern Hemisphere with expertise in a specific topic and comparative methods. Moreover, as each country typically has to raise its own funds, researchers from the Global South are often structurally disadvantaged due to poorer funding conditions (e.g., see Cohen, 2012; Lauerer & Hanitzsch, 2019). Finally, pragmatic considerations are also relevant for team composition, which are oftentimes influenced by the politically motivated interests of national or regional funding agencies (such as the EU) in finding collaborators from specific countries (e.g., see Hasebrink et al., 2009; Lauerer & Hanitzsch, 2019).

When team building is successful and teams begin working together, other problems often arise. The most common barrier faced by teams is probably the varying degrees of commitment or lack of commitment of individual members throughout a collaboration process, for example, individuals' unreliability, unpunctuality, or non-responsiveness (Volk, 2021). Another prominent challenge is diversity in the team, such as when members have different theoretical preferences (e.g., macro vs. micro level theories) or methodological preferences (e.g., qualitative vs. quantitative methods) and come from various disciplinary backgrounds (e.g., sociology, political science, psychology). While team diversity can be a source of creativity and innovation, it can also make building consensus throughout collaboration difficult (Hanitzsch et al., 2019; Wilke & Heimprecht, 2012). In addition, the goals of the collaborators can also differ and lead to conflicting goals or destructive internal competition (Cohen, 2012), ultimately even resulting in team member drop-out. In international teams, geographical distance and working across different time zones also pose a challenge, as face-to-face meetings and informal discussions are only possible to a limited extent (e.g., see Cummings & Kiesler, 2005; Walsh & Maloney, 2007). Moreover, especially for large teams, a team spirit or shared identity as a team can be difficult to achieve.

Task-related challenges

At the task level, the conceptual and methodological complexity of international research projects, in particular, is probably the most typical obstacle. For example, disagreements about a common theoretical framework that is suitable to different contexts (e.g., journalism ideals) can arise and lead to poor compromises that fall short of comparative goals (e.g., see Edelstein, 1982; Hanusch & Hanitzsch, 2017). In addition, difficulties typically arise in developing adequate instruments (e.g., context sensitive emic or standardized etic instruments) and achieving equivalent translations across different languages (Hanitzsch et al., 2019; Hasebrink et al., 2009). Furthermore, international teams may face difficulties in targeting hard-to-reach groups such as journalists or politicians, especially in non-democratic countries, accessing media archives and particularly non-digitized media, or dealing with varying database restrictions and legal situations in different countries. Furthermore, the division of tasks poses a typical problem because national sub-projects are often strongly interdependent, since equivalence is to be achieved, and require high levels of methodological expertise from all collaborators (e.g., see Mante-Meijer & Haddon, 2005; Wilke & Heimprecht, 2012). Other typical challenges include unclear responsibilities for tasks and coordinator roles, and insufficient or varying data quality between sub-projects, which can lead not only to conflicts but also to lack of comparability or even failure to achieve the project objectives (e.g., see Lauerer & Hanitzsch, 2019). Such challenges often surface in the planning and execution phases (i.e., during instrument development and data collection).

Project-related challenges

In terms of project-related challenges, the biggest obstacles appear to be in the areas of communication and leadership. Communication problems particularly include language barriers among non-native English speakers but also a lack of sufficient or informal communication, which can both lead to misunderstandings and loss of information. The fact that English is the lingua franca in academia, and is consequently used in most research collaborations, can be a disadvantage, especially for researchers from the Global South, where English may not be taught or spoken as frequently as in Western countries (Suzina, 2021). Especially in the initiation phase of a project, obtaining sufficient funding poses a typical challenge, ranging from general uncertainties related to insufficient funding in different national contexts to complicated administrative requirements of the funding organization (especially the EU), or to the unwanted influence of private sponsors on the project (e.g., see Esser & Hanitzsch, 2012). In addition, a lack of or unstable leadership and unplanned shifts in management tasks during the course of the project often pose a challenge, which may result in negative consequences, such as budget or schedule overrun. When resources are unequally distributed in teams and Western scholars have privileged positions in teams, this can possibly also result in acceptance problems for the project leadership (e.g., see Wilke & Heimprecht, 2012). Over time, project-related challenges are likely to surface across all phases.

Context-related challenges

At the contextual level, challenges are likely to arise in relation to different cultural norms, for example, when team members from different backgrounds have conflicting working styles (e.g., with regard to punctuality) or diverging expectations about academic hierarchies (e.g., authority, power distance) or gender roles. In larger international teams, national academic cultures in particular can create tension. For example, when positivist-empirical approaches typical of Anglophone cultures clash with hermeneutic epistemologies, which are more prevalent in the French or Latin American contexts (e.g., see Averbeck-Lietz & Löblich, 2017). Moreover, team members may hold conflicting assumptions about scientific quality and rigor (e.g., with regard to "messy and fuzzy" datasets, see Badr, 2023) or diverging views of research as a form of activism (Hasebrink et al., 2009; Mante-Meijer & Haddon, 2005). The sociopolitical environment can also become an obstacle, particularly in situations where political instability or war among countries hinders or complicates collaboration, for example, when researchers from regions in direct conflict with each other (e.g., Israel and the Middle East, Russia and Ukraine) are to work together. Different degrees of academic freedom in different countries can also pose challenges, especially for researchers from non-democratic countries with restricted freedom of expression who may have more difficulty carrying out certain research or even fear safety concerns or reprisals (Badr, 2023).

Typical solutions of international research collaborations

In addressing the multiple challenges, international research teams in communication and media studies can draw on different solutions at the team, task, project, and context levels, which can be combined in different ways and at different times to avoid or mitigate obstacles. There is no large-scale evidence on the prevalence of specific solutions in international research teams, but it is likely that the adoption of a solution depends on the project configurations and the project coordinator. In what follows, typical solutions identified in a qualitative study by Volk (2021) are described.

Team-related solutions

At the team level, a typical solution is to invest a significant amount of time in assembling a reliable and committed team, taking into account not only scientific but also "soft" criteria, such as pre-existing relationships (e.g., prior collaborations) and trust (e.g., reliability). For fruitful diversity in a team, members should have a common scientific perspective, but also bring different expertise, skills, and contextual knowledge to the table (e.g., see Hantrais, 2009) and have a genuine interest in the research subject. Differences in the goals of individual team members (e.g., regarding publications or authorships) should be discussed and reconciled at the beginning. To address team cohesion issues, a common solution is to invest time in team building and cultivating the imagination of a team as a "family" or "community" (Volk, 2021). Team building can rely on the creation of frequent team events (e.g., during international conferences) and opportunities for socializing and informal conversations, such as team lunches or dinners (e.g., Mante-Meijer & Haddon, 2005; Wöhlert, 2020). In addition, creating and maintaining a good working atmosphere in which team members trust each other and practice mutual respect, openness, and a willingness to compromise is a typical approach to prevent or deal with conflicts.

Task-related solutions

A typical solution to task-related challenges is to establish decentralized coordinators, typically at the country or regional level, to monitor data collection and coordinate work progress (e.g., Lauerer, 2023). To prevent shortfalls in task fulfillment, a common solution is to divide work packages based on interests and expertise and decouple tasks from each other so that collaborators can work less interdependently. A common approach to address the complexity of comparative research is to allow sufficient time for theoretical discussions about core concepts and definitions and for developing a common understanding of scientific principles that guide data collection and analysis. To support collaborators with varying methodological skills, written instructions (e.g., guidelines), method training (e.g., coding, data analysis), or support structures (e.g., statistical consultation) can be set up (Hanitzsch et al., 2019). For the problems posed by reintegrating separated work packages or datasets collected at the country level, one solution is to provide standardized templates and detailed guidelines for all methodological steps (e.g., instruments, sampling), facilitating the harmonization of datasets for the purpose of comparative analysis.

Teams can follow the proposed process model for comparative communication research (Figure 2) and discuss the requirements for comparability and equivalence step-by-step (Volk, 2021, p. 111). At the justification level, teams need to answer the initial question: What is the rationale for a comparison, that is, *Why compare?* At the conceptual level, teams need to answer the question *What to compare?* and agree on an appropriate theoretical approach for comparison, select comparable cases and equivalent objects, and conceptualize the role of contextual factors. At the methodological level, it is imperative that teams discuss *How to compare?* and establish equivalence of constructs, instruments, translations, samples, administration, methods, and measurements. Finally, at the interpretative level, teams need to reflect on *How comparable are the results?* and discuss the findings against the background of contextual conditions as well as in light of possible limitations or alternative explanations.

Figure 2: Process model of comparative communication research (source: Volk, 2021, p. 111)

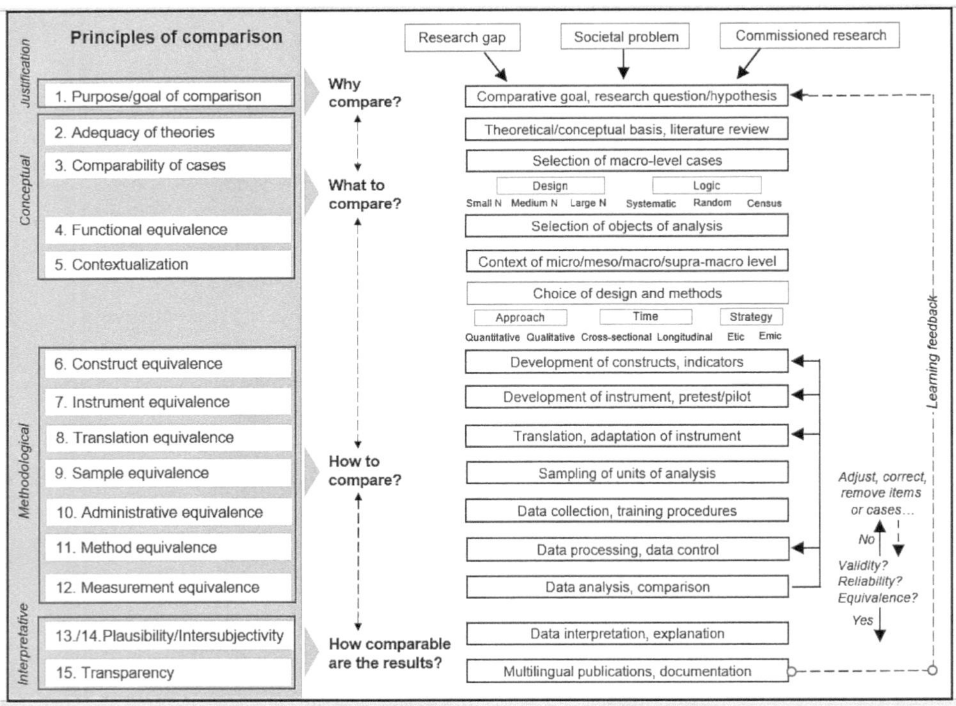

Project-related solutions

At the project level, various solutions can be adopted to tackle problems, not only with respect to project management issues but also team- or task-related problems. One of the most common solutions is to discuss and formalize agreements at the beginning of a research project, for example, regarding authorship, data access and ownership, or intellectual property rights (Cohen, 2012; Lauerer & Hanitzsch, 2019). Such formalized agreements should also define clear project goals, time plans, and work packages, as well as sanctions to be applied if team members do not fulfill their tasks (e.g., no access to full comparative datasets; Lauerer & Hanitzsch, 2019). Another common solution for avoiding difficulties in project management is to establish a clear leadership structure and self-governing bodies, such as an executive board, in the initiation phase. Appropriate digital communication infrastructures (e.g., videoconferencing), suitable tools for collaboration (e.g., Google Docs, Slack), and regular meetings can mitigate communication losses (e.g., Wilke & Heimprecht, 2012). An egalitarian discussion culture and clear decision-making rules (e.g., voting mechanisms) can facilitate efficient consensus. To solve funding problems, cross-financing of countries with fewer or no resources by well-resourced (mostly Western) universities or countries can be used (e.g., Hanitzsch et al., 2019). Also, pre-structured templates for national funding applications can be provided to improve the chances of success in obtaining funding.

Context-related solutions

At the contextual level, solutions often aim to make cultural diversity and differences between national academic cultures fruitful and to use potential points of friction in a productive way (Volk, 2021). Such solutions typically try to appeal to team members' sensitivity and tolerance toward cultural differences, to a mindset of openness to other epistemological perspectives, and to patience and reflexivity (e.g., Bozeman & Youtie, 2017; Haintrais, 2009). In the event of conflicts, a constructive or even humorous approach to dealing with differences (e.g., joking about cultural stereotypes) can help to seek a good compromise (Volk, 2021). In order to prevent sociopolitical aspects from having a negative impact on research, one solution is to leave political differences aside as far as possible and instead focus on common goals—namely, the joint research project.

Toward assembling truly international research teams

This chapter has examined the composition of international teams and shown that they are often characterized by power asymmetries and are not inclusive of scholars from the Southern hemisphere. It has described and systematized typical challenges and solutions to international collaboration, showing that assembling *truly* international teams poses a considerable challenge in itself. Yet, teams need to become more internationally inclusive—even if this means more potential for friction—because it can be assumed that increased diversity of disciplinary, epistemological, theoretical, and cultural backgrounds will be beneficial for the progress of the discipline. Looking ahead, more research is needed to examine the power asymmetries in international research teams, underlying structural causes, and consequences for what is (not) researched.

There have already been many statements and demands on how the "decentering," "de-Westernization," "deep internationalization," or "cosmopolitan transformation" of the discipline can and should be advanced (e.g., Badr et al., 2020; Curran & Park, 2000;

Waisbord, 2019). What they probably have in common is that further development of the discipline is to be understood as a joint responsibility of various actors in the academic community and science policy.

With regard to the question of how research teams can become more truly international in the future, the following demands can be made of *researchers, associations,* and *funding institutions*:

a) *Researchers in the Global North* can contribute to more inclusive teams by not only looking for collaborators in their existing networks but by specifically addressing calls for participation to researchers from the Global South and practicing openness toward different theoretical and epistemological approaches (Ganter & Badr, 2022). They could use AI-powered translation tools for identifying and translating relevant publications from journals published in languages other than English to engage with research written in Arabic, Chinese, French, Spanish, and so forth. When setting up project management structures, leadership or coordinator roles should be equally distributed across regions, rather than being dominated by Western scholars. To counteract structural inequalities within research teams, resources can be redistributed so that richer countries cross-fund data collection in less well-financed countries (Lauerer, 2023). Including colleagues from non-Western countries with contextual knowledge and language skills promises valuable insights, as data could be gained about regions that have so far been underrepresented in communication and media studies, which would be beneficial for future theorizing and context sensitive research (Hantrais, 2009). Since large heterogeneous teams may have more potential for conflict than small teams, it is advisable for project leaders to anticipate typical pitfalls and prevent them as best as possible in order to find a good balance between efficiency, scientific rigor, and inclusivity (Lauerer, 2023). For this purpose, they can draw on a growing number of toolkits and best practice guides for successful team research (e.g., Bennett et al., 2018; Facer & Enright, 2016).
b) *Researchers from the Global South* can participate in virtual conferences or workshops, use online platforms to make their research and expertise visible, and express interest in collaborative opportunities. They can gain collaboration experience in local settings and build regional networks within the Global South (e.g., Demeter et al., 2023), which may also lend them more visibility in other scientific communities. Since international professional associations are the main venue for scientific exchanges, becoming a member provides access to global networks and information on international projects seeking partners. To attend international conferences and network with other scholars in person, researchers can search for funding opportunities and conference grants specifically designed for researchers from underrepresented geographic regions. They can also take advantage of the recent internationalization efforts of international associations, such as making submissions in languages other than English, editing special issues with a regional focus on scholarship from the Global South, or hosting regional conferences (e.g., Takahashi, 2023). Moreover, they can take advantage of free AI-powered tools for translation or English copy editing of their manuscripts and submit them to high-impact journals, which are key to international visibility.

c) *Scientific associations* can take further steps to foster genuine internationalization, both at the central level and at the level of divisions (Badr et al., 2020). On the one hand, they should continue to support but also invest more in reducing conference fees or subsidizing travel grants for scholars from underrepresented countries, or provide competitive funding grants for innovative research ideas, especially from the Global South. On the other hand, they could establish mentorship networks or virtual workshops (e.g., providing advice for conference submissions, publishing, and first-time attendees) and offer virtual networking platforms that enable building more diverse and inclusive research networks (e.g., Esser, 2023; Lauerer, 2023; Mitchelstein, 2023). Conference planners should ensure that research from the Global North and South is not segregated, and that research is equally visible (e.g., Esser, 2023). Efforts for internationalization should also concern the structures of associations and affiliated journals, for example, by filling more leadership positions with researchers from non-Western regions.
d) *Funding agencies* need to create more suitable framework conditions for international research in the social sciences, for example, by developing appropriate internationalization strategies and cooperating more with non-Western national funding agencies to enable cross-border research. Targeted funding of smaller North–South collaborations is needed (Waisbord, 2023), as research shows that collaboration partners are often selected pragmatically on the basis of calls for proposals for third-party funding. In addition, more funding lines for comparative research on a small international scale are desirable.

References

Akkerman, S., Admiraal, W., & Simons, R. J. (2012). Unity and diversity in a collaborative research project. *Culture & Psychology*, *18*(2), 227–252. https://doi.org/10.1177/1354067X11434835

Averbeck-Lietz, S., & Löblich, M. (2017). Kommunikationswissenschaft vergleichend und transnational: Eine Einführung. In S. Averbeck-Lietz (Ed.), *Kommunikationswissenschaft im internationalen Vergleich: Transnationale Perspektiven* (pp. 1–29). Springer VS. https://doi.org/10.1007/978-3-531-18950-5_1

Badr, H. (2023). It is epistemic, folks! Why our knowledge from WEIRD contexts is limited and what we can learn from Arab contexts. *Political Communication Report*, *28*. http://dx.doi.org/10.17169/refubium-41237

Badr, H., Behmer, M., Fengler, S., Fiedler, A., Grüne, A., Hafez, K., Hahn, O., Hamidi, K., Hanitzsch, T., Horz, C., Illg, B., Litvinenko, A., Löffelholz, M., Radue, M., Richter, C., Thomaß, B., & Töpfl, F. (2020). Kosmopolitische Kommunikationswissenschaft: Plädoyer für eine "tiefe Internationalisierung" des Fachs in Deutschland: Ein wissenschaftliches Positionspapier. *Publizistik*, *65*(3), 295–303. https://doi.org/10.1007/s11616-020-00576-6

Beaver, D. D. (2004). Does collaborative research have greater epistemic authority? *Scientometrics*, *60*(3), 399–408. https://doi.org/10.1023/B:SCIE.0000034382.85360.cd

Bennett, L. M, Gadlin, H., & Marchand, C. (2018). *Collaboration and team science: A field guide* (2nd ed.). National Cancer Institute. https://www.cancer.gov/about-nci/organization/crs/research-initiatives/team-science-field-guide/collaboration-team-science-guide.pdf

Bozeman, B., & Youtie, J. (2017). *The strength in numbers: The new science of team science.* Princeton University Press. https://doi.org/10.2307/j.ctvc77bn7

Bozeman, B., Fay, D., & Slade, C. P. (2013). Research collaboration in universities and academic entrepreneurship: The-state-of-the-art. *Journal of Technology Transfer*, *38*(1), 1–67. https://doi.org/10.1007/s10961-012-9281-8

Chakravartty, P., Kuo, R., Grubbs, V., & McIlwain, C. (2018). #CommunicationSoWhite. *Journal of Communication*, *68*(2), 254–266. https://doi.org/10.1093/joc/jqy003

Chen, K., Zhang, Y., & Fu, X. (2019). International research collaboration: An emerging domain of innovation studies? *Research Policy*, *48*(1), 149–168. https://doi.org/10.1016/j.respol.2018.08.005

Chompalov, I., Genuth, J., & Shrum, W. (2002). The organization of scientific collaborations. *Research Policy*, *31*(5), 749–767. https://doi.org/10.1016/S0048-7333(01)00145-7

Cohen, A. A. (2012). Benefits and pitfalls of comparative research on news: Production, content and audiences. In I. Volkmer (Ed.), *The handbook of global media* research (pp. 533–546). Wiley Blackwell. https://doi.org/10.1002/9781118255278.ch31

Cohen, A. A., Hanitzsch, T., Stepinska, A., Porath, W., & Heimprecht, C. (2013). Rationale, design, and methodologies. In A. A. Cohen (Ed.), *Foreign news on television: Where in the world is the global village* (pp. 3–22). Peter Lang.

Cummings, J. N., & Kiesler, S. (2005). Collaborative research across disciplinary and organizational boundaries. *Social Studies of Science*, *35*(5), 703–722. https://doi.org/10.1177/0306312705055535

Curran, J., & Park, M.-J. (Eds.). (2000). *De-Westernizing media studies*. Routledge.

Demeter, M. (2018). Changing center and stagnant periphery in communication and media studies: National diversity of major international journals in the field of communication from 2013 to 2017. *International Journal of Communication*, *12*, 2893–2921. https://ijoc.org/index.php/ijoc/article/view/9122

Demeter, M., Vozab, D., & Boj, F. J. S. (2023). From Westernization to internationalization: Research collaboration networks of communication scholars from Central and Eastern Europe. *International Journal of Communication*, *17*, 1211–1231. https://ijoc.org/index.php/ijoc/article/view/20176

Edelstein, A. S. (1982). *Comparative communication research*. Sage.

Engelbrecht, L., Spolander, G., Martin, L., Strydom, M., Adaikalam, F., Marjanen, P., Pervova, I., Sicora, A., & Tani, P. (2016). Reflections on a process model for international research collaboration in social work. *International Social Work*, *59*(4), 438–451. https://doi.org/10.1177/0020872814531305

Esser, F. (2019). Comparative research. In H. Van den Bulck, M. Puppis, K. Donders, & L. van Audenhove (Eds.), *The Palgrave handbook of methods for media policy research* (pp. 85–101). Palgrave Macmillan. https://doi.org/10.1007/978-3-030-16065-4_5

Esser, F. (2023). Growing, evolving, engaging: The latest from ICA's political communication division. *Political Communication Report*, *28*. https://politicalcommunication.org/article/icas-political-communication-divisions-2023-2024-update/

Esser, F., & Hanitzsch, T. (2012). Organizing and managing comparative research projects across nations: Models and challenges of coordinated collaboration. In I. Volkmer (Ed.), *The handbook of global media research* (pp. 521–532). Wiley Blackwell. https://doi.org/10.1002/9781118255278.ch30

Facer, K., & Enright, B. (2016). *Creating living knowledge: The connected communities programme, community university relationships and the participatory turn in the production of knowledge.* University of Bristol. https://research-information.bris.ac.uk/ws/portalfiles/portal/75082783/FINAL_FINAL_CC_Creating_Living_Knowledge_Report.pdf

Ganter, S. A., & Badr, H. (2022). Conclusion: Cosmopolitan critique as a counterhegemonic methodology. In S. A. Ganter & H. Badr (Eds.), *Media governance: A cosmopolitan critique* (pp. 283–305). Palgrave Macmillan. https://doi.org/10.1007/978-3-031-05020-6_14

Gazni, A., Sugimoto, C. R., & Didegah, F. (2012). Mapping world scientific collaboration: Authors, institutions, and countries. *Journal of the American Society for Information Science and Technology*, *63*(2), 323–335. https://doi.org/10.1002/asi.21688

Goyanes, M., & de-Marcos, L. (2020). Academic influence and invisible colleges through editorial board interlocking in communication sciences: a social network analysis of leading journals. *Scientometrics*, *123*(2), 791–811. https://doi.org/10.1007/s11192-020-03401-z

Hanitzsch, T. (2008). Problemzonen kulturvergleichender Kommunikatorforschung: Methodologische Fallstudien. In G. Melischek, J. Seethaler, & J. Wilke (Eds.), *Medien & Kommunikationsforschung im Vergleich* (pp. 253–270). VS Verlag für Sozialwissenschaften. https://doi.org/10.1007/978-3-531-90808-3_14

Hanitzsch, T., Lauerer, C., & Steindl, N. (2019). Komplexität managen: Organisatorische, konzeptionelle und methodische Herausforderungen in der internationalen Verbundforschung am Beispiel der "Worlds of Journalism Study". In B. Dernbach, A. Godulla, & A. Sehl (Eds.), *Komplexität im Journalismus* (pp. 41–52). Springer. https://doi.org/10.1007/978-3-658-22860-6

Hantrais, L. (2009). *International comparative research: Theory, methods and practice*. Palgrave Macmillan.

Hanusch, F., & Hanitzsch, T. (2017). Comparing journalistic cultures across nations: What we can learn from the Worlds of Journalism Study. *Journalism Studies*, *18*(5), 525–535. https://doi.org/10.1080/1461670X.2017.1280229

Hanusch, F., & Vos, T. P. (2020). Charting the development of a field: A systematic review of comparative studies of journalism. *International Communication Gazette*, *82*(4), 319–341. https://doi.org/10.1177/1748048518822606

Hara, N., Solomon, P., Kim, S.-L., & Sonnenwald, D. H. (2003). An emerging view of scientific collaboration: Scientists' perspectives on collaboration and factors that impact collaboration. *Journal of the American Society for Information Science and Technology,* 54(10), 952–965. https://doi.org/10.1002/asi.10291

Hasebrink, U., & Herzog, A. (2002). Vergleichsweise schwierig: Internationale Kooperation. In K. Hafez (Ed.), *Die Zukunft der international vergleichenden Kommunikationswissenschaft in Deutschland* (pp. 155–169). Deutsches Übersee-Institut.

Hasebrink, U., Ólafsson, K., & Štětka, V. (2009). Opportunities and pitfalls of cross-national research. In S. Livingstone & L. Haddon (Eds.), *Kids online: Opportunities and risks for children* (pp. 41–54). Policy Press. https://doi.org/10.2307/j.ctt9qgvds.9

Henriksen, D. (2016). The rise in co-authorship in the social sciences (1980–2013). *Scientometrics, 107*(2), 455–476. https://doi.org/10.1007/s11192-016-1849-x

Hu, Z., Tian, W., Guo, J., & Wang, X. (2020). Mapping research collaborations in different countries and regions: 1980–2019. *Scientometrics, 124*(1), 729–745. https://doi.org/10.1007/s11192-020-03484-8

Katz, J. S., & Martin, B. R. (1997). What is research collaboration? *Research Policy, 26*(1), 1–18. https://doi.org/10.1016/S0048-7333(96)00917-1

Kosmuetzky, A. (2018). A two-sided medal: On the complexity of international comparative and collaborative team research. *Higher Education Quarterly, 72*(4), 314–331. https://doi.org/10.1111/hequ.12156

Kraidy, M. M. (2018). Global media studies: A critical agenda. *Journal of Communication, 68*(2), 337–346. https://doi.org/10.1093/joc/jqx024

Kwiek, M. (2020). What large-scale publication and citation data tell us about international research collaboration in Europe: Changing national patterns in global contexts. *Studies in Higher Education, 46*(12), 2629–2649. https://doi.org/10.1080/03075079.2020.1749254

Laudel, G. (2002). What do we measure by co-authorships? *Research Evaluation, 11*(1), 3–15. https://doi.org/10.3152/147154402781776961

Lauerer, C. (2023). Balancing efficiency and inclusivity: Open science principles and practices in international collaborative research projects. *Digital Journalism, 11*(2), 390–398. https://doi.org/10.1080/21670811.2022.2123020

Lauerer, C., & Hanitzsch, T. (2019). Surveying journalists around the world: A methodological framework. In T. Hanitzsch, F. Hanusch, J. Ramaprasad, & A. S. de Beer (Eds.), *Worlds of journalism: Journalistic cultures around the globe* (pp. 47–66). Columbia University Press. https://doi.org/10.7312/hani18642-004

Lind, F., Song, H., Boomgaarden, H. G., Kathirgamalngam, A., Syed Ali, K. P., & Vliegenthart, R. (2025). Research practices in comparative communication research: Visibility, topical and geographical disparities, and their longitudinal patterns. *International Journal of Communication, 19*, 1103–1128. https://ijoc.org/index.php/ijoc/article/view/23510

Lind, F., & Volk, S. C. (2025). Towards more truly international comparative research: Current opportunities and challenges of multilingual text analysis with computational methods. In H. Badr & K. G. Wilkins (Eds.), *Critical communication research with global inclusivity*. Routledge.

Mante-Meijer, E., & Haddon, L. (2005). Working in international research groups. In L. Haddon (Ed.), *International collaborative research: Cross-cultural differences and cultures of research* (pp. 117–152). EU Publications Office.

Meissner, F., Weinmann, C., & Vowe, G. (2022). Understanding and addressing problems in research collaboration: A qualitative interview study from a self-governance perspective. *Frontiers in Research Metrics and Analytics, 6*, Article 778176. https://www.frontiersin.org/article/10.3389/frma.2021.778176

Mitchelstein, E. (2023). Imagined academic communities: Three observations about the de-westernization of political communication. *Political Communication Report, 28*, 1–6. http://dx.doi.org/10.17169/refubium-41234

Mutsvairo, B., Borges-Rey, E., Bebawi, S., Márquez-Ramírez, M., Mellado, C., Mabweazara, H. M., Demeter, M., Głowacki, M., Badr, H., & Thussu, D. (2021). Ontologies of journalism in the Global South. *Journalism & Mass Communication Quarterly, 98*(4), 996–1016. https://doi.org/10.1177/10776990211048883

Olechnicka, A., Ploszaj, A., & Celińska-Janowicz, D. (2019). *The geography of scientific collaboration*. Routledge. https://doi.org/10.4324/9781315471938

Olson, J. S., Hofer, E. C., Bos, N., Zimmerman, A., Olson, G. M., Cooney, D., & Faniel, I. (2008). A theory of remote scientific collaboration. In G. M. Olson, A. Zimmerman, & N. Bos (Eds.), *Scientific collaboration on the Internet* (pp. 73–97). MIT Press. https://doi.org/10.7551/mitpress/9780262151207.003.0005

Powers, M., & Vera-Zambrano, S. (2018). The universal and the contextual of media systems: Research design, epistemology, and the production of comparative knowledge. *International Journal of Press/Politics, 23*(2), 143–160. https://doi.org/10.1177/1940161218771899

Richter, C. (2016). Area Studies: Regionalstudien in der Kommunikationswissenschaft. In S. Averbeck-Lietz & M. Meyen (Eds.), *Handbuch nicht standardisierte Methoden in der Kommunikationswissenschaft* (pp. 95–108). VS Verlag für Sozialwissenschaften. https://doi.org/10.1007/978-3-658-01656-2_26

Salas, E., Rico, R., & Passmore, J. (Eds.). (2017). *The Wiley Blackwell handbook of the psychology of team working and collaborative processes*. Wiley Blackwell. https://doi.org/10.1002/9781118909997

So, C. Y. K. (2017). Mapping comparative communication research: What the literature reveals. In J. M. Chan & F. L. F. Lee (Eds.), *Advancing comparative media and communication research* (pp. 12–35). Routledge.

Sonnenwald, D. H. (2007). Scientific collaboration: A synthesis of challenges and strategies. *Annual Review of Information Science and Technology, 41*(1), 643–681. https://doi.org/10.1002/aris.2007.1440410121

Stevenson, R. L. (2003). Mapping the news of the world. In B. Dervin & S. H. Chaffee (Eds.), *Communication, a different kind of horse race: Essays honoring Richard F. Carter* (pp. 149–165). Hampton Press.

Stokols, D., Misra, S., Moser, R. P., Hall, K. L., & Taylor, B. K. (2008). The ecology of team science: Understanding contextual influences on transdisciplinary collaboration. *American Journal of Preventive Medicine, 35*(2), 96–115. https://doi.org/10.1016/j.amepre.2008.05.003

Suzina, A. C. (2021). English as lingua franca. Or the sterilisation of scientific work. *Media, Culture & Society, 43*(1), 171–179. https://doi.org/10.1177/0163443720957906

Takahashi, B. (2023). Towards inclusive international environmental communication scholarship: The role of Latin America. *International Journal of Cultural Studies, 26*(4), 372–391. https://doi.org/10.1177/13678779221146302

Ulnicane, I. (2015). Why do international research collaborations last? Virtuous circle of feedback loops, continuity and renewal. *Science and Public Policy, 42*(4), 433–447. https://doi.org/10.1093/scipol/scu060

Volk, S. C. (2021). *Comparative communication research: A study of the conceptual, methodological, and social challenges of international collaborative studies in communication science.* Springer VS. https://doi.org/10.1007/978-3-658-36228-7

Volk, S. C., Hagelstein, J., Álvarez-Nobell, A., Buhmann, A., Macnamara, J., Meng, J., Molleda, J.-C., Moreno, Á., Tench, R., Verčič, D., & Wiesenberg, M. (2025). Internationalization of PR research: Evolution, impact, and learnings of the Global Communication Monitor study. In L. Dühring, F. Krebber, A. Melzer, E. Oliveira, & J. Seiffert-Brockmann (Eds.), *Kommunikation als Strategie*. Springer VS.

Wagner, C. S. (2018). *The collaborative era in science: Governing the network*. Palgrave Macmillan. https://doi.org/10.1007/978-3-319-94986-4

Wagner, C. S., Park, H. W., & Leydesdorff, L. (2015). The continuing growth of global cooperation networks in research: A conundrum for national governments. *PLoS ONE*, *10*(7), Article e0131816. https://doi.org/10.1371/journal.pone.0131816

Waisbord, S. (2019). *Communication: A post-discipline*. Polity Press.

Waisbord, S. (2023). De-Westernizing political communication: Why? How? *Political Communication Report*, *28*, 1–6. http://dx.doi.org/10.17169/refubium-41232

Waisbord, S., & Mellado, C. (2014). De-Westernizing communication studies: A reassessment. *Communication Theory*, *24*(4), 361–372. https://doi.org/10.1111/comt.12044

Walsh, J. P., & Maloney, N. G. (2007). Collaboration structure, communication media, and problems in scientific work teams. *Journal of Computer-Mediated Communication*, *12*(2), 712–732. https://doi.org/10.1111/j.1083-6101.2007.00346.x

Wang, J., & Hicks, D. (2015). Scientific teams: Self-assembly, fluidness, and interdependence. *Journal of Informetrics*, *9*(1), 197–207. https://doi.org/10.1016/j.joi.2014.12.006

Wilke, J., & Heimprecht, C. (2012). Vergleichen in der Forschungspraxis: Ein Erfahrungsbericht. In B. Hung, M. Magin, O. Jandura, & M. Maurer (Eds.), *Methodische Herausforderungen komparativer Forschungsansätze* (pp. 76–94). Herbert von Halem.

Wirth, W., & Kolb, S. (2004). Designs and methods of comparative political communication research. In F. Esser & B. Pfetsch (Eds.), *Comparing political communication: Theories, cases, and challenges* (pp. 87–111). Cambridge University Press. https://doi.org/10.1017/CBO9780511606991.006

Wöhlert, R. (2020). Communication in international collaborative research teams: A review of the state of the art and open research questions. *Studies in Communication and Media*, *9*(2), 151–217. https://doi.org/10.5771/2192-4007-2020-2-151

Wuchty, S., Jones, B. F., & Uzzi, B. (2007). The increasing dominance of teams in production of knowledge. *Science*, *316*(5827), 1036–1039. https://doi.org/10.1126/science.1136099

List of contributors

Stefanie Averbeck-Lietz, Dr. phil. habil., is a full professor of communication studies and communication ethics at the University of Greifswald, Germany. In her research, she focuses on communication and media ethics, communication theory, communication history and inter-/transcultural communication. She is a co-editor of *Communications. The European Journal of Communication Research* and headed a DFG-funded project on the communication history of the League of Nations.

Hanan Badr, Dr. phil., is full professor and head of the public spheres and inequalities unit at the Department of Communication Studies at the University of Salzburg, Austria. Her academic socialization has evolved between European and Arab universities where she studied journalism and international communication in Cairo and Erfurt. Her work focuses on global inequalities and communication, journalism and media coverage, comparing media systems, activism and transformation. Recent books include *Arab Berlin: Dynamics of Transformation* (2023), *Media Governance: A Cosmopolitan Critique* (2022) and *Critical Research Methods with Global Inclusivity* (2025). She is associate editor of *Journalism Practice*.

Lisa Bolz, Dr. phil, is an associate professor at the School for Communication and Journalism (CELSA) at Sorbonne University, France. Her research interests include digital journalism, digital media cultures, journalism history and transcultural communication. She is a founding member of the recently created French-speaking journalism scholar network GER "Journalismes".

Regina Cazzamatta, Dr. phil., is a postdoctoral researcher at the University of Erfurt and the principal investigator of the DFG-funded comparative research project "Disinformation environments and the emergence of fact-checking organisations in Europe and Latin America." Her research encompasses comparative media systems, disinformation, fact-checking, journalism cultures, platformization, public spheres, media discourse, and international news. She has published in leading journals, including *Journal-*

ism, *Journalism & Mass Communication Quarterly*, *New Media & Society*, *Journalism Practice*, and *International Journal of Press/Politics*.

Otávio Daros, Dr. phil., is a researcher in the postgraduate program in communication at the Pontifical Catholic University of Rio Grande do Sul, Brazil. His research focuses on the history of communication studies, the historiography of journalism, and the critical theory of media culture. In addition to his work published in journals such as *Communication Theory*, *Media, Culture & Society*, *Journalism Studies*, and *Philosophy & Social Criticism*, he is the author of the books *Writing Journalism History: The Press and Academia in Brazil* (2024) and *History of Brazilian Journalism: From Print to Digital* (2025), both published by Routledge.

Ines Drefs, Dr. phil., is a senior consultant with Deutsche Welle Akademie's research and evaluation unit. Her research interests include international media development, media and democratization as well as changing conditions for journalism. Formerly, Ines was a postdoc at the Erich Brost Institute for International Journalism and director of the Graduate Program "MEDAS 21 – Media Development in the 21st Century", funded by Volkswagen Foundation.

Thomas Eckerl is a PhD candidate at the chair of journalism at the University of Passau, Germany. His research topics mainly focus on international cross-border consortia such as the International Consortium of Investigative Journalists (ICIJ) and the Organized Crime and Corruption Reporting Project (OCCRP) as well as on their impact on investigative journalism.

Pauline Gidget Estella, Dr. phil., is a postdoctoral researcher and lecturer at Technische Universität Ilmenau in Germany and former assistant professor at the University of the Philippines. Currently, she is working on a project on journalism education in the era of artificial intelligence, funded by the Thüringen state ministry and Technische Universität Ilmenau. Pauline was also one of the researchers in the DECIPHER project, a comparative study that examined pandemic-related risk and crisis communication in Europe and USA. Her research interests lie in the areas of journalism education and competence, Global South studies, and critical theory.

Susanne Fengler, Dr. phil., is the academic director of the Erich Brost Institute for International Journalism (www.brost.org) at TU Dortmund University and is full professor for international journalism at the Institute of Journalism, TU Dortmund University, Germany. She has directed numerous comparative research projects in the field of media and migration, media accountability, and foreign coverage. Among her recent key publications are the *Global Handbook of Media Accountability* (2022) and the UNESCO Handbook *Reporting on Migrants and Refugees* (2021).

Anke Fiedler, Dr. phil., is a researcher and lecturer at the University of Greifswald, Germany. Her research focuses on media structures in (post-)conflict and transitional states, media and migration, and the role of media development in democratic transitions. Her research has taken her to countries in North and Sub-Saharan Africa and the Middle East. She is a founding member of the DFG network Cosmopolitan Communication Studies (KosmoKW).

Sarah Anne Ganter, Dr. phil., is a permanent associate professor in the School of Communication at Simon Fraser University in Canada. Her work has authoritatively addressed cosmopolitanism in media and communication studies for some while. She has (co)-authored several book chapters and journal articles addressing cosmopolitanism as a countering approach to the field. She is the lead author of *The Invisibility of Latin American Scholarship in European Media and Communication Studies* (2019) published in the *International Journal of Communication*, and the lead editor of *Media Governance: A Cosmopolitan Critique* (2022).

Anne Grüne, Dr. phil., is senior lecturer and researcher of comparative cultural, media and communication studies at the University of Erfurt, Germany. Her research and teaching focuses on global media and communication, popular culture, cosmopolitan communication aspects of the lifeworld as well as on qualitative and comparative methodology. She coordinates the master's program "Global Communication: Politics and Society". Recent publications include *Foundations of Global Communication* (2022, with K. Hafez) and the co-edited volume *Media and Transformation in Germany and Indonesia* (2019).

Kai Hafez, Dr. phil. habil., is chair and full professor of international and comparative media and communication studies at the University of Erfurt, Germany. His research focuses on theory of global communication, communication relations between the Islamic world and the West, media and political transformation in the Middle East, communication and migration/Islamophobia, comparative media ethics. He has been a frequent political advisor to German governments and he is an advisory board member of many international academic journals. Recent publications include *Foundations of Global Communication* (2022, with A. Grüne) and the co-edited volume *Media and Transformation in Germany and Indonesia* (2019).

Oliver Hahn, Dr. phil., is full professor of journalism at the University of Passau, Germany. He focuses, among others, on international comparative research on journalism cultures and media systems, foreign correspondents, media in transition, EU communication, pan-European broadcasting, and public diplomacy.

Kefa Hamidi, Dr. phil., is head of the Centre for Communication for Social Change (C4SC) at the Institute for Communication and Media Studies at the University of Leipzig, Germany. His research interests include communication for social change, participatory and transformative communication, transnational communication and journalism studies. For his work and social commitment, he was awarded the DAAD Prize in 2006. From 2010 to 2013, he was a doctoral fellow of the Ryoichi Sasakawa Young Leaders Fellowship Fund (SYLLF).

Christine Horz-Ishak, Dr. phil., is professor of transcultural media communication at the University of Applied Sciences, Cologne (TH Köln), Germany. Her research and teaching focus on media, migration and diaspora, as well as on the diversity and participation of minorities in traditional media and in the digital context. She is a scholarship holder of the "Initiativprogramm" of TH Köln. She is co-founder and -editor of the *Global Media Journal – German Edition*, an open access journal specialized on international and transcultural communication and founding member of the DFG-Network Cosmopolitan Communication Studies (KosmoKW).

Beate Illg, Dr. phil., is a professor of communication studies at Jade University of Applied Sciences in Wilhelmshaven, Germany. Her research projects deal with journalism in young democracies like Nepal and Bhutan. Beside this she focuses on gender, media reception and empirical methods, especially qualitative methods. She is a confidential professor for Heinrich-Böll-Stiftung e.V.

Mira Keßler, is a lecturer and research associate at the international master's program IMS at Deutsche Welle Akademie in Bonn, Germany. Her research and teaching focus on journalism education and media development, comparative journalism research, intercultural communication and approaches of postcolonialism and de-Westernization. She was a research fellow at "MEDAS 21 – Media Development in the 21st Century", funded by the Volkswagen Foundation. Mira has worked as a freelancer for the public broadcaster SWR, local online news and science communication. Beyond her academic work she is engaged in working groups dealing with racism and discrimination.

Michel Leroy, Dr. des, is a postdoctoral researcher at the Erich Brost Institute for International Journalism at TU Dortmund University, Germany. His research focuses on the sustainability and social impact of media development assistance interventions, with a particular emphasis on critical discourse analysis. The book based on his thesis is published by Palgrave in 2025. Apart from his academic work, he has been advising donors and implementers in the media and development fields for over 25 years, helping them to think strategically and manage transformational change. He is currently academic manager of an Erasmus+ cross-border project of an African online course on media coverage of migration.

Anna Litvinenko, Dr. phil. habil., is a researcher at the digitalization and participation department at the Institute for Media and Communication Studies, Freie Universität Berlin, Germany and an associated researcher at the Weizenbaum Institute for the Networked Society. Her research focuses on political communication in authoritarian regimes, Internet governance, as well as on comparing mediated communication across different contexts. She is a founding member of the DFG network Cosmopolitan Communication Studies (KosmoKW).

Martin Löffelholz, Dr. phil., is full professor of media studies and head of the International Crisis Communication Research Group at the Technische Universität Ilmenau, Germany, and former president of the Swiss German University in Jakarta, Indonesia.

He is (co-)author or (co-)editor of more than 400 publications, including 21 books, most recently, *Krieg der Narrative. Russland, die Ukraine und der Westen* (2024), *Handbuch Journalismustheorien* (2024), *Palgrave Handbook of Cross-border Journalism* (2023). His research has been published in journals such as *Brazilian Journalism Research*, *Equid Novi*, *Journalism Practice*, *Journalism Studies*, *Journalism Theory*, *International Journal of Communication*, *Vietnam Social Sciences*, among others.

Johanna Mack, is a research associate at the Erich Brost Institute for International Journalism at TU Dortmund University and Hochschule Bonn-Rhein-Sieg, Germany and a doctoral candidate in the graduate program "MEDAS 21 – Media Development in the 21st Century". She edits the German edition of the European Journalism Observatory. Her research interests include international journalism, media systems research, migration and media development.

Fabíola Ortiz dos Santos, is a PhD candidate at the graduate program "MEDAS 21 – Media Development in the 21st Century" at the Erich Brost Institute for International Journalism, TU Dortmund University, Germany and research associate at the Institute of Communication Studies at the University of Duisburg-Essen, Germany. She holds a MA from the Erasmus Mundus Journalism Program with a specialization in war and conflict from Aarhus University, Denmark, and Swansea University, United Kingdom. Her research focuses on media development/assistance in (post)conflict societies, media systems in Africa, communication for development/ social change.

Melanie Radue, Dr. rer. pol., is a research fellow at the chair of development politics at the University of Passau, Germany and works on topics of international communication with a specification in de-Westernization. She completed her doctorate at the Friedrich-Alexander University Erlangen-Nuremberg on the topic of *Media Systems in Non-Western Contexts. Freedom of the Media in Southeast Asia*. Her research interest is in media freedom, comparative media systems and social movements. Beyond Melanie's academic work she is engaged in the design, implementation, and management of media assistance projects.

Carola Richter, Dr. phil., is professor for international communication at Freie Universität Berlin, Germany. In her research, she focuses on media systems and communication cultures in the MENA region (Middle East and North Africa), media and migration, foreign news coverage as well as on public diplomacy. She is the co-founder of AREACORE, the Arab-European Association of Media and Communication Researchers, and director of the Center for Media and Information Literacy (CeMIL) at Freie Universität Berlin. She is co-founder and -editor of the *Global Media Journal – German Edition*, an open access journal for international and transcultural communication.

Aynur Sarısakaloğlu, Dr. phil., is a postdoctoral researcher at the Institute of Media and Communication Studies at Technische Universität Ilmenau, Germany. She earned her BA, MA, and PhD degrees in communication science from the University of Salzburg, Austria. She was an assistant professor at the Department of Communication at the

Turkish-German University in Istanbul. Her main research interests include artificial intelligence in communications, journalism research, and intercultural communication. Her recent research focuses on the transformation of journalism induced by artificial intelligence technologies. She is co-editor of the *Handbook of Artificial Intelligence and Journalism* (in press).

Kathrin Schleicher, M.A., is a PhD candidate at the Institute of Media and Communication Studies at Technische Universität Ilmenau, Germany and a member of the International Research Group on Crisis Communication. From 2021 to 2024, she worked as a research assistant in the MIRKKOMM research project, analyzing German risk and crisis communication at federal, state and municipal level in the context of the COVID-19 pandemic. Her research interests include organizational communication, crisis communication and war communication.

Viviane Schönbächler, Dr. des., is a postdoctoral researcher at the University of Hamburg, Germany. Her main fields of research are feminist media studies, media and conflict, intersectional methodologies, and journalists' safety and digital security. She completed her PhD at the Ruhr-University in Bochum on women journalists' contribution to conflict resolution processes in Burkina Faso, which will be published in 2025. Currently, she is working on digital security in journalism in Germany. She has a background in peace & conflict studies and worked in various humanitarian and peace projects in the Middle East and West Africa.

Barbara Thomass, Dr. phil., is professor em. for international comparison of media systems at the Institute for Media Studies, Ruhr-University in Bochum, Germany. Her main fields of research are political, economic and legal dimensions of media systems, media politics, international communication, public service media, media and journalism ethics. She is the second deputy head of the board of the ZDF, the national public service broadcaster in Germany. She has been working with international organizations for several years in courses on journalism standards and ethics in East and Southeast Europe, in West Africa and Southeast Asia.

Sophia Charlotte Volk, Dr. rer. pol., is a senior research and teaching associate at the Department of Communication and Media Research (IKMZ) at the University of Zurich, Switzerland. She received her PhD from Leipzig University, Germany, with a dissertation on *Comparative Communication Research* (2021). Her research interests include strategic communication, science communication, evaluation and impact measurement, digital media environments and technologies, international comparative research and team science. She currently serves as speaker of the PR/Organizational Communication Division of the German Communication Association (DGPuK).

Stefan Wollnik, M.Sc., M.A., is a doctoral candidate at the Institute of Communication Studies at the University of Duisburg-Essen in Germany and a member of the graduate school "MEDAS 21 – Media Development in the 21st Century". In his dissertation, he deals with the social importance of health journalism in sub-Saharan Africa. In this context, his research focuses in particular on the role of global media development assistance. His research interests include political communication, health communication, intercultural communication, and concepts and theories of communication studies and media sociology.

Yi Xu, Dr. phil., is a postdoctoral researcher at the Institute of Communication Studies (IfKW) at Friedrich Schiller University Jena, Germany. She received her PhD degree from Technische Universität Ilmenau. Her research interests include multimodal framing, visual communication, risk and crisis communication, journalism studies, and public diplomacy. From 2021 to 2024, she worked for the DFG-funded cross-national comparative research project "Deciphering the 'pandemic public sphere': Government communication, (social) media discourses on and citizens' responses to COVID-19 in Europe and the USA".

Roja Zaitoonie, M.A., LL.M. (Human Rights), is a social scientist, doctoral candidate at TU Dortmund University, and associated researcher at the Erich Brost Institute for International Journalism in Dortmund, Germany. Her PhD project is about the UN's media strategies in peacekeeping operations with a special focus on UN radios in Sierra Leone, Liberia, and Côte d'Ivoire. Her research interests include media development, peace and conflict studies, human rights, and the United Nations. Currently, she also works for the University of Applied Sciences for Police and Public Administration in Duisburg, Germany.